Understanding

Biotechnology

An Integrated and Cyber-Based Approach

Understanding
Biotechnology

An Integrated and Cyber-Based Approach

George Acquaah
Langston University

PEARSON
Prentice
Hall

Upper Saddle River, New Jersey 07458

Library of Congress Cataloging-in-Publication Data

Acquaah, George.
  Understanding biotechnology: an integrated and cyber-based approach / by George
Acquaah.—1st ed.
     p. cm.
  Includes bibliographical references and index.
  ISBN 0-13-094500-5
    1. Biotechnology. I. Title.

TP248.2 .A275 2004
660.6—dc21   2002190731

Editor-in-Chief: Stephen Helba
Executive Editor: Debbie Yarnell
Development Editor: Kate Linsner
Managing Editor: Mary Carnis
Production Editor: Amy Hackett, Carlisle Publishing Services
Production Liaison: Janice Stangel
Director of Manufacturing and Production: Bruce Johnson
Manufacturing Buyer: Cathleen Peterson
Creative Director: Cheryl Asherman
Cover Design Coordinator: Miguel Ortiz
Marketing Manager: Jimmy Stephens
Cover Design: Amy Rosen
Cover Illustration: Genetic Transformation of Cassava. Courtesy of Nigel J. Taylor.

Pearson Education LTD.
Pearson Education Australia PTY, Limited
Pearson Education Singapore, Pte. Ltd
Pearson Education North Asia Ltd
Pearson Education Canada, Ltd.
Pearson Educatión de Mexico, S. A. de C. V.
Pearson Education—Japan
Pearson Education Malaysia, Pte. Ltd

10 9 8 7 6 5 4 3 2 1
ISBN: 0-13-094500-5

To Theresa. . . . He who finds a good wife finds a good thing!

Contents

11   Genome Mapping and DNA Sequencing   134

12   Storage and Retrieval of Genetic Information   145

Part III   Approaches of Biotechnology   151

15   Modifying Protein Production and Function   200

Part IV   Specific Applications   215

16   Food Biotechnology   216

17 Human Health and Diagnostics   240

18   Industrial Applications   263

Preface

*B*iotechnology is such a rapidly evolving field that perhaps the only way it makes sense to have a textbook on the subject is for the author and publisher to plan to annually update it. Unless, of course, a strategy such as the one adopted in this textbook is followed, whereby the textbook is linked to the very wheels of the vehicle for modern information dissemination—the *Internet*. In other words, a *cyber-based approach* is critical to continually keeping a biotechnology textbook current. Without question, there is a tremendous need for textbooks in biotechnology to inform and to instruct. Even though the Internet is currently the most important Information Age technology, we do not carry computers with us all the time, nor would we want to stare at a computer screen all the time even if we did carry one. There is a role for the written word in the Information Age that cannot be replaced by computers.

Notwithstanding the rapid rate of its evolution, biotechnology evolves around certain core principles and practices. A textbook on the subject should, therefore, first emphasize these core principles and practices, which are relatively immutable and time honored, and then allow room for growth through planned revisions. Methodologies that are deemed to be "standard" today may be obsolete tomorrow. Even so, it is informative and instructive to understand the evolution of the technologies of biotechnology. Such knowledge guides and stimulates the development of newer technologies. Furthermore, new technology does not become assimilated into the tradition of science overnight. Some older technologies, the proverbial "old reliables," are still depended upon in certain cases for a variety of reasons. For example, isozyme technology is an older technology that is still used in research and industry as a quick and inexpensive means of authenticating the hybridity of a cross, among other uses. In situations where resources are limited, as well as in small-scale studies, older technologies may sometimes be more economical and convenient to use than newer ones.

The Internet currently provides the most expansive forum for the exchange of ideas. It is a playground for both amateurs and professionals, for "sowing wheat and tares," to borrow from the Bible. Without specific guidance, an Internet user has to spend considerable time to sort through the tremendous volume of information available to find useful data. Obviously, a textbook that is supported by a quick guide to supplemental material on the Internet would be very attractive and desirable. This book is supported by such a website guide to important and informative sites.

Another challenge in writing a biotechnology textbook is determining the audience, which determines the depth and scope of coverage as well as the style of presentation of the material. Biotechnology is a household word that has permeated instructional curricula even at the kindergarten level. High school students engage in sophisticated biotechnology projects, as evidenced by the high quality of projects exhibited at science fairs. This textbook is designed to be useful to a fairly broad audience, but especially to *early college-level* students. Hence, the presentation in this text is graduated, whereby a topic is first introduced in a general way, providing an overview of the subject so that the casual reader who wants to know about the topic for self-edification can understand it. The discussion then proceeds to provide further details.

An introductory textbook should not be bogged down with laboratory protocols. However, whenever appropriate, simple methodologies are described in this textbook. Also, standard protocols are provided for the illustration of certain key techniques. In practice, in academia and industry, the trend is for various research laboratories to customize laboratory protocols according to their needs and system of operation. Furthermore, there are good sites on the Internet to which a reader can be referred to access numerous methodologies.

The Internet also provides opportunities for the reader to view animated clips to help explain the principles and techniques discussed in books. There are also numerous graphics that can be accessed to supplement what is provided in the textbook. By adopting a cyber-based approach, this book provides students with an enhanced learning environment in which additional materials can be accessed as needed, thereby making it unnecessary to include such bulky material in the textbook. This is appropriate as the Internet has become an integral part of the instructional delivery process of modern education.

Biotechnology is ubiquitous in society. It impacts the environment, agriculture, nutrition, industry, and health. Biotechnology depends on principles from a variety of science disciplines, especially biochemistry, microbiology, molecular biology, physiology, and genetics. Practitioners with good familiarity in these disciplines are more successful. Biological materials share certain biomolecules in common, with a major group being nucleic acids. The major nucleic acid that links all forms of life (with minor exceptions) is DNA. Most analyses and methodologies of biotechnology are generally applicable across species, with slight modifications in certain cases. Furthermore, these technologies and methodologies are developed by scientists working in various disciplines of science and on various organisms, and then adapted for application by workers in other disciplines. For this reason, "agricultural biotechnology" and "medical biotechnology," and for that matter any category of biotechnology, differ in the applications but not the essential principles of these technologies. The author knows persons trained in plant molecular biology who were hired from a highly competitive pool of applicants to work on the Human Genome Project. However, it should be pointed out that certain fundamental biological differences between plants and animals make the application of certain biotechnological techniques more challenging in one area than another.

Another reason for an integrated approach in this text is that the products and use of biotechnology in plants directly or indirectly affect animals, humans, and the environment. Genetically modified plants fed to animals reach humans as meat and other animal products. The nature of these modified plants affects how they are cultivated, and hence the impact of crop culture on the environment.

In view of the foregoing, an introductory textbook on biotechnology should, to some extent, integrate all areas of application. It is informative and useful for a student of agriculture to know how biotechnology is applied to solve problems in industry, medicine, and the environment, and to appreciate the underlying relationships among these areas. This book adopts such an integrated approach, pointing out, where applicable, the differences in the application of techniques between plants and animals. Examples of applications in each field are presented and discussed, with significant emphasis on agriculture where biotechnology application is most visible and most controversial. Since the discipline is rapidly evolving, newer ways of using biotechnology will continue to emerge.

Another aspect of importance that should be addressed in an introductory biotechnology textbook is the matter relating to the acceptance of the processes and products by the beneficiary of research in biotechnology—the general public. Because of the nature of some of the component technologies, the development and application of biotechnology is embroiled in significant controversy. There is a fair amount of public apprehension, some of it rooted in fear from lack of information or misinformation. Some criticisms leveled against biotechnology are also based on personal ethics. Such controversies have economic and political implications. A section in this textbook is devoted to discussing such issues associated with biotechnology.

Finally, biotechnology at an introductory level should be presented in a captivating fashion without watering down the material. Throughout this textbook, the student is engaged through questions and suggestions of things to do to facilitate his or her understanding of the concepts. Frequent reference to the Internet for alternative and colorful graphics and animation breaks the monotony of staring at white pages. The textbook also provides a glossary of terms commonly encountered in biotechnology. This book may be used as a primary text or supplementary text for an undergraduate or graduate course. It may also be utilized for instruction at the pre-college level.

Acknowledgments

*T*he author offers deep appreciation to Janet Rogers, Manager of the DNA/Protein Core Facility at Oklahoma State University, Noble Research Center, for arranging for the photos of most of the equipment used in this text. The exceptional technical assistance provided by Kate Linsner of Prentice Hall during the preparation and production stages of this textbook is acknowledged with many thanks. Also, the assistance with library search provided by Njambi Kamoche, Director of Libraries at Langston University, is deeply appreciated. The author extends thanks to Dr. Kanyand Matand, Coordinator of the Biotechnology Research and Education Center at Langston University for the variety of literature materials provided for preparing this text. Dr. Marvin Burns, Dean of the School of Agriculture is deeply appreciated for his support and encouragement throughout this project. In addition, several individuals provided helpful insight in the developmental stages of the manuscript. I am indebted to the following reviewers for their valuable assistance with this project: Joseph Gindhart, University of Massachusetts-Boston; Robert L. Houtz, University of Kentucky; Stephen Moose, University of Illinois at Urbana-Champaign.

1 What Is Biotechnology?

PURPOSE AND EXPECTED OUTCOMES

When was the first time you heard about biotechnology? What does biotechnology mean to you? Do you think biotechnology has received too much press, either good or bad? Do you think you know enough to weigh in on the current debate about biotechnology's role in modern society? The purpose of this chapter is to help you understand what biotechnology is all about; why it deserves all the press, good or bad; and why it is an important subject to study. To this end, you will be briefly introduced to its impact on society so far, and to some of its anticipated impact. It is important to realize that technologies, like the tools they are manifested in, can be used "for better or for worse." For this reason, there is much controversy surrounding biotechnology; it has the potential for good as well as evil. But is it the process or the product that is questionable? By the end of this textbook, you will learn enough to make informed judgments on the issues surrounding the development and application of biotechnology.

The purpose of this introductory chapter is to present an overview of the nature and applications of biotechnology, as well as its impact on society. In this chapter, you will learn:

1. The definition of biotechnology.
2. The overview of the impact of biotechnology on society.
3. Specific milestones in the development and application of biotechnology.

INTRODUCTION

Every once in awhile, the world experiences a revolution. Revolutions come in different forms, and in different fields. They also have different degrees of impact; some are localized, while others have regional and even worldwide impact. Some revolutions are political, while others are cultural, industrial, or religious in nature. In addition, others are scientific or technical in nature. The 1960s witnessed a biological revolution, dubbed the "Green Revolution," in which genetically improved seed was developed for use in selected economically disadvantaged parts of the world, especially the tropical regions. The new and improved wheat and rice cultivars transformed food production in the targeted developing countries and earned the originator of this revolution, Norman Borlaug, the Nobel Peace Prize in 1970.

Periodically, society can be transformed in a dramatic fashion by some ideological, philosophical, or technological innovation. The world of the twenty-first century is witnessing the unfolding of an event that some believe to be another biological revolution. Called **biotechnology,** this revolution began in the late twentieth century and is currently impacting food, health, and the environment in very dramatic ways, the full extent of which is, as yet, unpredictable. In the process, biotechnology is also stirring up significant controversy in the areas of ethics and religion. Biotechnology is big business, and it promises to be even bigger with time. The growth of biotechnology will continue to impact world trade and global economics, and in the process will also have sociological and political ramifications. There is a need for the consuming public to be properly informed about biotechnology in order to create an atmosphere in which objective discussion can occur. The continued success and growth of biotechnology depends to a great extent on consumer perception and acceptance of the technology and its application.

WHAT'S IN A NAME?

There is no clear consensus among the scientific community on the definition of biotechnology. Criticisms run rampant in literature about proposed definitions by courageous individuals. Some definitions are criticized as being "too broad to be useful," while others are said to be "too narrow to be informative." Someone more creatively stated that biotechnology is "a lexicographic amoeba," while others have noted that some definitions are "politically charged."

How and when did the term "biotechnology" originate? Robert Bud of the Science Museum in London, UK, is credited with the first use of the term "biotechnology" in about 1917. During the first World War, the term was used to refer to industrial fermentation processes used to produce industrial feedstock (e.g., acetone that was used to manufacture the explosive, cordite). A Hungarian engineer, Karl Ereky, is credited with coining the term in 1919, to refer to all the lines of work by which products are produced from raw materials with the aid of living organisms. However, in certain accounts, a Danish microbiologist is credited with coining the term in 1941, to refer to a technique for precise selection of yeast strains.

The definition of biotechnology may be inferred from an etymological analysis of the term, whose root is in the ancient Greek language. *Bios* is Greek for "life"; *teuchos* means "tool"; while *logos* is Greek for "word," "study of," or "essence." Therefore, biotechnology may be defined as *the study of tools from living things.* Better still, it may be defined as *the use of techniques based on living systems (plants, animals, or microbes) to make products or improve other species.*

BIOTECHNOLOGY TAKES ACADEMIA BY STORM

In the early years of the biotechnology revolution, when it was starting to gain momentum in academia, there appeared to be some "academic turf wars" waged over which scientific discipline was the appropriate one to claim biotechnology under its umbrella. The author recalls hearing a description of a molecular geneticist by a biochemist as "a geneticist practicing biochemistry without a license." Currently, molecular genetics or molecular biology has evolved into a full-fledged discipline of science (William Astbury is credited with the first use of the term "molecular biology" in 1945). Some scientists hailed biotechnology as a radically new technology while others argued for its historical origins. The technology was presented initially as manipulating the genetics of organisms in ways that transcended natural barriers, which was conducted *directly* at the molecular level (directly involving the DNA). In contrast, traditional plant and animal breeders also manipulate organisms, although they do it *indirectly* and from the whole organism level (genes are manipulated through the sexual process). Microbiologists, similarly, can manipulate microbes to perform useful functions such as fermentation. Before long, the definition of biotechnology began to accommodate all these "interest groups." Furthermore, there was a scramble to incorporate biotechnology into all biological disciplines in academia. Classical breeders were "pressured" to retool (through some kicking and screaming). To get around the issue, some established "old school" professors resorted to hiring recent graduates from the "new school" to broaden the scope of existing research programs.

THE PUBLIC WEIGHS IN ON THE BIOTECH DEBATE

Apart from academia, the consuming public would soon throw a monkey wrench into the discussion, compelling scientists to take note and accommodate them in the still-evolving definition of the term. Presenting biotechnology as "radically new" might have been glamorous, but it only raised the eyebrows of consumers who, thanks to activists, soon began to see biotechnology as radical and alien. The result was a perception that early biotechnology products were "unnatural" and "freakish." References to "new biotechnology" as "genetic engineering" did not help the situation, since they connoted the idea of humans redesigning and reconstructing God's creation, a no-no in the eyes of many. Such an unfavorable public perception was damaging to the biotechnology industry right from the start of the revolution. Even though products of traditional plant and animal breeding had been accepted for centuries, suddenly consumers, especially those in Europe, were finding it hard to embrace biotechnology food products. The current trend is to present biotechnology (agricultural biotechnology, that is) as a more efficient and effective extension of the techniques of classical plant and animal breeding. But one may ask, "Is it the *process* or the *product* that scares people?" What can and should be done to correct this situation?

Some of the techniques of biotechnology do not involve genetic manipulation. For example, bacteria and yeast are used in food processing without the need to first modify their genomes. Other techniques involve the manipulation of the genome. This latter group of techniques is the one that is most readily identified with biotechnology and thus receives all the positive as well as negative feedback. Consequently, an emerging trend is to define biotechnology in two ways: broad and narrow. A narrow definition equates biotechnology with recombinant DNA technology or genetic engineering. Biotechnology may be narrowly defined as the use of gene transfer methods to improve individuals or make products. The definition previously given will suffice as a broad definition (i.e., the use of techniques based on living systems (plants, animals, or microbes) to make products or improve other species).

THE "OLD" VERSUS THE "NEW" BIOTECHNOLOGY

The cell is the basic unit of organization of living things. There are two basic classes of cells—prokaryotic and eukaryotic. Prokaryotes (bacteria) have cells that lack distinct nuclei and compartmentalization into discrete bodies called organelles. Eukaryotes (e.g., corn, cows, and humans) have cells with distinct nuclei and compartmentalization into organelles with different functions. Some organisms consist of just one cell (unicellular, like bacteria) while others are multicellular (e.g., corn). Multicellular organisms are structurally organized in a certain order of increasing complexity, from macromolecules to whole organisms.

Biotechnology research and applications may involve whole organisms or parts of them. Classical or traditional ("older") biotechnological manipulation of organisms is done at the whole organism level. Typically, breeders orchestrate gene mixing through planned crosses or mating, followed by selection to identify individuals that have the desired combination of genes. In a way, this genetic manipulation is relatively unprecise, hence the long duration of breeding programs. It might take up to 10 years or even more for plant breeders, for example, to develop a breeding program and release a new cultivar. The newer tools of biotechnology, embodied in **recombinant DNA technology,** enable scientists to extract and transfer specific genes from one parent and insert them into others, irrespective of the species. That is, they circumvent sexual propagation as a means of mixing genes. The newer technologies are applied at the molecular (DNA) level, and are more precise in terms of what genetic material is transferred. Research is ongoing to discover newer techniques for inserting target genes into specific locations in the genome of the host. Specific techniques are discussed in detail later in this textbook. It should be made clear that just because a technique is described as "older" does not mean it is no longer useful. In fact, classical breeding methodologies remain the way by which improved plants and animals, irrespective of the method of development, are finally put together for release to the public. That is, the "old" and "new" biotechnologies are used in complementary fashion.

BIOTECHNOLOGY TIMELINE

The timeline presented in Table 1–1 shows the primitive use of biotechnology by humans, primarily in food processing, as early as 1750 B.C. (some accounts place the date as early as 10,000 B.C.). Given the technology and scientific knowledge of that period, it is not likely the people were aware that they were manipulating microorganisms to make products. In this modern era, the proliferation of scientific knowledge makes it possible for us, in many cases, to know exactly what we are doing, and to nudge nature in the direction we desire for higher efficiency. As such, the timeline shows landmark discoveries of technologies that are key to the development and application of biotechnology. A variety of timelines may be consulted on the Internet, including the one provided in the reference at the end of this chapter.

WHY IS BIOTECHNOLOGY SUCH A BIG DEAL?

Biotechnology is important because of the extent to which it has already impacted modern society. In terms of its benefits to society, many believe that more is yet to come. New techniques and applications are continually discovered. An overview of the benefits of biotechnology is discussed in this section. Selected applications of biotechnology are discussed in later chapters of the book. The purpose of this brief overview of benefits is to justify the great attention biotechnology commands in society.

The benefits of biotechnology may be discussed under six major categories—agriculture, industry, health (or medicine), environment, forensic, and advancement of knowledge.

1. **Agriculture**

   Biotechnology provides a more efficient means of crop and animal improvement. Instead of extensive mixing of genes, as occurs in conventional breeding, biotechnology enables targeted gene transfer to occur. The genome of the recipient individual remains intact, except for the introduced gene, thus accelerating breeding programs. Furthermore, biotechnology enables gene transfer across natural barriers, breaking down mating barriers and creating a sort of "universal gene pool" or "universal breeding population" accessible to all organisms. Biotechnology is used to improve the yield of crop and animal products, their quality, the flavor of foods, and also the shelf life of products. In addition to these benefits, biotechnology reduces the need for agrochemicals through disease resistance breeding, thereby reducing environmental pollution from chemical runoff. Increased yields and higher food quality reduce world hunger and malnutrition.

TABLE 1–1
Selected milestones in the timeline of biotechnology.

Date	Event
1750 B.C.	Sumerians brew beer
1830	Protein discovered
1833	First enzymes isolated
1863	Mendel's discoveries
1919	Term "biotechnology" first used by Hungarian scientist
1938	Term "molecular biology" coined
1941	Term "genetic engineering" first used by Danish scientist
1944	DNA confirmed as genetic material
1954	Cell culturing techniques developed
1967	First protein synthesized by automation
1973	First recombinant DNA experiment by Cohen and Boyer
1974	NIH form rDNA Advisory Panel to oversee rDNA research
1975	Colony hybridization and Southern blotting developed
1980	U.S. Supreme Court approves the principle of patenting of genetically modified life forms, in case of *Diamond v. Chakrabarty*
1983	Polymerase chain reaction (PCR) technique developed; first genetic transformation of plant cells by Ti plasmids accomplished
1984	DNA fingerprinting technique developed
1985	First unauthorized field test of a genetically engineered plant (tobacco); first biotech-engineered interferon drug (Biogen's Intron A and Genentech's Roferon A approved)
1987	First authorized field test of a genetically modified organism—Frostban by Advanced Genetic Sciences
1988	Funds for Human Genome Project approved by U.S. Congress
1994	Calgene produced Flavr Savr tomato
1997	Dolly the sheep cloned by Scottish scientists
2000	Human genome sequence draft produced

2. **Industry**

Biotechnology is used to develop alternative fuels. Cornstarch is converted by yeast into ethanol, which is used to produce gasohol (a gasoline-ethanol mix). Bacteria are also used to decompose sludge and landfill wastes. While cleaning the environment, methane (natural gas) is produced for fuel. Through biotechnology, microbes or their enzymes are used to convert biomass into feedstocks, which are used for manufacturing biodegradable plastics, industrial solvents, and lubricants. Plants are being genetically manipulated to produce plastics (bioplastics) in their tissues. Organisms (microbes and mammals) are used as pharmaceutical factories for producing chemical compounds that are extracted from their products and processed as drugs and other products. Plant and animal fibers are used in making a variety of fabrics, threads, and cordage. Biotechnology is used to improve the quality and quantity of these products. Biopulping is a technique whereby a fungus is used to convert wood chips into pulp for papermaking.

3. **Health/medicine**

   In the area of health/medicine, biotechnology is used to develop diagnostic tools for identifying heritable diseases. The results of such diagnoses are used in genetic counseling to aid in making informed choices by parents who are predisposed to the birth of children with genetic abnormalities. Diagnostic tools for pregnancy tests, as well as other tests, have also been developed for early detection. Biotechnology is used to produce more effective and efficient vaccines, antibiotics, and other therapeutics. The famous drug, penicillin, is a microbial product. Through the biotechnology of gene therapy, scientists are taking a crack at curing genetic diseases by attempting to replace defective genes with healthy ones. A revolutionary strategy is being developed whereby staple foods such as potatoes, bananas, and others are used as delivery vehicles for vaccines to facilitate the immunization of people in economically depressed regions of the world.

4. **Environmental**

   Developing and using alternative fuels that burn cleaner improves air quality through reduced pollution of the environment. Microbes are used to decompose and clean up contaminated sites by the technology of bioremediation. The use of disease-resistant cultivars makes crop production less environmentally intrusive by reducing the use of agrochemicals.

5. **Forensic**

   Forensic application of biotechnology has entered deliberations in the judicial system in many countries. The high-profile O.J. Simpson murder trial in the 1990s was perhaps the most dramatic evidentiary use of DNA technologies in the history of the U.S. courts. DNA evidence is used in cases involving paternity disputes and family relationships.

6. **Advancement of knowledge**

   This benefit of biotechnology is often unheralded. Biotechnology provides tools for more in-depth probing of nature to understand how things function. As one author puts it, biotechnology is a revealer of knowledge through "unlocking nature's black box."

BIOTECHNOLOGY CAN BE ABUSED

It is important to emphasize that, while the application of a technology can be evil, the technology *per se* is not. The biotechnology applications cited so far paint a picture of a technology that can do no wrong. However, just as splitting the atom brought about innumerable benefits to society, it is also the reason the world has the atomic bomb and other destructive armaments. Similarly, one of the deadliest forms of warfare in modern times is biological warfare, in which deadly microbial organisms are deliberately unleashed into the environment to inflict diseases on victims. Fortunately, such deliberately destructive and objectionable use of microorganisms is not among the mainstream biotechnology applications. However, it is instructive to mention the potential danger such a powerful technology poses when in the hands of desperately wicked individuals. In 1952 the Protocol for the Prohibition of the Use in War of Asphyxiating, Poisonous or Other Gases, and of Bacterial Methods of Warfare was signed. In 1969, the United States unilaterally renounced the first use of chemical weapons and all methods of biological warfare. Later in 1975, the United States signed an updated version of the 1952 convention prohibiting the production, possession, stockpiling, transfer, and use of such weapons as well as providing for the destruction of existing stock. The UN Security Council is involved in this international effort. A Verification Expert Group has been formed to provide ongoing monitoring for compliance of the prohibitions, and to determine the appropriate exemptions on the grounds of legitimate use of the technology.

Biotechnology can also be used to resurrect the deplorable eugenics movement. Anti-biotechnology activists persistently protest certain biotechnology applications, citing potential damage to the environment and humans, as well as ethical grounds (e.g., the cloning of humans).

CYBER-BASED INTRODUCTORY MATERIALS

There are numerous sites on the Internet that may be consulted by the reader for additional materials presenting an overview of biotechnology. Selected examples are as follows.

1. **Video**

    It may be helpful to view an audiovisual presentation on biotechnology that gives an overview of the discipline. One such presentation, produced by Norvatis corporation, can be accessed at *http://whybiotech.com/* (Understanding Biotechnology).

2. **Quiz: biotech IQ**

    What do you know about biotechnology products in the food chain? What is their impact on the environment? Take a couple of quizzes to test your "biotech IQ" at *http://www.hort.purdue.edu/hort/people/faculty/goldsbrough.html.* Answers are provided at the end of the quiz.

3. **Timeline**

    Additional milestones in the development and application of biotechnology may be obtained from *http://www.bio.org/timeline/timeline.html.*

4. **Applications**

    Further overview of biotechnology applications may be obtained from:

    http://www.whybiotech.com/en/default.asp
    http://www.accessexcellence.org/AB/BA/
    http://www.accessexcellence.org/AB/GG/biotechnology.html (summary of biotechnology—present and future)
    http://www.accessexcellence.org/AB/WYW/fink/fink_1.html (the history, methods, and promise of plant biotechnology—overview)

USING THE INTERNET MATERIAL IN THIS TEXTBOOK

One of the major features of this book is the extensive use of Internet links to provide additional information, or present information in an alternative format or from a different perspective. Unlike textbooks, which usually are peer-reviewed for accuracy of information prior to publication, Internet sites are often maintained by individuals who post materials that have not been independently checked for accuracy. Because of this, it is possible to encounter inaccurate materials. To protect against this, multiple sites have been provided on the same topic for comparison. One reader may find the presentation at one site more useful than at another. Some of these websites provide colorful graphics and animation that make it much easier to understand the principle or concept being presented. Frequently, the selected sites provide the reader with links to additional resources.

KEY CONCEPTS

1. Biotechnology entails the use of living organisms to make products or to improve other organisms.
2. Ancient civilizations used organisms in a basic fashion; the modern applications are more refined, more efficient, and relatively more radical.
3. Some techniques utilize organisms in whole or in part without modification of the genomes. More powerful modern technologies enable the manipulation of the genome of the organism to varying extents, even disregarding natural genetic barriers.

4. There is considerable public apprehension regarding the development and application of biotechnology for various reasons, among which is the perception that biotechnology is an "unnatural" means of developing products.
5. The realized and anticipated benefits of biotechnology are astonishing. Benefits accrue to society through agricultural, medical, environmental, and industrial applications. Also, genetic-based diagnostic tools are applied in the justice system.
6. The development and application of biotechnology contributes to the advancement of knowledge.
7. Like almost everything, biotechnology has some risks.
8. Biotechnology can be used for deliberately destructive purposes.
9. Biotechnology depends on advances in disciplines such as biochemistry, genetics, molecular biology, microbiology, physiology, and physics.

OUTCOMES ASSESSMENT

1. Define "biotechnology" to a layperson.
2. How do you think the development and application of biotechnology should proceed? Explain your answer.
3. Do you think biotechnology could be the answer to world hunger and malnutrition? Explain.
4. Do you think the world is better or worse off because of biotechnology? Explain.
5. How has this introductory chapter affected your previous perception of biotechnology?

ADDITIONAL QUESTIONS AND ACTIVITIES

1. Search the literature/Internet for different opinions in favor of and against biotechnology (five of each).
2. Conduct a public opinion survey regarding biotechnology on your campus.
3. What is the most recent news item on biotechnology in your local newspaper or other news media?
4. Visit your grocery store and make a list of biotechnology products on display.

REFERENCES AND SUGGESTED READING

Abelson, P. H. 1994. Continuing evolution of U.S. agriculture. *Science, 264:*1383.

Baumgardt, B. R., and M. A. Martin (eds). 1991. *Agricultural biotechnology: Issues and choices.* West Lafayette, IN: Purdue University Agricultural Experiment Station.

Davis, J., and D. L. Ritter. 1989. How genetic engineering got a bad name. *Imprints, 18(2):*1–5.

Krimsky, S., and R. P. Wrubel. 1996. *Agricultural biotechnology and the environment: Science, policy, and social issues.* Chicago: University of Illinois Press.

Lee, T. F. 1993. *Gene future: The promise and perils of the new biology.* New York Plenum Press.

Miller, H. 1993. Perceptions of biotechnology risks: The emotional debate. *Bio/Technology, 11:*1075–1076.

Miller, M. 1991. The promise of biotechnology. *Journal of Environmental Health, 54:*13–14.

Roberts, L. R. 1992. Science in court: A culture clash. *Science, 253:*732–736.

Rochelle, G. 1990. Tinkering with the secrets of life. *Health, 22:*46–86.

Sharpe, F. 1987. Regulation of products from biotechnology. *Science, 235:*1329–1332.

Tokar, B. (ed). 2001. *Redesigning life? The worldwide challenge to genetic engineering.* New York: Zed Books.

Verma, I. M. 1990. Gene therapy. *Scientific American, 263:*68–84.

PART I

Brief Review of the Underlying Science

The Nature of Living Things: How They Are Organized

PURPOSE AND EXPECTED OUTCOMES

Biotechnology includes tools that enable scientists to transfer genes freely among living things, irrespective of biological barriers. This activity involves the manipulation of organisms at the molecular level.

In this chapter, you will learn:

1. The levels of eukaryotic organization.
2. The structural organization of living things at the cellular level.
3. The structure and properties of key cellular macromolecules (DNA, RNA, and proteins).
4. The roles of cellular structures and biomolecules in biotechnology.

BRIEF TAXONOMY OF LIVING THINGS

There are five major groups (kingdoms) of all organisms: **Plantae, Animalia, Fungi, Protista,** and **Monera** (see Table 2–1). The kingdom Monera, comprised solely of bacteria, is very critical to genetic engineering technologies, as it provides enzymes (restriction endonucleases) and **plasmids** (circular DNA molecules) for cloning. Bacteria also provide valuable genes that are used to enhance animals and plants (e.g., the *Bt* gene). Bacteria are very useful in applications such as **bioreactors** (for making pharmaceutical products) and food technology applications (e.g., fermented food products such as cheese and wine). Fungi are also used in various food applications as well as genetic engineering methodologies. These applications are discussed in detail in other chapters. A key piece of information to note at this point is that, in spite of the genetic barriers and enormous biological variation in nature, biotechnology enables scientists to circumvent natural reproductive barriers and move genes within and between kingdoms. In a sense, biotechnology has created a "universal gene pool" to allow unrestricted mixing of genes by scientists.

LEVELS OF EUKARYOTIC ORGANIZATION

The **cell** is the fundamental unit of organization in organisms. Some organisms consist entirely of one cell (**unicellular**). Other organisms consist of many cells working together (**multicellular**). Except for bacterium, which lacks compartmentalization into **organelles**

TABLE 2–1
The key features of the five kingdoms of life. Organisms and/or their products, from each of these kingdoms, are utilized in biotechnology for making products or improving other species.

Monera	• Have prokaryotic cells • Unicellular organisms • Bacteria belong here
Protista	• Eukaryotes that do not fit other kingdoms • Most are unicellular • Slime molds, algae
Fungi	• Filamentous eukaryotes that lack plastids and photosynthetic pigments
Plantae	• Photosynthetic • Multicellular • Have cell walls
Animalia	• Multicellular • Non-photosynthetic • Complex sensory and neuromotor systems

with specific functions (called a **prokaryote**), all other cells consist of a membrane-bound nucleus and other membrane-enclosed organelles. These organisms are called **eukaryotes.** A single eukaryotic, higher organism may also be structurally organized at various levels of complexity: whole organism, organs, tissues, cells, organelles, and molecules (see Figure 2–1). Genetic manipulation by conventional methods (plant and animal breeding) normally occurs at the whole organism level, and involves sexual processes (crossing or hybridization). In the new biotechnology, genetic manipulation focuses on the macromolecules directly (i.e., direct DNA manipulation). Other techniques of biotechnology (tissue culture, cell culture, protoplast culture) are also conducted below the whole organism level.

THE CELL

Understanding cell structure and function is critical to biotechnology. Genetic engineering for stable transformation of plants targets single cells. Scientists must know how to manipulate a single cell and culture it back into a full organism. Called **regeneration,** this capability must of necessity precede transformation.

The eukaryotic cell consists of many organelles and structures with distinct as well as interrelated functions (see Table 2–2). Some organelles occur only in plants while others occur only in animals. For example, an animal cell lacks a cell wall, which affects how animal and plant cells are manipulated in biotechnology. Examples of methodologies that differ between plant and animal research include DNA extraction, transformation, and tissue culture. For example, the sample of plant tissue needs to be ground in a mortar to break down the cell wall in order to facilitate the DNA extraction process. It is important to understand the structure of the plasma membrane because many critical physiological processes are associated with it. For example, the histocompatibility antigens that play a role in tissue and organ transplants are part of the cell coat in animals. However, such antigens are not known to occur in plants.

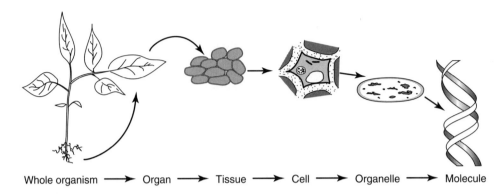

Whole organism ——→ Organ ——→ Tissue ——→ Cell ——→ Organelle ——→ Molecule

FIGURE 2–1
Levels of structural organization in eukaryotes. Multicellular organisms have a
hierarchical structural organization that ranges from visible features to invisible
molecules. Genetic engineering often involves the manipulation of the organism at the
cellular and molecular levels.

The major organelles of special interest in biotechnology include the **nucleus** that
houses the DNA (the hereditary material that is the primary target of genetic manipulation). The **mitochondria** and **chloroplasts** (present only in green plants) have their
unique DNA and are used in unique ways in biotechnology (e.g., role of mitochondrial
DNA in establishing kinship).

SUBCELLULAR ORGANIZATION

Did you know that a typical cell contains about 10^4 to 10^5 different kinds of molecules?
Did you know that a simple bacterium like *E. coli* has over 1,000 different proteins involved in its cellular metabolism? It is important to know the structure of these cellular
molecules because their three-dimensional structures determine their biological properties and, consequently, their roles in the function and development of the cell and the
whole organism in multicellular organisms. Further, biotechnology research involves the
use and manipulation of cellular macromolecules, specifically proteins and nucleic acids
(DNA and RNA). Genes are expressed as proteins. An understanding of protein and nucleic acid structure and function is hence critical to understanding biotechnology.

■ TYPES OF CELLULAR MOLECULES

Cells contain an astonishing number of organic and inorganic molecules of varying sizes
and complexities. There are **small organic molecules** of molecular weights ranging between 100 and 1,000 that occur free in solution in the cytoplasm, and form a pool of in-
termediates from which larger molecules are formed. These small molecules may be
placed into four broad families: **simple sugars, fatty acids, amino acids,** and **nucleotides.** It should be pointed out that some cellular compounds do not fit into any of
these four categories.

Larger molecules are synthesized from these small molecules. About 50 percent of
cellular organic molecules are described as **macromolecules** because of their relatively enormous molecular weights that typically range in order of magnitude between 10^4 and 10^{12}. Macromolecules are chains (polymers) of simpler and smaller
molecules that act as their building blocks (monomers). The three major classes of cellular macromolecules are **proteins, nucleic acids,** and **polysaccharides.** These

TABLE 2–2
Structure of plants and animal cells. Certain structures occur either only in animals or plants.

Present in Both Plant and Animal Cells

Plasma membrane	Differentially permeable cell boundary; delimits cell from immediate external environment; membrane surface may contain specific receptor molecules and may elicit an immune response.
Nucleus	Contains genetic material, DNA; DNA and proteins in condensed form exist as strands called chromosomes; in uncoiled dispersed form, DNA/protein complex exists at chromatin.
Cytoplasm	The part of the cell excluding the nucleus and enclosed by the plasma membrane; consists of colloidal material called cytosol; contains organelles.
Endoplasmic reticulum	Membranous structure that increases the surface area for biochemical synthesis; two kinds exist—smooth (no ribosomes) or rough (with ribosomes).
Ribosome	Contains RNA; site of protein synthesis.
Mitochondrion	Contains DNA; site of aerobic respiration.
Golgi apparatus	Or dictyosome; has role in cell wall formation.

Present Only in Plant Cells

Chloroplast	A plastid containing chlorophyll; site of photosynthesis; contains DNA.
Vacuoles	Storage region structure for useful and undesirable compounds; helps to regulate water pressure within the cell and maintain cell rigidity.
Cell wall	Rigid boundary outside of the plasma membrane.

Present Only in Animal Cells

Centrioles	Occur in the centrosome; associated with the organization of the spindle fibers during mitosis and meiosis.
Lysosomes	Contain hydrolytic enzymes (e.g., lysozyme); enzymes are inactive until released during intracellular digestion.

three classes of macromolecules primarily determine the structure and function of cells. The building blocks of these macromolecules are polymerized into linear chains. Whole molecules or portions of molecules interact through the formation of a variety of noncovalent bonds. This results in the formation of characteristic three-dimensional configurations that confer certain biological properties on these molecules. Macromolecules will be reviewed in greater detail because they are the key targets of biotechnological manipulation.

Proteins

Proteins are very complicated macromolecules. Unlike DNA, which has a universal double-helical structure, each species of protein molecule has a unique three-dimensional structure. Proteins exhibit tremendous variation in size, shape, and composition. There are 20 monomers that can be polymerized in any order, hence the tremendous potential for variation in structure, and consequently, in function. Some proteins have structural roles in the cell as part of cellular components and membrane components. Proteins may be substrates in cellular metabolism or catalyze cellular reactions as enzymes. The diversity in protein structure confers specificity to enzymes regarding the chemical reactions they speed up. Proteins may regulate the expression of other proteins.

Primary Structure

The monomers of proteins are called amino acids, 20 of which occur in biological systems. All amino acids have the same basic structure comprising the following:

   a. An amino (NH_2) group.
   b. A carboxyl (COOH).
   c. A side chain (designated by R), attached to the carbon atom (see Figure 2–2).

Amino acids are distinguished by their side chains, the smallest being a hydrogen atom as found in the amino acid glycine. Polymers of amino acids are synthesized through repeated dehydration reactions in which the carboxyl group of one amino acid is linked to the amino group of the next, with the loss of water. The resulting chemical bond is called a **peptide bond**, and the product a **dipeptide** (see Figure 2–3). The peptide bond is planar. Amino acids may be linked in this fashion to form a polypeptide chain that may contain hundreds of amino acids, and stretch to a length of about 1,000 to 5,000Å. The end of the protein chain with the free NH_2 group is called the **amino terminus,** while the other end with the free COOH is called the **carboxyl terminus.**

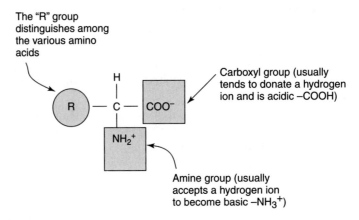

The "R" group distinguishes among the various amino acids

Carboxyl group (usually tends to donate a hydrogen ion and is acidic –COOH)

Amine group (usually accepts a hydrogen ion to become basic –NH_3^+)

FIGURE 2–2
The basic amino acid structure consists of an amino group, a carboxyl group, and a side chain (R) that distinguishes among amino acids.

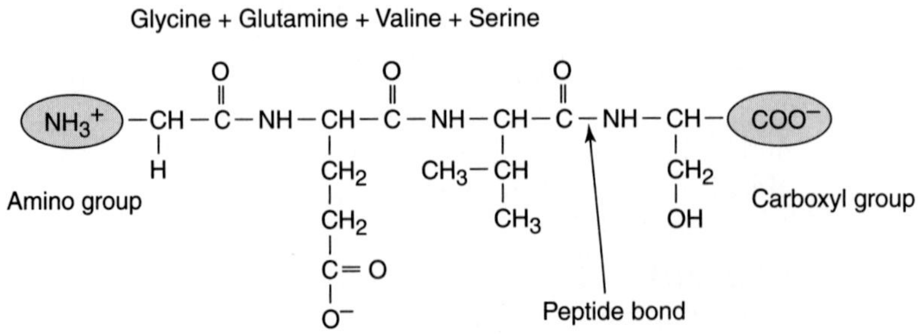

Glycine + Glutamine + Valine + Serine

FIGURE 2–3
A polypeptide showing a peptide bond between two adjacent amino acids to produce a dipeptide.

Secondary Structure

The sequence of the subunits (the amino acids) of a polypeptide chain is preserved by strong covalent bonds. However, in its linear and stretched-out form, the polypeptide (protein) is biologically inactive. The information carried by a macromolecule is expressed primarily by means of weak noncovalent bonds that determine the unique three-dimensional structure of the molecule. The linear polypeptide chain (primary structure) undergoes a series of folding into a large number of shapes or **conformations,** determined by the physical properties of the peptide bond and the kinds of amino acids in the chain.

During the folding of the polypeptide, side chains of amino acids cannot overlap. Similarly, two charged groups with the same sign will not occur near each other, leading to extension of the chain. In proteins containing sulfur-containing amino acids (e.g., cysteine), the –SH (sulfhydryl) groups of nearby amino acids interact to form a covalent bond called a **disulfide.** When folding is completed, hydrophilic (polar) amino acids tend to reside on the exterior of the protein, while hydrophobic (non-polar) amino acids occur mainly on the interior. Common folding patterns recur in different protein chains, although the overall conformation of each protein is unique. Two patterns are most commonly produced by H-bonding interactions. Of these secondary structures α-**helix** (see Figure 2–4) is the most preferred form of a polypeptide chain, in the absence of all other interactions. This conformation is generated when a single polypeptide chain turns regularly about itself. Alpha helices can be coiled about themselves in two, three, or four coil conformations. They can occur internally in proteins, or on protein surfaces, or in membranes (e.g., α-keratin of skin, hair, and feathers). Hydrogen bonds may also create the beta (β) configuration (**beta strands** or **beta sheets**). Beta sheets are formed when an extended polypeptide chain folds back and forth upon itself. The conformation makes up extensive regions of most globular protein.

Tertiary Structure

The **tertiary structure** of proteins results from the folding of secondary structures (α-helices and β-sheets) into three-dimensional structures (see Figure 2–5). Proteins seldom consist entirely of one kind of tertiary structure. Where one kind of folding predominates, long and thin proteins, called **fibrous proteins,** are formed. On the other hand, when the forms are short, the proteins formed tend to be somewhat spherical in shape and are called **globular proteins.** Fibrous proteins occur where structural roles are required. Proteins with catalytic functions (enzymes) are globular proteins.

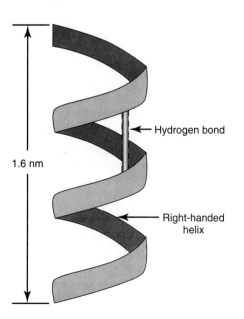

FIGURE 2–4
The secondary structure of proteins is a right-handed helix.

1.6 nm

Hydrogen bond

Right-handed helix

FIGURE 2–5
The tertiary structure of
proteins.

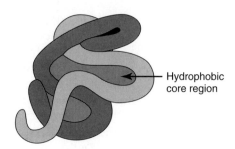

Hydrophobic
core region

FIGURE 2–6
The quaternary structure of
proteins consists of
aggregates of tertiary
structures.

Quaternary Structure

Individual globular proteins may noncovalently assemble into larger protein aggregates. An aggregate may comprise varying numbers of these individual proteins (or subunits). Aggregates of two (called a dimer), three (trimer), four (tetramer), or even six subunits (hexamer) occur in nature (see Figure 2–6). This clustering is common because even though polypeptide chains typically fold such that non-polar side chains are internal, some non-polar groups usually remain on the outside. These exposed hydrophobic regions can be covered up through the aggregation of subunits.

Protein Function

Proteins may be characterized as the workhorses of living systems. They respond to changing environments and other conditions impacting the organism's physiology by, for example, relocating within the cell or changing the molecules they bind to. Proteins are precisely constructed such that an alteration as slight as a change in a few atoms in one amino acid can have dramatic consequences in structure and, hence, function. The chemical properties of a protein molecule are primarily dependent upon its conformation and the exposed surface residues. A protein interacts with another molecule by forming weak noncovalent bonds between them. To be effective, the molecule (called a **ligand**) must fit exactly into a specific region on the protein called its **binding site.** The chemical reactivity of selected amino acid side chains can be changed as a result of the interaction between neighboring residues on the surface of a protein. This affects access to a ligand. When side chain reactivities are diminished below what is required for a specific task, proteins depend on a special group of non-polypeptide molecules called **coenzymes** for assistance. Coenzymes are very complex organic molecules that are frequently tightly bound to the protein they are assisting. Examples of coenzymes are hemes of cytochromes and hemoglobin, and thiamine phyrophosphate.

Enzymes

Enzymes are a special group of proteins that catalyze chemical reactions (accelerate reaction rates). This activity is highly specific, in that an enzyme may catalyze only one or a set of closely related reactions. The first step in enzyme catalysis is the binding to its substrate (the ligand) to form an **enzyme–substrate complex.** This event is highly specific and is responsible for the high selectivity in enzyme catalysis. One theory describes this specific binding as the **lock-and-key model,** in which the shape of the binding site (or active site) is complementary to the shape of the substrate. The second model, the **induced fit,** is believed to explain the binding phenomenon much better. According to this model, the enzyme alters its shape after binding to the substrate so that the shape of the binding site complements that of the substrate. The strength of the binding between the enzyme and its substrate is described as the **affinity** the enzyme has for the substrate.

Nucleic Acids

Nucleic acids are polymers of nucleotides. There are two kinds of nucleic acids: **deoxyribonucleic acid (DNA)** and **ribonucleic acid (RNA).** A **nucleotide** consists of three basic components: pentose sugar, nitrogenous base, and a phosphate group (see Figure 2–7). The cyclic five-carbon sugar is ribose in RNA and deoxyribose in DNA. Similarly, there are two kinds of bases: **purines** and **pyrimidines.** There are two purines—**adenine (A)** and **guanine (G),** and three pyrimidines—**cytosine (C), thymine (T),** and **uracil (U).** Thymine occurs only in DNA, while uracil occurs only in RNA. The letters A, C, T, and G are commonly referred to as the alphabets of life.

When a base is linked to a sugar, the product is called a **nucleoside.** A nucleoside linked to a phosphate forms a nucleotide. Two nucleotides may be linked by a phosphodiester group to form a **dinucleotide** (see Figure 2–8). Shorter chains (consisting of less than 20 nucleotides) are called an **oligonucleotide** while longer chains are called **polynucleotides.** A single nucleoside is also called **nucleoside monophosphate (NMP),** while two nucleosides form a **nucleoside diphosphate (NDP).** Triphosphates are important in cellular bioenergetics, especially **adenosine triphosphate (ATP)** and **guanosine triphosphate (GTP).** When these compounds are hydrolyzed, inorganic phosphate is produced, accompanied by the release of energy (e.g., ATP → ADP + energy).

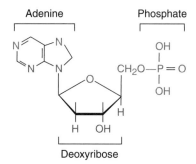

FIGURE 2–7
Structure of a nucleotide.

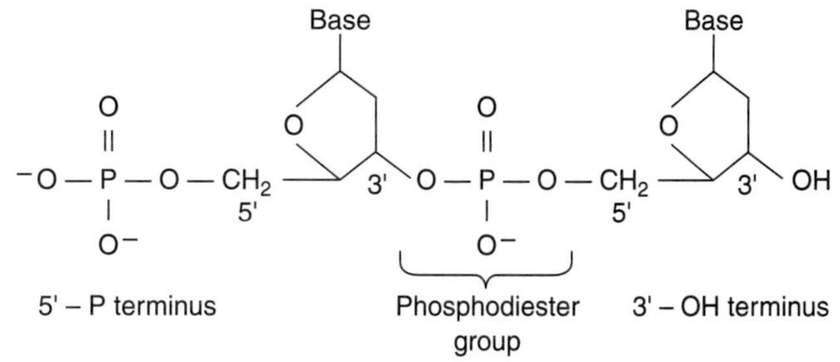

FIGURE 2–8
A dinucleotide structure showing the phosphodiester bond that links two nucleotides.

Structure of DNA

DNA is the universal, hereditary material (except in certain viruses). The most powerful direct evidence for DNA being the hereditary material is currently provided by the cutting edge technology of **recombinant DNA,** the principal subject of this book. The structure of the DNA molecule is a **double helix** (see Figure 2–9), and its key features are:

a. Two polynucleotide chains coiled around a central axis in a spiral fashion such that a right-handed double helix is produced of diameter 2.0 nm.

b. The polynucleotide chains are **antiparallel;** one chain runs in the 5′ to 3′ orientation and the other 3′ to 5′ (carbon atoms of a sugar are conventionally numbered from the end closest to the aldehyde or ketone) (see Figure 2–10).

c. The two bases in each base pair lie in the same plane. Each plane is perpendicular to the axis of the helix. There are 10 base pairs per helical turn.

d. The helix has two kinds of alternating external grooves; a deep groove (called the **major groove**) and a shallow groove (called the **minor groove**).

FIGURE 2–9
The double-stranded helix is the predominant form of DNA structure. This right-handed molecule consists of two complementary strands that run in antiparallel orientation, with alternating minor and major grooves.

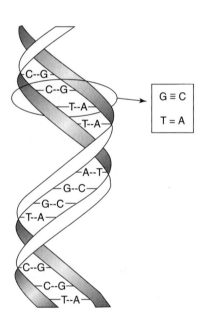

FIGURE 2–10

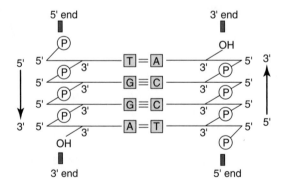

A DNA duplex consists of two strands in antiparallel orientation such that the sugars and phosphates form a backbone on the outside of the helix. The phosphates (P) join the 3′ carbon of one sugar to the 5′ carbon of the adjacent sugar. The two strands are joined on the inside by hydrogen bonds between the nitrogenous bases.

e. The nitrogenous bases on one strand pair with those on the other strand in complementary fashion (A always pairs with T, while G pairs with C).

In addition to the features previously described, certain implications deserve emphasis.

1. Complementary base pairing means that the replicate of each strand is given the base sequence of its complementary strand when DNA replicates.
2. Because the strands are antiparallel, when two nucleotides are paired, the sugar portions of these molecules lie in opposite directions (one upward and the other downward along the chain).
3. Because the strands are antiparallel, the convention for writing the sequence of bases in a strand is to start from the 5′ – P terminus at the left (e.g., GAC refers to a trinucleotide P′-5′-GAC-3′-OH).
4. The conventional way of expressing the base composition of an organism is by the percentage of [G] + [C]. This value is approximately 50 percent for most eukaryotes with only minor variations among species. In simpler organisms, there are significant variations (e.g., 27 percent *Clostridium,* 50 percent for *E. coli,* and 76 percent for *Sarcina,* all of these organisms being bacteria).
5. The chains of the double helix are held together by hydrogen bonds between base pairs in opposite strands. The bond between A and T is a double bond, while the bond between G and C is a triple hydrogen bond.

Structure of RNA

Ribonucleic acids (RNA) are similar in structure to DNA. However, there are significant differences, the key ones being:

1. RNA consists of ribose sugar (in place of deoxyribose) and uracil in place of thymine.
2. Most RNA is predominantly single-stranded (except in some viruses). Sometimes the molecule folds back on itself to form double-stranded regions.
3. Certain animal and plant viruses use RNA as their genetic material.
4. A typical cell contains about 10 times more RNA than DNA.
5. Whereas DNA stores genetic information, RNA most often functions in the expression of the genetic information.
6. There are three major classes of RNA known to be involved in gene expression: **ribosomal RNA (rRNA), messenger RNA (mRNA),** and **transfer RNA (tRNA).** The site of protein synthesis, the ribosome, contains rRNA.

Messenger RNA Structure

Messenger RNA is the molecular carrier of genetic information from the DNA to ribosomes, where the DNA transcript or template (copy of the nuclear DNA, the mRNA) is translated (the genetic information of DNA transcript is expressed) into proteins. Because genes vary in size (number of nucleotides) the mRNA species are variable in length. The structure of mRNA is discussed further in Chapter 3.

Transfer RNA (tRNA) Structure

The structure of transfer RNA (tRNA) is very unique among the three key RNA molecules in the cell. These molecules are small in size and very stable. tRNA molecules range in size from 75 to 90 nucleotides. A single-stranded tRNA molecule is able to fold back onto itself and undergo complementary base pairing in short stretches to form double strands. This folding also creates four characteristic loops and a cloverleaf two-dimensional structure (see Figure 2–11). Of the four loops, three are involved in translating the message of the mRNA. The **anticodon loop** (or simply **anticodon**) consists of a sequence of three bases that are complementary to the sequence of a codon on the mRNA. The stop codons do not have tRNA with anticodons for them. Another feature of the tRNA molecule is the occurrence of the sequence **pCpCpA-3′** at the 3′ end. The

FIGURE 2–11
The tRNA molecule has a cloverleaf structure and consists of a total of 76 bases. Bases 34, 35, and 36 constitute the anticodon (GAA). Bases 74, 75, and 76 (CCA) constitute the amino acid attachment site that occurs at the 3' end.

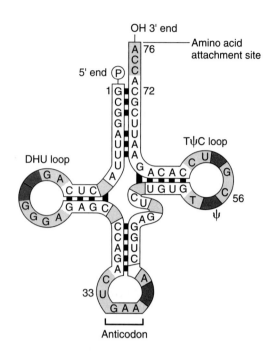

terminal adenine residue is the point of attachment for an amino acid and hence is called the **amino acid attachment** (or **binding) site.** During protein synthesis, the amino acid corresponding to a particular mRNA codon (which base pairs with the tRNA anticodon) is attached to this terminal and transported to the appropriate segment of the mRNA.

Ribosomal Structure

Ribosomes are the sites ("factories") of polypeptide synthesis. A bacterial cell may contain about 1,000 ribosomes. A ribosome consists of two subunits, which together form the **monosome.** The ribosomal particles are classified according to their sedimentation coefficient or rate (S). Monosomes of bacteria are **70S (70S ribosomes)** whereas eukaryotic monosomes are about **80S.** Because sedimentation coefficients are not additive, a 70S monosome in actuality comprises two subunits that are **50S** and **30S,** while an 80S monosome consists of **60S** and **40S** subunits. A ribosome subunit consists of a molecule's rRNA and proteins. For example, the 50S subunit contains one 55 rRNA molecule, one 235 rRNA molecule, and 32 different ribosomal proteins.

KEY CONCEPTS

1. The cell consists of a large number of organic molecules that can be grouped into four generalized categories: simple sugars, fatty acids, amino acids, and nucleotides.
2. The molecules may also be categorized according to size as either micromolecules or macromolecules. There are three groups of macromolecules: proteins, nucleic acids, and polysaccharides.
3. The structure and function of cells are determined primarily by macromolecules.
4. Proteins may have structural roles. They may be substrates, or have catalytic roles in the cell.
5. Genes are expressed as proteins (some code for regulatory factors).
6. Amino acids are the building blocks of proteins. There are 20 commonly occurring amino acids in the cell. They are distinguished by their side chains (R).

7. Proteins have four basic levels of structural organization: primary, secondary, tertiary, and quaternary. The primary structure is a linear polypeptide chain that undergoes various kinds of folding to produce complex conformations.

8. Enzymes are a special group of proteins that catalyze chemical reactions (accelerate reaction rates). This activity is highly specific, in that an enzyme may catalyze only one or a set of closely related reactions.

9. There are two kinds of nucleic acids: DNA and RNA. They are characterized by three basic components: pentose sugar, nitrogenous base, and phosphate group. The difference between the two kinds lies in the sugar, being deoxyribose in DNA and ribose in RNA.

10. There are five nitrogenous bases: adenine (A), guanine (G), cytosine (C), thymine (T), and uracil (U). A and G are called purines, while C and G are pyrimidines. Uracil is the fifth base; it occurs only in RNA in place of thymine.

11. DNA has a double-helical structure with a sugar-phosphate backbone and nitrogenous bases pairing in complementary fashion between the two antiparallel polynucleotide chains. The pairing of bases is highly restrictive and specific, with A pairing only to T, and C to G.

12. Whereas DNA stores genetic information, RNA most often functions in the expression of genetic information.

OUTCOMES ASSESSMENT

Internet based
1. Problems associated with enzymes: *http://web.mit.edu/esgbio/www/chapters.html*
2. Problems with large molecules: *http://esg-www.mit.edu:8001/esgbio/lm/problems.html*
3. Problems on nucleic acids: *http://www.biology.arizona.edu/molecular_bio/problem_sets/nucleic_acids/nucleic_acids_1.html*
4. Large molecules: *http://www.biology.arizona.edu/biochemistry/problem_sets/large_molecules/large_molecules_problems.html*

ADDITIONAL QUESTIONS AND ACTIVITIES

1. Explain why it is at least theoretically possible to undertake horizontal gene transfer among living things.
2. Why are proteins important to living things?
3. Give four major characteristics of DNA.
4. Give all the sources of cellular DNA.
5. Give the major structural difference between plant and animal cells.
6. Discuss the concept of a "universal gene pool."
7. Discuss the importance of the kingdom Monera in biotechnology.
8. Discuss why an understanding of protein and DNA structure is important to biotechnology.

INTERNET RESOURCES

General biology
1. Life hierarchy: *http://web.mit.edu/esgbio/www/chapters.html*
2. Details of taxonomy: *http://www.ultranet.com/~jkimball/BiologyPages/T/Taxonomy.html*

3. Weak bonds: *http://www.accessexcellence.org/AB/GG/weakBonds1.html*
4. Weak bonds: *http://www.accessexcellence.org/AB/GG/weakBonds2.html*
5. Enzyme kinetics: *http://www.ultranet.com/~jkimball/BiologyPages/E/ EnzymeKinetics.html*
6. Mechanism of enzyme activity: *http://www.accessexcellence.org/AB/GG/enzyme.html*
7. Enzyme structure and function: *http://www.ultranet.com/~jkimball/BiologyPages/E/ Enzymes.html*

The cell—structure and function

1. Cell basics: *http://web.mit.edu/esgbio/www/chapters.html*
2. Animal cells, organelles, and their functions: *http://www.ultranet.com/~jkimball/ BiologyPages/A/AnimalCells.html*
3. Plant cells, organelles, and their functions: *http://www.ultranet.com/~jkimball/ BiologyPages/P/PlantCell.html*
4. Cell membrane: *http://www.ultranet.com/~jkimball/BiologyPages/C/CellMembranes.html*
5. Virtual cell: *http://www.life.uiuc.edu/plantbio/cell/*
6. The nucleus—detailed structure and function: *http://www.ultranet.com/~jkimball/ BiologyPages/N/Nucleus.html*
7. Details of chloroplast structure and function: *http://www.ultranet.com/~jkimball/ BiologyPages/C/Chloroplasts.html*
8. Organellar chromosomes: *http://vector.cshl.org/dnaftb/30/concept/index.html*

Cell molecules—proteins

1. Macromolecules in the cell: *http://www.accessexcellence.org/AB/GG/macroMols.html*
2. Families of small organic molecules: *http://www.accessexcellence.org/AB/GG/ small_orgMols.html*
3. Proteins—structure and function: *http://www.ultranet.com/~jkimball/ BiologyPages/P/Proteins.html*
4. Protein structure: *http://www.accessexcellence.org/AB/GG/protein.html*
5. Protein structure: *http://www.accessexcellence.org/AB/GG/prot_Struct.html*
6. Protein structure: *http://web.mit.edu/esgbio/www/chapters.html*
7. Amino acid structure—general structure and individual structures: *http://www.ultranet.com/~jkimball/BiologyPages/A/AminoAcids.html*
8. Polypeptide: *http://www.ultranet.com/~jkimball/BiologyPages/P/Polypeptides.html*
9. Amino acid structure and grouping: *http://www.accessexcellence.org/AB/GG/aminoAcids2.html*
10. Formation of peptide bonds: *http://web.mit.edu/esgbio/www/chapters.html*
11. Amino acids: *http://esg-www.mit.edu:8001/esgbio/lm/proteins/aa/aminoacids.html*
12. Noncovalent bonds: *http://www.ultranet.com/~jkimball/BiologyPages/N/ Noncovalent.html*
13. Rules of protein structure: *http://www.ultranet.com/~jkimball/BiologyPages/D/ DenaturingProtein.html*
14. Binding site of proteins: *http://www.accessexcellence.org/AB/GG/prot_Bindg.html*

Cell molecules—sugars and polysaccharides

1. Sugars and polysaccharides: *http://www.ultranet.com/~jkimball/BiologyPages/C/ Carbohydrates.html*
2. Sugars and polysaccharides: *http://www.accessexcellence.org/AB/GG/sugarTypes2.html*
3. Monosaccharides: *http://www.accessexcellence.org/AB/GG/sugarTypes1.html*
4. Sugars: *http://web.mit.edu/esgbio/www/chapters.html*
5. Fats: *http://www.ultranet.com/~jkimball/BiologyPages/F/Fats.html*
6. Fatty acids, lipids, and steroids: *http://www.accessexcellence.org/AB/GG/fattyAcids2.html*
7. Fatty acids, lipids: *http://www.accessexcellence.org/AB/GG/fattyAcids1.html*
8. Phospholipid structure: *http://www.ultranet.com/~jkimball/BiologyPages/P/ Phospholipids.html*

Cell molecules—DNA and RNA

1. Nucleotides: *http://www.ultranet.com/~jkimball/BiologyPages/N/Nucleotides.html*
2. DNA is the genetic material: *http://vector.cshl.org/dnaftb/17/concept/index.html*
3. Bacterial DNA: *http://vector.cshl.org/dnaftb/18/concept/index.html*
4. DNA and proteins are key cellular molecules: *http://vector.cshl.org/dnaftb/15/concept/index.html*
5. DNA structure: *http://vector.cshl.org/dnaftb/19/concept/index.html*
6. Hydrogen bonds: *http://www.ultranet.com/~jkimball/BiologyPages/H/HydrogenBonds.html*
7. DNA double-helical structure: *http://www.ultranet.com/~jkimball/BiologyPages/D/DoubleHelix.html*
8. Base pairing in DNA: *http://www.ultranet.com/~jkimball/BiologyPages/B/BasePairing.html*
9. DNA is packaged in chromosomes: *http://vector.cshl.org/dnaftb/29/concept/index.html*
10. Nucleotides: *http://www.accessexcellence.org/AB/GG/nucleotide.html*
11. Nucleotides . . . details: *http://www.accessexcellence.org/AB/GG/nucleotide2.html*
12. RNA molecule and its role in heredity: *http://vector.cshl.org/dnaftb/27/concept/index.html*
13. RNA as genetic material of some viruses: *http://vector.cshl.org/dnaftb/25/concept/index.html*

REFERENCES AND SUGGESTED READING

Alberts, B., D. Bray, J. Lewis, M. Raff, K. Roberts, and J. D. Watson. 1993. *Molecular biology of the cell,* 3rd ed. New York: Garland Publishing, Inc.

Darnell, J. E., Jr. 1985. RNA. *Scientific American, 243* (Nov.): 68–78.

Dickerson, R. E. 1983. The DNA helix and how it is read. *Scientific American, 249* (June): 94–111.

Doolittle, R. F. 1985. Proteins. *Scientific American, 253* (Oct.): 88–89.

Felsenfeld, G. 1985. DNA. *Scientific American, 253* (Oct.): 58–78.

Freifelder, D., and G. M. Malacinski. 1993. *Essentials of molecular biology,* 2nd ed. Boston, MA: Jones and Bartlett Publishers.

Sharon, N. 1980. Carbohydrates. *Scientific American, 243* (Nov.): 90–116.

Watson, J. D. 1968. *The double helix.* New York: Atheneum.

Watson, J. D., and F. C. Crick. 1953a. Genetic implications of the structure of deoxyribose nucleic acid. *Nature, 171*:984.

Watson, J. D., and F. C. Crick. 1953b. Molecular structure of deoxyribose nucleic acid. *Nature, 171:* 737–738.

Weinberg, R. A. 1985. The molecules of life. *Scientific American, 253* (Oct.): 48–57.

The Nature of Living Things: How They Function

PURPOSE AND EXPECTED OUTCOMES

Have you ever wondered how a gene (DNA) directs the cell to perform certain functions? In other words, what is the message of the gene and how is it coded and decoded? Before scientists can manipulate genes, they have to understand gene structure and function. They have to understand the "language of heredity" thoroughly; they have to know the "alphabets" of the language of heredity, how genetic "words" and "sentences" are formed, and how instructions are carried out by the cell. The language of heredity is universal. This is the reason that recombinant DNA technology is possible. Scientists are able to transfer genes from animals to plants because the genetic instruction from an animal can be acted upon by a plant and vice versa. In Chapter 2, we learned that proteins are the workhorses of living systems. In this chapter, we will learn how proteins are synthesized.

In this chapter, you will learn:

1. The central dogma of molecular biology.
2. The genetic code.
3. How the genetic code is translated into proteins.
4. How organisms metabolize food.
5. The opportunities in metabolic pathways for genetic manipulation of organisms.

GENETIC BASIS OF FUNCTION

One of the most important concepts to understand regarding how an organism functions is that its function is an interplay of genetics and the environment. Genes are not expressed in a vacuum but in an environment. We shall learn later in this chapter that genes are expressed as proteins. Furthermore, more than one protein can be expressed from one gene. The particular protein expressed is determined by factors that include the cell's environmental conditions (e.g., temperature, humidity) and the presence of signal transducers and activators of transcription (referenced by the acronym STATs, to be discussed further later in the text).

■ WHAT IS THE CENTRAL DOGMA?

As previously stated, DNA is the genetic material of nearly all life, except RNA-viruses. But just how does the information in the DNA develop into visible traits? In other words, how does the DNA function? The **central dogma of molecular biology** is the concept

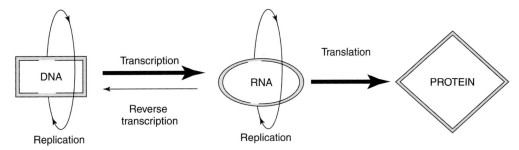

FIGURE 3–1
The central dogma of molecular biology as originally proposed by Watson and Crick.

that information flow progresses from DNA to RNA to protein but not the reverse (see Figure 3–1). Watson and Crick (who discovered the physical structure of DNA) first stated this dogma. However, current knowledge necessitates a modification of this "old" dogma. The "new" dogma should take into account the fact that an organism's environment impacts when and how some of its genes are expressed, and also that more than one protein can be produced by a single gene.

Whereas the central dogma holds true in nature, one of the astonishing facts of modern science is the ability of scientists to create a gene by working backward from its protein product. The technique of reversing the central dogma is the synthesis of a **complementary DNA (cDNA)** using an enzyme called reverse transcriptase. This technique is discussed in detail later in the textbook.

Three types of processes are responsible for the inheritance of genetic information and for its conversion from one form to another. These are **replication, transcription,** and **translation.** But are there significant advantages to the cell for having an intermediate between DNA and the protein it encodes? Certainly. By copying the message and taking it away to the cytoplasm where it is interpreted, the original DNA remains pristine. It is also shielded from chemicals of the cytoplasm. Another advantage to the multi-step flow of genetic information is that the genetic information can be amplified through the production of numerous copies from just one template. Furthermore, by having multiple steps, there are more opportunities for controlling the expression of the gene under different conditions.

■ *DNA Replication*

How does DNA retain its original constitution as a cell divides and increases in number? Replication is DNA's mode of perpetuation, whereby daughter molecules that are identical to the parent DNA molecule are created. This process of duplication fulfills the property of the genetic material to transmit information from parent to progeny. The entire genome of a cell must be replicated precisely once each time a cell undergoes **cell division.** Replication must be executed to a high degree of fidelity in order to preserve genetic continuity between cells following cell division. To this end, the replication process incorporates editing mechanisms to correct errors. In spite of this safeguard, errors do occur, but at a low rate of 1 in 10^9 to 1 in 10^{10} base pairs.

What is the exact mechanism by which replication occurs? To duplicate a double-stranded DNA molecule, the double helix first unwinds so that each strand serves as a template for the synthesis of its complement. This means each replicated DNA molecule would comprise one new strand and one parent strand. This mode of replication is called **semiconservative replication.**

DNA replication starts at an **origin of replication (ori)**. Once started, it continues until the entire genome has been duplicated. The unit or length of DNA that is replicated following one initiation event at one origin of replication is called a **replicon.** Whereas the bacterial chromosome constitutes one replicon, eukaryotic chromosomes contain numerous

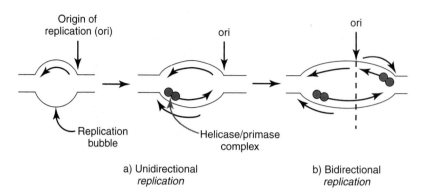

a) Unidirectional
replication   b) Bidirectional
                                    replication

FIGURE 3–2
DNA replication starts at an origin of replication and may proceed in one or two directions on opposite sides of the origin of replication. A bubble is formed as replication continues.

replicons (e.g., 35,000 in faba bean (*Vicia faba*)). The point on the DNA at which replication is occurring is called the **replication fork.** Once replication begins, it may proceed either in one direction (**unidirectional**) or two directions (**bidirectional**) (see Figure 3–2).

DNA replication is underlain by complex enzymatic processes involving a battery of bioactive proteins, notably **DNA polymerases I and III.** It is important to mention that DNA polymerase I and III cannot initiate synthesis of a new strand on a bare single strand. They require a **primer** (an oligonucleotide that is hydrogen bonded to the template strand) with a 3′ -OH group onto which a dNTP (a precursor molecule) can attach. Chain elongation occurs only in the 5′ to 3′ direction by the addition of nucleotides to the growing 3′ terminus.

DNA polymerase I has both 3′ - 5′ and 5′ - 3′ exonuclease activity, which is the ability to pause in the middle of polymerization and remove or excise nucleotides just added to the growing strand. This proofreading or editing function ensures fidelity of replication. In the event an incorrect nucleotide is added that cannot base-pair with the nucleotide in the template, polymerase I is able to remove the unpaired base. Conversely, nucleotides can be removed one at a time, from the 5′ -P terminus and replaced by a process called **nick translation.**

Replication of the duplex DNA may be divided into three stages: **initiation, elongation,** and **termination.** First, the origin of replication must be recognized. The origin of replication of *E. coli* (*ori C*) consists of 245 base pairs. Before replication can start, the double helix will have to unwind and separate. The complex of proteins associated with the elongation stage of replication is called a **replisome.** As this complex moves along the DNA, the two parental strands unwind while new (daughter) strands are synthesized. As already mentioned, the DNA strands are antiparallel. Consequently, as the replication fork moves from 5′ to 3′ on one strand, it moves in the opposite direction (3′ to 5′) on the other. However, nucleic acids are synthesized in one direction only: 5′ to 3′. Therefore one strand, called the **leading strand,** is synthesized continuously from 5′ to 3′ while the other strand, **lagging strand,** is synthesized discontinuously (see Figure 3–3). Discontinuous synthesis involves short fragments that are synthesized in the reverse direction (relative to the fork movement). These fragments are eventually connected to form a continuous strand. The fragments are called **Okazaki fragments,** after their discoverer. This mode of DNA replication is called **discontinuous replication.** Each Okazaki fragment starts with a primer. The primers are removed and replaced by appropriate nucleotides before linking. The nicks between adjacent fragments are sealed by the enzyme DNA ligase.

The mode of DNA replication in prokaryotes is generally applicable to eukaryotes. However, there are significant differences stemming largely from the relatively high

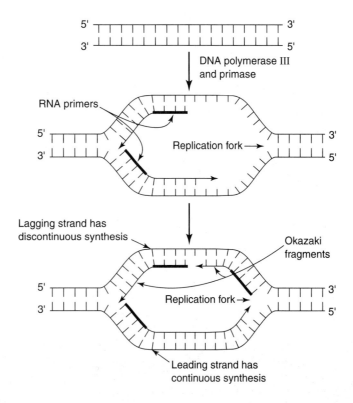

FIGURE 3–3
Replication of the two strands in each bubble proceeds in opposite directions. The leading strand is continuously extended in the direction of the replication fork, while the lagging strand is replicated in a discontinuous fashion, with each fragment (Okazaki fragment) being initiated by a primer.

complexity of eukaryotic DNA organization in the chromosome, and the enormous amount of DNA per cell (50 times that in prokaryotes). As previously stated, because of the enormous amount of DNA, eukaryotic cells have numerous initiation sites. *Drosophila* has about 3,500 replicons per genome, while mammals have about 25,000 replicons per genome, with sizes ranging between 100 to 200 kb. Another distinction between prokaryotic and eukaryotic DNA replication is that, sometimes, the replication of the leading strand is not continuous. This mode is called **semidiscontinuous replication.**

■ *DNA Transcription*

DNA replication is the duplication of a sequence or strand of DNA. Transcription is the process by which the genetic information stored or encoded in the DNA is copied, with the copy being in the form of an RNA molecule.

Which Strand Is Transcribed?

To accomplish transcription, the RNA molecule is synthesized on a DNA template by an enzymatic process that involves several distinct events. First, transcription starts at a specific site to which the appropriate enzyme binds. After this, polymerization is initiated, followed by chain elongation. Finally, the synthesis of the chain comes to an end (chain termination), after which the product (RNA molecule) is released. Binding sites on the DNA molecule are called **promoters.** These are base sequences of about 40 bp in length. The DNA, as described previously, is double stranded. However, only one strand is a

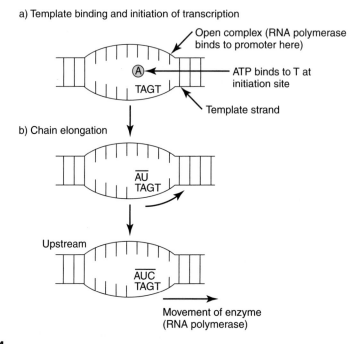

a) Template binding and initiation of transcription

Open complex (RNA polymerase binds to promoter here)

ATP binds to T at initiation site

Template strand

TAGT

b) Chain elongation

AU
TAGT

Upstream

AUC
TAGT

Movement of enzyme (RNA polymerase)

FIGURE 3–4
Transcription starts with the binding of RNA polymerase to the promoter, subsequently creating an open complex. Only one of the two strands (the template strand) is transcribed. ATP binds to T on the template strand. RNA synthesis proceeds in the direction of movement of the RNA polymerase, with complementary nucleotide triphosphates being added as the enzyme moves.

template strand. Nonetheless, both strands are involved in the recognition site interactions. In order for the promoter to function properly, the RNA polymerase must recognize and bind to it properly. The first nucleotide triphosphate to be positioned is usually a purine (ATP or GTP). Consequently, the first base to be transcribed is usually T (thymine) or C (cytosine) (see Figure 3–4). Since only one of the two DNA strands is transcribed, the strand that has base pairs complementary to the first nucleotide triphosphate to reach the template strand (also called the **antisense strand**) is selected for transcription. The other strand is the **sense** or **coding strand.**

The Product of Transcription

Messenger RNA is the product of transcription, that is, the RNA synthesized from the template DNA. It is the mRNA that is decoded to determine the amino acid sequence of a DNA strand. The term **cistron** is used to refer to a segment of DNA corresponding to a polypeptide, including the start and stop sequences. When an mRNA codes for one polypeptide, it is called a **monocistronic mRNA.** Frequently, prokaryotic mRNA is **polycistronic,** encoding several different polypeptide chains or different genes. When this happens, the polycistronic mRNA may be interspersed by sequences called **spacers.** Furthermore, prokaryotic mRNA is short lived, degrading within a few minutes after synthesis.

Transcription in eukaryotes is complicated by their genetic organization. Eukaryotic cells have compartmentalized and membrane-bound subunits called organelles, one of the most distinct being the nucleus. The DNA in the chromosomes is tightly bound to nucleoproteins (histones), forming a complex structure called **chromatin.** This structure is altered to permit the transcription of specific segments. Another significant difference is that transcription in eukaryotes occurs in the nucleus, and then the mRNA is transported out of the nucleus into the cytoplasm for translation.

Regulation of Transcription

Two types of regulatory sequences located upstream from the point of initiation are involved in the stimulation and initiation of gene transcription in eukaryotes. Specific proteins called **transcriptional factors** that facilitate the binding of RNA polymerase II recognize these sequences (promoters and enhancers). These proteins are indispensable because, unlike prokaryotes, RNA polymerase cannot bind directly to eukaryotic promoters. Most promoters (sequences that literally promote and regulate DNA transcription or expression of the gene) have a sequence comprised of repeats of thymine (T) and adenine (A) nucleotides (called the **TATA box**) that is usually located about −25 base pairs upstream of the start point. However, many housekeeping genes do not always contain TATA boxes (or **Goldberg–Hogness boxes,** after their discoverers). There are other promoter modules such as the CCAAT and GC boxes (which contain the sequence GGGCGG). It should be mentioned that none of these sequences is uniquely essential for promoter function. However, the choice of starting point depends on the TATA box, and promoters that lack TATA boxes usually lack unique start points. Promoters and their associated transcriptional factors control the degree of transcription initiation and consequently the amount of transcription of the corresponding gene. **Enhancers** (sequences that increase the transcriptional activity of genes) interact with promoters to increase the rate of transcription initiation. Enhancers vary in their location, and may be found upstream, downstream, or even within the gene. A promoter, however, may be operationally defined as a sequence (or sequences) of DNA that must be in a relatively fixed location with regard to the start point of transcription.

Post-transcriptional Processing

Are the products of transcription conceptually (regarding readiness for translation) the same in prokaryotes and eukaryotes? No, they are not. Whereas the mRNA produced from DNA transcription in prokaryotes may be submitted directly for translation into amino acids, the initial product of transcription in eukaryotes is a **pre-mRNA.** It requires considerable processing to produce a native mRNA. This is because of the lack of a direct relationship or colinearity between a gene and its product. Most eukaryotic genes are interspersed by non-coding sequences. The coding sequences are called **exons** while the non-coding intervening sequences are called **introns.** Introns are removed by the process called **splicing.** Introns present in a single pre-mRNA derived from a single gene may be spliced in more than one way, consequently yielding different collections of exons in the mature mRNA. This mechanism is called **alternative splicing** and yields different but related proteins (called **isoforms**) upon translation (see Figure 3–5). Isoforms occur widely in nature, including in humans.

You will later learn that this phenomenon makes scientific attempts to identify and characterize all the proteins produced by genes very challenging (see **proteomics**). Does one gene encode one protein as previously thought? It appears this is no longer true. In terms of application, several key terminologies have been introduced in this section that are commonly used in biotechnology research. These are primers and promoters. Scientists artificially synthesize primers for specific purposes (e.g., in PCR technology). Likewise, promoters may be extracted from certain organisms and redesigned to modify the expression of a gene as desired by the researcher (e.g., strong promoters can induce overexpression of a gene product that is normally present in small amounts).

■ TRANSLATION OF mRNA (PEPTIDE SYNTHESIS)

How is a chain of nucleotides (an mRNA transcript) interpreted to produce gene products? Translation is the process by which genetic information, in the form of base sequences that were transcribed from the DNA as mRNA, is converted into an amino acid sequence and then polymerized into a polypeptide chain and, eventually, protein. In short, translation converts nucleic acid language to protein language.

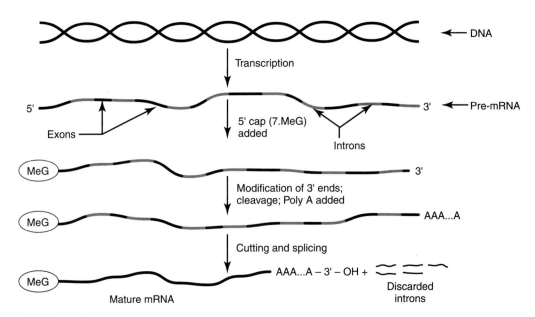

FIGURE 3–5

Eukaryotic DNA is interspersed with non-coding regions (introns). Introns are removed from the primary transcript (pre-mRNA) during RNA processing and coding pieces (exons) are joined to produce the mature mRNA. The primary transcript is capped prior to being released for post-transcriptional modification.

In translation, the nucleic acid language (based on only four "alphabets"—A, T, C, and G—representing the four nitrogenous bases of DNA) is converted into the polypeptide language (based on 20 amino acids as building blocks). A triplet of bases called a **codon** encodes an amino acid. The collection of triplet codons that specify specific amino acids is called the **genetic code.** There are 20 amino acids; hence, there must be more than 20 codons (counting the signals for starting and terminating the synthesis of a protein). On the basis of a triplet code, there are 64 possible codons (see Figure 3–6). In translating mRNAs, the codons do not overlap, but are read in a sequence.

The reading of the genetic code begins at a fixed point in the gene and proceeds sequentially without any interruption by "punctuation." The **reading frame,** as it is called, is set by an initiator codon. When there is an interruption in the sequence by a mutation (either through insertion or deletion of a nucleotide pair(s), the consequence is a shift in the reading frame (causing a **frameshift mutation**) (see Figure 3–7). A frameshift mutation creates new codons that are read to specify a new polypeptide, if the new codons are not nonsense codons (which signal the termination of translation). The consequence of frameshift mutations depends on the portion of the polypeptide affected by the alteration. If the affected portion is relatively nonfunctional, the polypeptide may retain a relatively normal function.

There is redundancy or degeneracy in the code (i.e., most amino acids, except methionine and tryptophan, have more than one codon). However, there is no ambiguity (i.e., none of the multiple codons for a specific amino acid represent any other amino acid). Furthermore, three codons, UAA, UAG, and UGA, are signal codes for termination of translation (nonsense codons) called **stop codons.** However, only one codon, AUG, can signal for the initiation of a polypeptide chain (thereby called a **start codon**). It should be pointed out that this codon doubles as a signal for initiation as well as for the encoding of the amino acid methionine. Furthermore, the codon GUG is used to initiate synthesis of certain polypeptides in certain organisms. The genetic code is universal in the sense that a particular codon is translated as the same amino acid in all forms

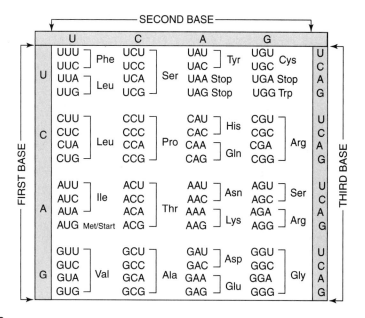

FIGURE 3–6

The genetic code showing the triplet combinations and their corresponding amino acids. The triplets UAA, UAG, and UGA are stop codons, while AUG (and rarely GUG) is a start codon. The genetic code is degenerate; hence, several different triplets may code for the same amino acid.

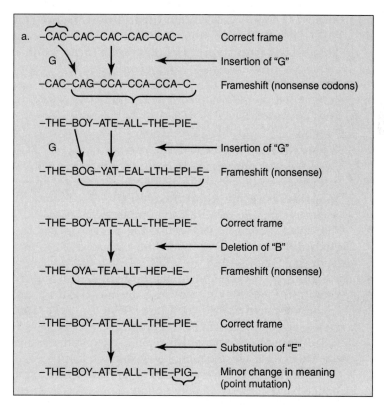

FIGURE 3–7

Frameshifts occur as a result of mutation by way of substitution, deletion, or insertion of a nucleotide. Such events cause a new set of triplet codes to be created with consequences that differ based on the role of the altered gene in the cell. Even a small change in the original property of the gene product, as occurs in point mutation, may be disastrous to the organism (e.g., sickle cell anemia).

of life. However, there are minor deviations such as those that occur in mitochondria and chloroplasts.

Peptide synthesis is akin to putting words into sentences. It entails the linking up of monomers (amino acids) into polymers, and involves three steps: initiation, elongation, and termination. The process in prokaryotes is as follows.

Initiation

The initiation of translation involves only a 30S (small ribosome subunit), an mRNA molecule, a specific charged initiator tRNA, guanosine 5'-triphosphate (GTP), Mg^{2+}, and at least three proteins called **initiator factors.** The linking of a tRNA to its respective amino acid is called **charging.** The tRNA so linked is said to be **charged** or **acylated** (because the enzyme involved in charging is **aminoacyl tRNA synthetase**). Each charging event is catalyzed by its unique enzyme (i.e., there are at least 20 different enzymes and 20 different tRNA molecules). The exception is where the process involves the conversion of the amino acid to an activated form by reacting with ATP to form an **aminoacyladenylic acid.** Even though theoretically there are 61 codons, there are not 61 tRNAs and enzymes. It has been determined that because of the redundancy in the genetic code, only the first two bases are critical to coding for a specific amino acid. The third base is not critical to defining the amino acid (i.e., ABC, ABD, and ABE all would normally correspond to the same amino acid). Francis Crick proposed the **wobble hypothesis** to explain this situation. Because of wobbling of the third amino acid, there are now believed to be at least 32 different tRNAs and 20 enzymes involved in the charging of a tRNA.

The initiation codon (**AUG**) binds to an initiator tRNA that is changed into a formylated amino acid called **formylmethionine** (formylated methionine) to form **fMet-tRNA.** The binding involves a unique base sequence of AGGAGGU (called the **Shine-Dalgarno sequence**) that occurs near the initiation codon. This purine-rich sequence forms base pairs with a complementary region of the 16S rRNA of the small ribosome. The fMet-tRNA binding is facilitated by another initiation factor. The resulting 30S preinitiation complex is joined by a 50S subunit to produce a **70S initiation complex** (see Figure 3–8).

Elongation

Once the initiation complex sets the reading frame, the stage is set for translation of the mRNA. The 50S subunit contains two sites for charged tRNA molecules, designated **P (peptidyl) site** and **A (aminoacyl) site.** The initiation tRNA (fMet-tRNA or tRNA^{fmet}) occupies the P site, while an appropriate tRNA (one whose anticodon can base pairs with the adjacent codon) may occupy the A site. Once both sites are occupied, the enzyme **pedtidyl transferase** catalyzes the formation of a peptide bond between the two amino acids (see Figure 3–9).

Once the bond is formed, the P-site tRNA leaves the ribosome while the A site tRNA, carrying the growing polypeptide chain, moves to the P site (called **translocation**). Translocation frees up the A site for a new tRNA. The release and shifting of the entire **mRNA-tRNA-aa$_2$-aa$_1$ complex** by precisely three nucleotides is facilitated by a number of proteins called **elongation factors.** The series of events are repeated, producing a growing polypeptide chain. In *E. coli*, the rate of elongation is about 15 amino acids per second at 37° C. In terms of efficiency, the error rate is about 10^{-4} (i.e., an incorrect amino acid will be incorporated once in every 20 polypeptide chains of an average length of 500 amino acids).

Termination

Polypeptide elongation is brought to an end when a stop codon (UAG, UAA, or UGA) is encountered. These codons (also called **termination codons** or **nonsense codons**) prevent the A site from being occupied by a tRNA. They rather cause GTP-dependent proteins called **release factors** to act to cleave the polypeptide chain at its point of attachment to

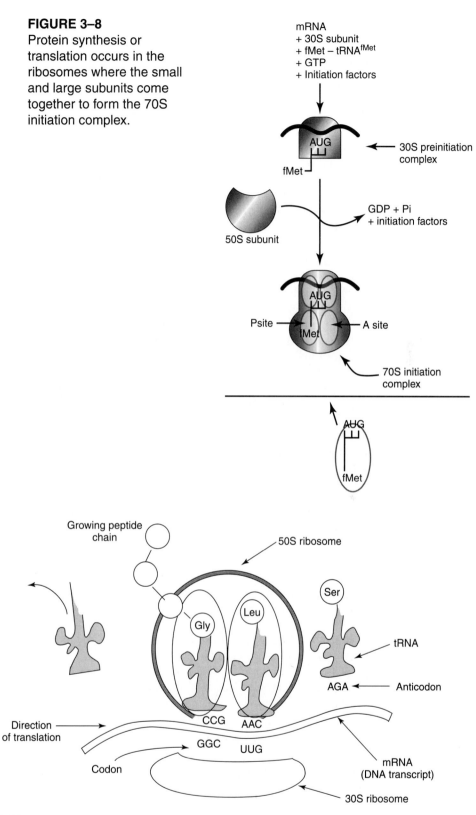

FIGURE 3–8
Protein synthesis or translation occurs in the ribosomes where the small and large subunits come together to form the 70S initiation complex.

mRNA
+ 30S subunit
+ fMet – tRNA^{fMet}
+ GTP
+ Initiation factors

30S preinitiation complex

AUG

fMet

50S subunit

GDP + Pi
+ initiation factors

AUG

Psite

fMet

A site

70S initiation complex

AUG

fMet

Growing peptide chain

50S ribosome

Ser

Gly

Leu

tRNA

AGA

Anticodon

Direction of translation

CCG

AAC

GGC   UUG

mRNA (DNA transcript)

Codon

30S ribosome

FIGURE 3–9
The ribosomes move along the mRNA in the 3' direction. The tRNA is charged with the correct amino acid and proceeds to the A site of the ribosome complex where the amino acid is peptide-bonded to the adjacent amino acid in the P site. The tRNA is translocated after delivering its amino acid. The process is repeated, leading to the elongation of the polypeptide chain.

the tRNA occupying the P site. The polypeptide chain is released, and the ribosome machinery is dissociated into its subunits. The mRNA is also degraded. Should a stop codon occur prematurely (i.e., in the middle of an mRNA as a result of a mutation), the translation is likewise terminated prematurely.

Polyribosome

Once a portion of mRNA has been translated and has exited the ribosome, it is free to be subjected to a new round of translation. This means that various parts of the mRNA may be translated simultaneously, each with its ribosome complex. The resulting structure is called a **polyribosome** or simply **polysome.** This mechanism of translation, in which a number of ribosomes are activated simultaneously, is the norm in all cells.

Translation in Eukaryotes

The general mechanism of translation is the same for all organisms. However, there are certain key differences. A mature eukaryotic mRNA is capped at the 5′ end. This cap is implicated in a role similar to that of the Shine-Dalgarno sequence in prokaryotes. Poor binding, and consequently poor initiation, results in mRNA being uncapped. The second difference is that there is no need for the amino acid formylmethionine. However, the recognition codon for initiation, AUG (which encodes the amino acid methionine), is still needed. Similarly, the starting tRNA is the unique **tRNA^{met}.** Just like prokaryotes, a host of proteins facilitate the three stages of translation (initiation, elongation, and termination). The numbers and complexity of these factors are greater in eukaryotes.

CELLULAR METABOLISM

The term **anabolism** is used to describe the synthesizing reactions that occur in the cell. The energy locked in anabolic products is made available through **respiration,** a destructive process that breaks down complex food substances (proteins, carbohydrates, and fats) into simpler compounds to release stored energy in the chemical bonds. The term **catabolism** is used to describe the destructive reactions in the organism's physiology. The two sets of reactions, anabolism and catabolism, are collectively called **metabolism.** The cell is thus a metabolic machine. Respiration produces energy and reducing agents used by the organism for constructing new materials and for maintaining existing structures that result in growth and development. Metabolic pathways offer tremendous opportunities for scientists to use biotechnology tools to influence the growth and development of an organism. Blocking a critical step in a biochemical pathway is a strategy used by scientists in developing pesticides and certain medicines. As you will learn later, scientists created a new metabolic pathway in rice to enable it to produce pro-vitamin A (refer to the discussion of golden rice).

■ *PHOTOSYNTHESIS*

Except for a few bacteria that derive their energy from sulfur and other inorganic compounds, life at any level depends on photosynthesis, the process by which plants use light energy to synthesize food molecules from carbon dioxide and water. The sugars produced may be further converted to structural materials and other cell constituents. Over 90 percent of dry matter of crop plants depends on photosynthetic activity. Crop yield depends on the rate or duration of photosynthesis. In the ecosystem, plants are called **primary producers** because they are the ultimate source of food for all life on earth.

Can a plant manufacture food in any other part, apart from the leaf? The leaf is the principal plant organ of eukaryotes in which photosynthesis occurs. The assimilation process occurs in the chloroplasts, which are found mainly in leaves. These are plastids that contain the green pigment **chlorophyll.** Chlorophyll is an enzyme and a protein, making it the most abundant enzyme or protein on earth. There are different types of chlorophyll. **Chlorophyll *a*** occurs in all photosynthesizing eukaryotes. However, vascular plants and bryophytes also require **chlorophyll *b*** for photosynthesis.

The Light Phase of Photosynthesis

What materials do plants require for photosynthesis? How do these materials combine to produce food products usable by plants? In what ways do humans involuntarily contribute to photosynthesis? The general reaction for photosynthesis is:

$$6CO_2 + 12H_2O \rightarrow C_6H_{12}O_6 + 6O_2 + 6H_2O$$

Carbon dioxide is obtained from the air while plants obtain water from the soil. The ultimate result of photosynthesis is the reduction of carbon dioxide to carbohydrate. In this regard, photosynthesis can be discussed with reference to the fate of carbon dioxide in three critical steps:

1. Entry of carbon dioxide into plant organs (especially leaves) by the process of diffusion. This is a physical activity.
2. Harvesting light by photochemical processes.
3. Reduction of carbon dioxide (assimilation) by biochemical processes.

Furthermore, the processes in these steps may be classified into two categories or phases, one set involving light (**light-dependent reactions**) and the other not needing it (**light-independent reactions**).

Harvesting light is a photochemical process. The ultimate source of energy in the ecosystem is sunlight. Visible light constitutes only a small proportion of the **electromagnetic spectrum** (the range of wavelengths emanating from the sun). The plant pigments associated with photosynthesis absorb light energy. Each of these pigments has a specific **action spectrum** (the relative effectiveness of different wavelengths of light for a specific light-dependent process such as photosynthesis or flowering). The various pigments absorb a range of wavelengths of light (**absorption spectrum**).

The light-absorbing pigments are packed in the thylakoids in discrete units called **photosystems (PS).** There are two photosystems, **photosystem I (PS I)** and **photosystem II (PS II).** In a photosystem, only a pair of chlorophyll molecules (called the **reaction center chlorophyll**) out of all the pigments present is capable of utilizing the light energy absorbed. The other pigments form the antenna pigments. The reaction center chlorophyll in PS I is a special molecule of chlorophyll *a* called **P_{700}** because it has a peak absorption at 700 nanometers. Photosystem II, also called **P_{680}** (for similar reasons), is more efficient in light harvesting than PS I. The high energy absorbed is transferred in a relay fashion from one photosynthetic pigment to another until it reaches the reaction center molecule (see Figure 3–10). The light energy excites electrons of the photosystem. The energy differential as the electrons return to a lower energy level is utilized in forming chemical bonds. The process of accomplishing this is called **photophosphorylation** (the reaction by which ATP (adenosine triphosphate) is formed from ADP (adenosine diphosphate) and phosphorus). Electrons are excited from P_{700} simultaneously as from P_{680}. As the P_{700} electrons return to lower energy via a photon gradient, they reduce coenzyme $NADP^+$ to NADPH.

The light reaction can be summarized by the following chemical equation:

Light quanta + $2H_2O$ + $2NADP^+$ + $3ADP$ + $3Pi \rightarrow 2NADPH + 2H + 3ATP + 3H_2O + O_2$

The ATP is the usable energy. Along with the reducing potential (NADPH), these two products enter the next phase of the photosynthetic cycle—light-independent reaction or dark reaction.

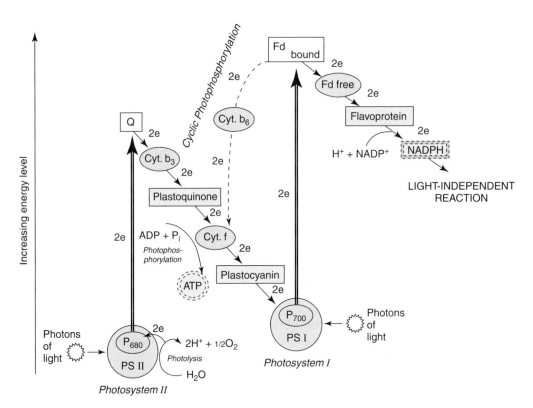

FIGURE 3–10

A summary of the light reactions of photosynthesis. The light-absorbing pigments are organized into two discrete units, photosystems I and II.

The Dark Phase of Photosynthesis

The final step in photosynthesis involves biochemical processes that result in the **fixation** or **assimilation of carbon dioxide,** using the energy (ATP) from the light-dependent reaction (photochemical) and the reducing agent (NADPH). These processes occur by one of two basic pathways, depending on the plant species. Based upon the first stage product of the pathway, the two processes are called **C_3** and **C_4 pathways.** As will be discussed later, one of the major strategies of herbicide design is to target one of these pathways.

C_3 Pathway

The first stable product of the C_3 pathway is a three-carbon compound called **phosphoglycerate** (see Figure 3–11). The C_3 pathway is called the **Calvin Cycle,** after its discoverer, and plants using this pathway for carbon dioxide assimilation are called **C_3 plants.** C_3 plants include both monocots (e.g., barley, wheat, rice, and oats) and dicots (e.g., soybean, pea, peanuts, and sunflower). Tuber crops such as potatoes also utilize this route for carbon dioxide assimilation. The assimilation process is catalyzed by an enzyme called **ribulose 1,5 bisphosphate carboxylase** (or simply **rubisco,** or **RuBP**). The overall reaction is represented by the following chemical equation:

$$6CO_2 + 12NADPH + 12H^+ + 18ATP \rightarrow 1 \text{ glucose} + 12NADP^+ + 18ADP + 18 \text{ Pi} + 6H_2O$$

C_4 Pathway

The first stable product resulting from this pathway is a four-carbon molecule called **oxaloacetate** (see Figure 3–12). **C_4 plants** include corn, sorghum, sugar cane, and millet. These plants produce two types of chloroplasts. One type occurs in the parenchymatous

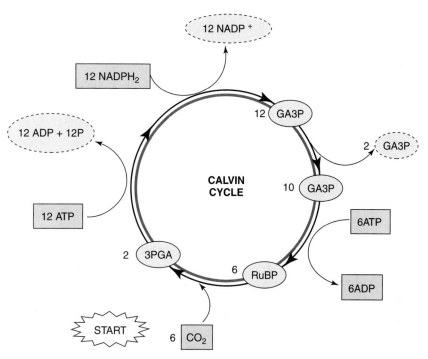

FIGURE 3–11
The Calvin Cycle or the C_3 pathway of carbon fixation. The first stable product in this pathway is GA3P (glyceraldehydes-3-phosphate).

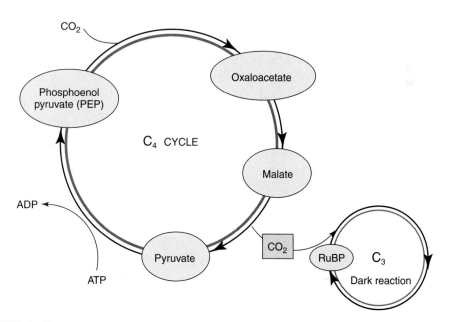

FIGURE 3–12
The C_4 pathway of carbon fixation. The first stable product of this pathway is a four-carbon substance.

tissue around the bundle sheath, and are large in size. Small chloroplasts occur in the mesophyll cells where phosphoenol pyruvate (PEP) carboxylase catalyzes the assimilation of carbon dioxide into PEP, which then is transported to the chloroplasts in the bundle sheaths. Malate is converted to pyruvate, releasing carbon dioxide in the process. The carbon dioxide enters the dark reaction cycle.

A third pathway in carbon dioxide assimilation, the **Crassulacean acid metabolism (CAM),** operates by fixing carbon dioxide in the dark. It depends on carbon dioxide that accumulates in the leaf at night. This pathway is utilized by many houseplants.

■ CELLULAR RESPIRATION

Cellular respiration is the process by which active cells obtain energy from food. All active cells undergo respiration. Respiration occurs continuously, 24 hours a day, in active cells, whether or not photosynthesis occurs simultaneously in the same cell. The harvested energy is stored in the chemical bonds of special energy molecules called ATP. Cellular respiration, like photosynthesis, is a reduction-oxidation (redox) reaction.

Photosynthesis occurs in chloroplasts; respiration occurs in mitochondria. Respiration may be described as the reverse of photosynthesis. It usually requires oxygen to occur and is thus called **aerobic respiration.** However, under certain conditions, respiration may occur in an oxygen-deficient environment and is called **anaerobic respiration.** In crop production, the producer should strive to avoid conditions that cause plants to respire anaerobically. However, silage-making is dependent upon anaerobic respiration. Similarly, in biotechnology, many food applications are based on the anaerobic respiration of microbes (e.g., wine, yogurt production).

Stages of Aerobic Respiration

Just how do cells extract the energy locked up in organic molecules? The general reaction involved in aerobic respiration may be summarized by the equation:

$$C_6H_{12}O_6 + 6O_2 \rightarrow 6CO_2 + 6H_2O + energy$$

There are three stages in aerobic respiration, **glycolysis, Kreb's (tricarboxylic acid) cycle,** and **electron transport chain.**

Glycolysis

Glycolysis or "sugar splitting" involves the breakdown of glucose into **pyruvic acid** (see Figure 3–13). It is a 10-step process that occurs in the cytosol. This stage in aerobic respiration may be summarized by the equation:

glucose + $2NADH^+$ + 2ADP + 2Pi $\rightarrow$ 2pyruvic acid + 2NADH + $2H^+$ + 2ATP + $2H_2O$

It should be mentioned that most steps in the glycolytic pathway are reversible (e.g., glucose can be synthesized from fructose-6-phosphate that has been dephosphorylated). The first half of glycolysis uses ATP. ATP and NADH are made in the second half of glycolysis. Most of the energy of glucose remains locked up in pyruvic acid until the next stage.

Krebs Cycle

The pyruvic acid enters the Krebs cycle (or TCA cycle) in the mitochondrion (see Figure 3–14). The summary of the reactions (called **oxidative decarboxylation**) is as follows:

Oxaloacetic acid + acetyl CoA + ADP + Pi + $3AND^+$ + FAD $\rightarrow$ oxaloacetic acid + $2CO_2$ + CoA + ATP + 3NADH + $3H^+$ + $FADH_2$

In both glycolysis and Krebs cycle, chemical energy from different substrates is used to bond phosphate to ADP to make ATP by the process called **substrate-level phosphorylation** that does not involve a membrane.

Electron Transport

The electron transport chain is the final step in the respiration process. It involves the transfer of electrons generated in glycolysis and Krebs cycle downhill in an electron gradient. The energy released is used to form ATP from ADP in a process called **oxidative phosphorylation** (see Figure 3–15). In oxidative phosphorylation, a series of oxidation-reduction reactions of the electron transport chain makes the cell use the energy in

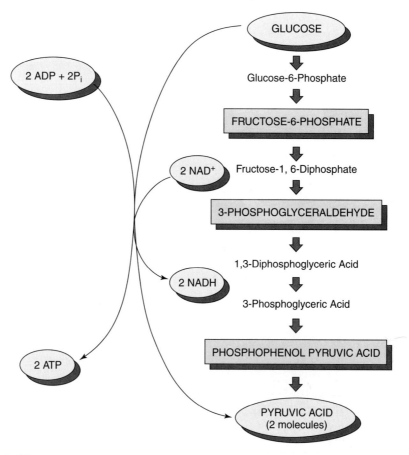

FIGURE 3–13

A summary of glycolysis, the initial phase of all types of respiration. Glucose is converted to pyruvic acid without involving free oxygen.

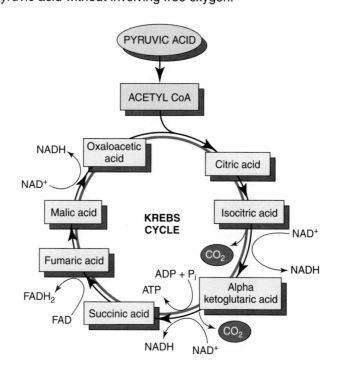

FIGURE 3–14

A summary of Krebs cycle or the tricarboxylic acid cycle. The cycle always begins with acetyl CoA, which is its only real substrate.

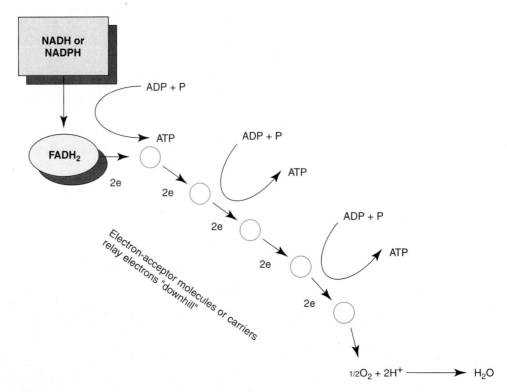

FIGURE 3–15
Electron transport involves the flow of electrons in an energetically downhill fashion, resulting in energy release for the formation of ATP.

NADPH and ubiquinol to phosphorylate ADP to ATP. Oxygen is required only at the end of the electron transport chain as an electron acceptor. It is the strongest acceptor in the chain. Lack of oxygen inhibits both the electron transport chain and the TCA cycle in glycolysis. At the end of one cycle of aerobic respiration, one molecule of glucose yields a net of 36 ATPs.

Anaerobic Respiration

Under conditions of oxygen deficiency, the end product of glycolysis is not pyruvate but **ethyl alcohol** (ethanol), a breakdown product of pyruvate. This process is also called **fermentation** (see Figure 3–16). Unless they have aerenchyma cells to adapt them, many plants ferment when grown in mud or water depleted of oxygen. The anaerobic pathway is less efficient than the aerobic, yielding only two ATP molecules per molecule of glucose.

GROWTH AND DEVELOPMENT

Growth is a progressive and irreversible process that involves three activities—cell division, **enlargement,** and **differentiation.** Cell division occurs by the process of **mitosis** followed by cytoplasmic division. Cells enlarge when they take in water by **osmosis** (the diffusion of water through a differentially permeable membrane from a region of higher concentration to one of a lower concentration). The water status of an organism is critical to its growth and development.

   Development is the term used to describe the continuing change in the form and function of the organism as it responds to environmental factors. It involves the coordination, growth, and longevity of new somatic (body or nonreproductive tissue) and reproductive

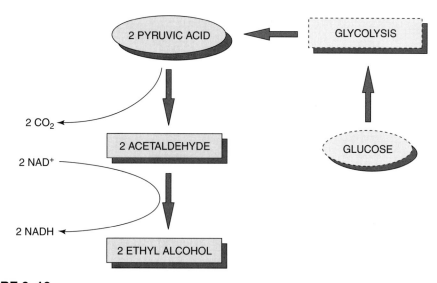

FIGURE 3–16
A summary of the process of fermentation. This anaerobic respiratory pathway is inefficient in yielding useful energy for the cell.

parts. Growth in animals occurs throughout the organism. However, growth in plants occurs at growth points where certain specialized cells (called **meristems**) occur. Meristematic cells have the capacity (called **totipotency**) to form any tissue of the organism. Meristematic cells also occur in animals (see stem cells). Some cells are not completely totipotent and cannot form all tissues of the organism. Meristematic cells are very important in biotechnology. They provide scientists with a tool to study the development of the organism.

Newly produced cells from the division of meristematic cells undergo change and specialization through differentiation (cells and tissues are assigned their adult structure and function), a process called **morphogenesis.** In tissue culture adult cells are dedifferentiated (lose their assigned structure and function) and then cultured to redifferentiate into the new structures of the new plant. The ability to regenerate a new plant from dedifferentiated (deprogrammed) cells is critical to plant transformation research in biotechnology (refer to the discussion of cell/tissue culture).

In tissue culture, adult cells are deprogrammed through the production of **callus** (a mass of undifferentiated cells). Through appropriate modifications of the culture environment (nutrients, hormones, and temperature), cells in the callus phase are reprogrammed to produce the entire plant or parts of it. Callus is the equivalent of stem cells in animals. These two technologies are discussed further in later chapters.

KEY CONCEPTS

1. The DNA is the genetic material of nearly all life, except RNA-viruses.
2. The central dogma of molecular biology is the concept that information flow progresses from DNA to RNA to protein.
3. Three types of processes are responsible for the inheritance of genetic information and for its conversion from one form to another. These are replication, transcription, and translation.
4. Replication is the mode of perpetuation of DNA whereby daughter molecules that are identical to the parent DNA molecule are created. The process is semiconservative, with each replicated molecule consisting of an old strand and a newly synthesized one.

5. Replication is an enzymatic process, involving DNA polymerases (I, III). These enzymes require a primer to initiate replication. The unit or length of DNA that is replicated following one initiation event at one origin of replication is called a replicon.

6. Replication of the duplex DNA may be divided into three stages: initiation, elongation, and termination.

7. Though similar to prokaryotes, the mode of DNA replication in eukaryotes is generally more complex.

8. The mode of DNA replication is discontinuous in eukaryotes. Each Okazaki fragment starts with a primer.

9. Transcription is the process by which the genetic information encoded in the DNA is copied in the form of an RNA molecule.

10. Only one of the two DNA strands is transcribed. The strand that has base pairs complementary to the first nucleotide triphosphate to reach the template strand (or the antisense strand) is selected for transcription. The other is the sense or coding strand.

11. Three classes of RNAs (mRNA, rRNA, and tRNA) are involved in transcription. Messenger RNA (mRNA) is the product of transcription.

12. The enzymes involved in transcription bind to promoters to initiate the process.

13. Mature mRNAs are not products of eukaryotic transcription. Instead, the pre-mRNAs are processed by splicing to remove introns. There are alternative ways of splicing.

14. Translation is the process by which genetic information, in the form of base sequences that were transcribed from the DNA as mRNA, is converted into an amino acid sequence and then polymerized into a polypeptide chain and eventually, protein.

15. A triplet of bases called a codon encodes an amino acid. The collection of triplet codons that specify specific amino acids is called the genetic code. There are 20 amino acids; hence, there must be more than 20 codons (counting the signals for starting and terminating the synthesis of a protein). On the basis of a triplet code, there are 64 possible codons.

16. Ribosomes are the sites ("factories") of polypeptide synthesis.

17. Peptide synthesis is akin to putting words into sentences. It entails the linking up of monomers (amino acids) into polymers, and involves three steps: initiation, elongation, and termination.

18. Transcription in eukaryotes is accompanied by three different RNA polymerases (unlike only one in prokaryotes).

19. Cells are maintained through the operation of several metabolic cycles, the key ones being photosynthesis (in plants) and respiration.

20. Certain steps in cellular metabolic pathways provide opportunities for scientists to manipulate organisms and to develop biotechnology products (e.g., in response to herbicides).

21. The cell is a metabolic machine. Metabolism consists of two processes: anabolism (synthesis) and catabolism (breakdown).

22. Except for a few bacteria that derive their energy from sulfur and other inorganic compounds, life at any level depends on photosynthesis, the process by which plants use light energy to synthesize food molecules from carbon dioxide and water.

23. The final step in photosynthesis involves biochemical processes that result in the fixation or assimilation of carbon dioxide. These processes occur by one of two basic pathways, depending on the plant species. Based upon the first stage product of the pathway, the two processes are called C_3 and C_4 pathways.

24. Cellular respiration is the process by which active cells obtain energy from food.

25. Not all cells in plants photosynthesize, but all active cells undergo respiration. Respiration occurs continuously, 24 hours a day, in active cells, regardless of whether photosynthesis occurs simultaneously in the same cell.

26. There are three phases in aerobic respiration: glycolysis, Krebs (tricarboxylic acid) cycle, and the electron transport chain.
27. Under conditions of oxygen deficiency, the end product of glycolysis is not pyruvate but ethyl alcohol (ethanol), a breakdown product of pyruvate. This process is also called fermentation.
28. Growth is a progressive and irreversible process that involves three activities—cellular division, cell enlargement, and cell differentiation.
29. Development is the term used to describe the continuing change in the form and function of the organism as it responds to environmental factors.
30. Newly produced cells from the division of meristematic cells undergo change and specialization through differentiation (cells and tissues are assigned their adult structure and function), a process called morphogenesis.

OUTCOMES ASSESSMENT

Internet Resources
1. Problems on central dogma of molecular biology: *http://web.mit.edu/egsbio/www/ chapters.html*
2. Problems associated with replication, translation, transcription, and genetic code: *http://www.biology.arizona.edu/molecular_bio/problem_sets/nucleic_acids/03q.html*
3. Problems associated with recombinant DNA: *http://www.biology.arizona.edu/ molecular_bio/problem_sets/Recombinant_DNA_Technology/recombinant_dna.html*
4. Molecular biology problems: *http://www.life.uiuc.edu/molbio/*
5. Problems on Mendelian genetics: *http://web.mit.edu/esgbio/www/chapters.html*
6. Solving pedigree problems: *http://web.mit.edu/esgbio/www/chapters.html*
7. Problems on glycolysis: *http://web.mit.edu/esgbio/www/chapters.html*
8. Problems on general cell biology: *http://web.mit.edu/esgbio/www/chapters.html*
9. Problems on Mendelian genetics: *http://www.biology.arizona.edu/mendelian_genetics/ mendelian_genetics.html*
10. Cell division, cellular organization: *http://www.biology.arizona.edu/cell_bio/tutorials/ cell_cycle/main.html*

ADDITIONAL QUESTIONS AND ACTIVITIES

1. Compare and contrast gene expression in prokaryotes and eukaryotes.
2. One gene does not always produce one protein. Discuss.
3. Suggest how metabolic pathways can be a basis for developing specific biotechnology products.
4. Suggest how researchers may manipulate gene expression.
5. Discuss the concept of "reading frame."
6. Describe how DNA is transcribed. How is transcription regulated?
7. Describe a typical structural gene.
8. Discuss the "gene code" and the implication of its degeneracy.
9. All codons do not code for amino acids. Explain.
10. Discuss the role of promoters in translation.
11. Describe briefly the process by which green plants manufacture food.
12. What is carbon fixation? Distinguish between C_3 and C_4 plants.
13. Give examples of C_3 and C_4 plants.
14. Discuss the importance of anaerobic respiration in biotechnology.

15. What is totipotency? What is its importance to biotechnology?
16. Discuss the importance of meristematic cells in biotechnology.

INTERNET RESOURCES

Metabolism
1. Cellular metabolism: *http://www.accessexcellence.org/AB/GG/cell_Metab.html*
2. Metabolism: *http://www.ultranet.com/~jkimball/BiologyPages/M/Metabolism.html*

Growth and Development
1. Differentiation: *http://vector.cshl.org/dnaftb/36/concept/index.html*

Photosynthesis
1. Photosynthesis and respiration relationship: *http://www.accessexcellence.org/AB/GG/photo_Resp.html*
2. Introduction to photosynthesis: *http://web.mit.edu/egsbio/www/chapter.html*
3. Light reactions: *http://web.mit.edu/egsbio/www/chapter.html*
4. Light reactions of photosynthesis: *http://www.ultranet.com/~jkimball/BiologyPages/L/LightReactions.html*
5. Dark reactions: *http://web.mit.edu/egsbio/www/chapter.html*
6. Carbon fixation cycle: *http://www.accessexcellence.org/AB/GG/carbFix_cyc.html*
7. The Calvin Cycle: *http://www.ultranet.com/~jkimball/BiologyPages/C/CalvinCycle.html*
8. C_3 versus C_4 plants: *http://web.mit.edu/egsbio/www/chapter.html*
9. Photorespiration and C_4 plants: *http://www.ultranet.com/~jkimball/BiologyPages/C/C4plants.html*
10. CAM plants: *http://www.ultranet.com/~jkimball/BiologyPages/C/ C4plants.html#CAMplants*

Respiration
1. Cellular respiration: *http://www.ultranet.com/~jkimball/BiologyPages/C/CellularRespiration.html*
2. Glycolysis and citric acid precursors: *http://www.accessexcellence.org/AB/GG/glycoCit_prec.html*
3. Cellular energy metabolism: *http://www.people.virginia.edu/~rjh9u/metab1.html*
4. Krebs cycle: *http://www.accessexcellence.org/AB/GG/citric_Cyc.html*
5. Glycolysis: *http://www.ultranet.com/~jkimball/BiologyPages/G/Glycolysis.html*
6. Glycolysis: *http://www.accessexcellence.org/AB/GG/out_Glycol.html*
7. Energy relationships in photosynthesis and respiration: *http://www.ultranet.com/~jkimball/BiologyPages/B/BalanceSheet.html*

REFERENCES AND SUGGESTED READING

Albersheim, P. 1975. The walls of growing plant cells. *Scientific American, 232* (April): 81–95.

Alberts, B., D. Bray, J. Lewis, M. Raff, K. Roberts, and J. D. Watson. 1993. *Molecular biology of the cell,* 3rd ed. New York: Garland Publishing, Inc.

Becker, W. M., and D. W. Deamer. 1991. *The world of the cell,* 2nd ed. Redwood City, CA: Benjamin Cummings.

Beckwith, J., and P. Rossow. 1974. Analysis of genetic regulatory mechanisms. *Annu. Rev. Genet., 8:*1–13.

Bertrand, K. 1975. New features of the regulation of the tryptophan operon. *Science, 189:*22–26.

Blixt, S. 1975. Why didn't Gregor Mendel find linkage? *Nature, 256:*206.

Buratowski, S. 1995. Mechanisms of gene activation. *Science, 270:*1773–1774.

Crick, F. H. C. 1966. Codon-anticodon pairing: The wobble hypothesis. *J. Mol. Biol., 19:*548–555.

Crick. H. H. C. 1962. The genetic code. *Scientific American, 207* (Oct): 66–77.

Cummings, M. R. 1994. *Human heredity: Principles and issues,* 3rd ed. St. Paul, MN: West Publishing.

Denhardt, D. T., and E. A. Faust. 1985. Eukaryotic DNA replication. *Bioessays, 2:*148–153.

Englesberg, E., and G. Wilcox. 1974. Regulation: Positive control. *Annu. Rev. Genet., 8:*219–224.

Freifelder, D., and G. M. Malacinski. 1993. *Essentials of molecular biology,* 2nd ed. Boston, MA: Jones and Bartlett Publishers.

Glick, B. R., and J. J. Pasternak. 1994. *Molecular biotechnology, principles and applications of recombinant DNA.* Washington, DC: ASM Press.

Gregory, R. P. F. 1989. *Photosynthesis.* Glasgow, UK: Blackie.

Klug, W. S., and M. R. Cummings. 1997. *Concepts of genetics,* 5th ed. Upper Saddle River, NJ: Prentice Hall.

Kornberg, A., and T. Baker. 1992. *DNA replication,* 2nd ed. New York: W.H. Freeman.

Lake, J. A. 1981. The ribosome. *Scientific American, 245* (Aug.): 84–97.

Mazia, D. 1961. How cells divide. *Scientific American, 205* (Jan.): 101–120.

Moore, R., and W. D. Clark. 1995. *Botany: Plant form and function.* Boston, MA: Wm. C. Brown Communications, Inc.

Nirenberg, M. 1963. The genetic code. *Scientific American,* (March): 80–94.

Noggle, G. R., and G. J. Fritz. 1983. *Introductory plant physiology,* 2nd ed. Englewood Cliffs, NJ: Prentice Hall.

Olby, R. C. 1985. *Origins of Mendelism,* 2nd ed. London, UK: Constable.

Platt, T. 1981. Termination of transcription and its regulation in the tryptophan operon of *E. coli. Cell, 24:*10–23.

Radman, M., and R. Wagner. 1988. The high fidelity of DNA replication. *Scientific American, 259* (Aug.): 40–60.

Salisbury, F. B., and C. W. Ross. 1992. *Plant physiology,* 4th ed. Belmont, CA: Wadsworth.

Sharp, P. A. 1987. Splicing of messenger RNA precursors. *Science, 235:*766–771.

Umbarger, H. E. 1978. Amino acid biosynthesis and its regulation. *Annu. Rev. Biochem., 47:*533–606.

Watson, J. D. 1963. Involvement of RNA in the synthesis of proteins. *Science, 140:*17–26.

4 The Nature of Living Things: Genetic Behavior

PURPOSE AND EXPECTED OUTCOMES

The structure and function of living things are prescribed and controlled by genetic factors. In Chapter 3, you learned how the message encoded by the DNA (genes) is translated into instructions to make proteins. One of the major activities in genetic engineering is to modify the expression of the gene, increasing it (over expression), or decreasing and even eliminating it (gene silencing).

In this chapter, you will learn:

1. How the nature of genes can be altered.
2. The genetic basis of observed traits.
3. How the expression of a gene can be regulated.

GENOTYPE VERSUS PHENOTYPE

The totality of the genes in an individual is called its **genome.** Scientists also use the term **genotype,** but usually use it to refer to a subset of the genes in an individual. Genes condition the traits of organisms. The alternative forms of a gene are called **alleles.** A **dominant allele** is that which suppresses the expression of the other allele (**recessive allele**) at the same locus (specific location of a gene on a chromosome). Another term, **phenotype,** is used to represent the visible manifestation of a genotype. Because genes are not expressed in a vacuum, a phenotype is the product of the interaction between a gene and its environment (i.e., phenotype = genotype + environment).

NUMBER OF GENES ENCODING A TRAIT

Traits vary in the number of genes by which they are controlled. Traits controlled by one or a few genes are classified as **qualitative traits** or **simply inherited traits,** while traits controlled by many genes are called **quantitative traits** or **polygenic traits.** Many hereditary diseases in humans (e.g., sickle cell anemia) and other organisms are conditioned by single genes. Many traits of importance to plant and animal breeders are polygenic or quantitative traits (e.g., crop yield). They vary more or less continuously among individuals, and consequently are placed in categories according to measured values of the trait (they are hence also called metric traits). Optimal expression of quantitative traits depends on the presence of optimal environmental factors for all the genes that are involved, a condition that is difficult to satisfy. Consequently, changing the growth environment (e.g., through irrigation, fertilization, and temperature control) can change the degree of expression of the trait during plant or animal production, sometimes dramatically. Because of the effect of the environment on their expression, quantitative traits are difficult to breed. To facilitate breeding, scientists search for genetic markers that are linked (tend to be inherited together) to **quantitative trait loci (QTLs).** Techniques of biotechnology have been used to develop more effective genetic marker systems (e.g., RFLP and AFLP, discussed later in the text).

Much of the success of biotechnology in genome manipulation so far has been with single gene traits. Qualitative traits discretely vary among individuals in the population and can be placed in non-overlapping categories by counting (e.g., purple flowers versus white flowers, presence or absence of an effect). Their expression does not vary from one environment to another (e.g., a white flower will always be white, regardless of the environment). Such traits are relatively easier to manipulate genetically.

GENOME VARIATIONS

Each species has a characteristic number of chromosomes (e.g., man has $2n = 46$, corn $2n = 20$). No two individuals are genetically identical (except identical twins or clones). However, genomic variations occur frequently. Agents of such variations include mutations and transposable elements.

■ MUTATIONS

Genetic recombination (gene mixing that occurs primarily through sexual reproduction) is the principal phenomenon by which genes or DNA sequences in the chromosomes are reshuffled. Genetic recombination is the reason that no two individuals in a sexually breeding population are identical. The process of **meiosis** (nuclear division that results in the production of a haploid (n) number of chromosomes) is accompanied by genetic recombination. Recombination entails two basic events: reshuffling of chromosomes and crossing over (physical exchange of parts between adjacent chromosomes) (see Figure 4–1). Genetic recombination is the primary reason why no two individuals are identical (except clones), and provides the genetic variation needed for conventional breeding (discussed later in the text).

Certain events in nature, called **mutations,** can alter the genetic content of an individual. The consequence of a heritable mutation is a change in the meaning of the affected gene (by changing the DNA sequence) and consequently the production of an abnormal phenotype. An organism or gene that is different from the normal type (or **wild type**) is called a **mutant.** Gene mutations are the source of most alleles and hence the origin of much of the genetic variability that occurs within populations.

Mutation is the ultimate source of biological variation, without which natural selection would be hampered. Mutations are indispensable in genetic analysis; they provide the basis for genetic studies. Mutant phenotypes allow scientists to study the genes that control the traits that have been altered. By being aware that the genome can tolerate variations, and by understanding the modes by which they arise, scientists are able to artificially induce genetic changes on purpose. Another important reason that scientists study these phenomena is that, knowing the roles of altered genes, they are able to study the function of the normal forms of genes.

Mutations that arise naturally are called **spontaneous mutations.** However, scientists can also deliberately induce them (called **induced mutations**) by **mutagenesis**

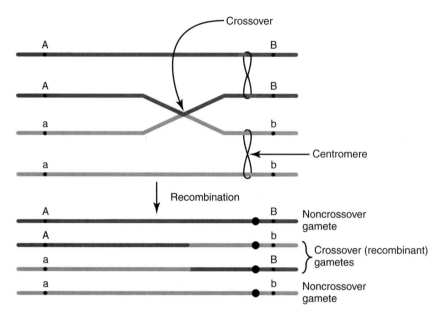

FIGURE 4–1
Following the synapsis (lateral association) of homologous chromosomes, contact is maintained in a region called the chiasma (zone of contact). Subsequently, crossing over occurs whereby homologous chromatids exchange parts at the chiasma. The process of physical genetic mixing is called recombination. The chromatids that exchange parts are called recombinants, whereas the others that remain intact are called nonrecombinants.

TABLE 4–1
Selected physical and chemical mutagens used in inducing mutations for the improvement of organisms. Mutagens differ in their depth of penetration, the type of mutations they induce, and the safety or risk to the operator, among other characteristics.

Physical Mutagens
1. X-rays
2. Gamma rays
3. Beta particles
4. Alpha particles
5. Neutrons
6. Protons

Chemical Mutagens
1. Ethyl methane sulfonate
2. Dimethyl methane sulfonate
3. Colchicine

(the process of inducing mutations). Agents that induce mutations are called **mutagens** and may be physical or chemical in nature (see Table 4–1). It should be pointed out that errors in DNA transcription and translation are not hereditary. However, mutant genes may produce defective proteins that may have adverse physiological consequences on the cell. Spontaneous mutations are assumed to be random events. Mutations, whether spontaneous or induced, are primarily random events. However, the technique of **site-directed mutagenesis** enables researchers to design a mutation and introduce it into the genome at a predetermined site.

Mutations may occur in **gametic** (or **germline**) or in **somatic cells.** Mutations in somatic cells are not heritable. Furthermore, somatic cell mutations that produce recessive **autosomal** alleles (an autosome is a chromosome other than a sex chromosome) have little or no impact on the organism since their expression is likely to be masked by dominant alleles. These mutations may have more impact if they occur early in development before cells become differentiated. However, gametic mutations are heritable. Dominant autosomal mutations in the germline will be expressed in the first generation. However, autosomal recessive mutations may be unexpressed through many generations because of heterozygosity (the deleterious recessive allele is unexpressed because of the presence of a dominant allele at the same locus). Currently, genetic manipulations that involve the germline in humans using U.S. federal funds is prohibited (refer to the discussion of gene therapy).

■ *TRANSPOSABLE GENETIC ELEMENTS*

Genomes are relatively static. However, they evolve, albeit slowly, by either acquiring new sequences or rearranging existing sequences. Genomes acquire new sequences either by the mutation of existing sequences or through introduction (e.g., by vectors or hybridization). Rearrangements occur by certain processes, chiefly genetic recombination and transposable genetic elements.

Transposable genetic elements (**transposable elements, transposons,** or "**jumping genes**") are known to be nearly universal in occurrence. How do genes spontaneously relocate themselves within the genome? These mobile genetic units relocate within the genome by the process called **transposition.** The presence of transposable elements indicates that genetic information is not fixed within the genome of an organism. The Human Genome Project has revealed that about 45 percent of all human genes are derived from transposable elements.

Barbara McClintock, working with corn in the 1940s, was the first to detect transposable elements, which she initially identified as **controlling elements.** This discovery was about 20 years ahead of the discovery of transposable elements in prokaryotes. Controlling elements may be grouped into families. The members of each family may be divided into two classes: **autonomous elements** or **nonautonomous elements.** Autonomous elements have the ability to transpose, whereas the nonautonomous elements are stable (but can transpose with the aid of an autonomous element through *trans*-activation).

McClintock studied two mutations: **dissociation (Ds)** and **activator (Ac).** The *Ds* element is located on chromosome 9. *Ac* is capable of autonomous movement, but *Ds* moves only in the presence of *Ac. Ds* has the effect of causing chromosome breakage at a point on the chromosome adjacent to its location (see Figure 4–2). The *Ac* element has an open reading frame (which has initiation and termination sequences and encodes genetic products). The activities of corn's transposable elements are developmentally regulated. That is, the transposable elements transpose and promote genetic rearrangements at only certain specific times and frequencies during plant development. Transposition involving the *Ac–Ds* system is observed in corn as spots of colored aleurone. A gene required for the synthesis of anthocyanin pigment is inactivated in some cells whereas other cells have normal genes, resulting in spots of pigment in the kernel (genetic mosaicism).

Transposable elements have been discovered in humans and are called the **Alu family** of short interspersed elements. They are so called because these 200 to 300 bp sequences are cleaved by the restriction enzyme (enzyme capable of cleaving DNA sequence), *AluI.* In *Drosophila* (fruitfly), the transposable elements called **P elements** are known to cause hybrid dysgenesis (a syndrome of abnormalities, including sterility, mutations, and distorted meiotic segregation, resulting from a cross of a male carrier of P elements and a female lacking it). The P element transposon has been a valuable tool for molecular geneticists in studying the function and regulation of many genes in the *Drosophila.* P elements are used as insertional mutagens and in germline transformation as vectors.

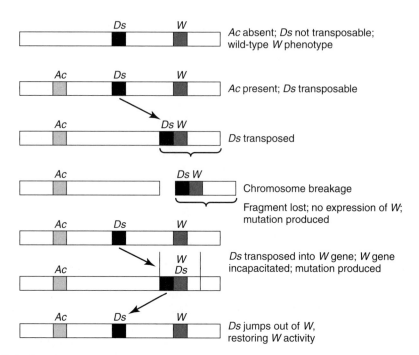

FIGURE 4–2

The effect of the *activator* (*Ac*) element on the *dissociation* (*Ds*) element in the *Ac–Ds* system in maize. In one scenario, transposition of the *Ds* to *W* causes chromosome breakage and subsequently the production of a mutant effect. In the second scenario, the *Ds* is transposed into the *W* gene, causing a mutant to form. It may also relocate out of the *W* to another region, restoring the wild-type phenotype.

REGULATION OF GENE EXPRESSION

Apart from transferring desirable genes into an organism, one of the key ways in which scientists genetically engineer organisms is by manipulating the gene expression. It is important first to understand how natural regulation of gene expression occurs so that scientists may know the opportunities that exist for manipulation. It is not only important for a gene to be expressed, but it is critical that its expression be regulated such that it can be "turned on" or "turned off" as needed. Some genes need to be expressed all the time, while others require expression only some of the time. It is through the regulation of gene expression that cellular adaptation, variation, differentiation, and development occur. As cells are exposed to varying environmental conditions, the kinds and amounts of proteins present are critical to cellular function. An organism should be able to switch from metabolizing one substrate to another.

The underlying principle of gene regulation is that there are regulatory macromolecules that interact with nucleic acid sequences to control the rate of transcription or translation. These macromolecules are usually proteins but can be RNA. One of the earliest successes in the application of biotechnology at the commercial level is the manipulation of gene expression to delay fruit ripening (refer to the discussion of antisense RNA).

■ REGULATION OF GENE EXPRESSION IN PROKARYOTES

How does a cell regulate the process of gene expression? Gene expression can be controlled at different stages that may be grouped into three general categories: transcription, processing, and translation. Transcriptional control is usually effected at the stage of transcription initiation, but not often at elongation. Splicing the RNA primary transcript may be subject to regulation. Finally, gene expression may be regulated at both the initiation and termination stages of translation.

There are several mechanisms for the regulation of transcription. Because enzymes can act in either a synthetic or degradative metabolic pathway, the mechanism of gene regulation used depends on the pathway. The specific molecular events involved in regulation are variable, but generally, there are two basic types: **negative regulation** or **positive regulation.** In negative regulation, the genes are transcribed unless turned off by the regulator protein or the **repressor protein.** For transcription to occur, a molecule called an **inducer** (an antagonist of the repressor) is required. However, in positive regulation, transcription does not occur unless a regulator protein, called an **activator**, directly stimulates RNA production (see Figure 4–3). A system may be regulated positively or negatively. Furthermore, enzyme activity rather than enzyme synthesis may be regulated by a mechanism called **feedback inhibition.** In this mode of regulation an enzyme, though present, is rendered inactive by the products of its catalytic process.

Genes that encode the primary structure of proteins required by the cell for enzymatic or structural functions are called **structural genes.** Most bacterial genes are structural genes, and include rRNA and tRNA. Structural genes tend to be organized into closely linked clusters that are transcribed as a single unit (i.e., coordinately controlled). The mRNA is thus called a **polycistronic mRNA.** An example of such a gene cluster is that involved in the control of lactose metabolism in *E. coli* (see Figure 4–4). The *lacZ* gene codes for the enzyme β-galactosidase, while the *lacY* gene codes for β-galactoside permease. The third gene, *lacA*, codes for β-galactoside transacetylase. When a mutation occurs in either *lacZ* or *lacY*, a genotype is produced in which the individuals are unable to utilize lactose.

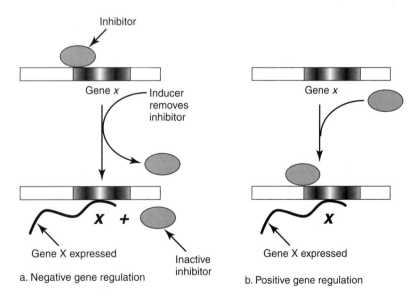

FIGURE 4–3

A comparison of the two basic categories of gene regulation: negative or positive regulation. In negative regulation, an inhibitor that is bound to the DNA must be removed for transcription to occur. In positive regulation, gene transcription occurs when an activator binds to the DNA.

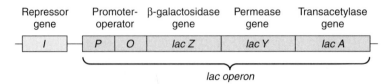

FIGURE 4–4
A sketch of the genetic map of *lac* operon (not to scale), showing the organization of the gene cluster and regulatory units involved in the control of lactose metabolism.

◼ THE OPERON AND OPERON MODEL

Is there a mechanism for the transcription of all genes? The **operon** is a unit of gene expression including structural and regulatory genes. Francois Jacob and Jacques Monod first proposed the concept of the **operon model** of negative control of gene expression in 1961. They proposed that the *lacI* gene regulates transcription by producing a **repressor molecule** that binds to a sequence of DNA called the **operator** (O_{lac}) located between the promoter (P_{lac}) and the cluster of structural genes (*lacZYA*) (see Figure 4–5). The binding of the repressor inhibits RNA polymerase from initiating transcription at the promoter. Furthermore, the repressor molecule is believed to be **allosteric,** or capable of conformational changes that alter both its shape and chemical activity. It has two binding sites, one for the inducer and the other for the operator. When a molecule binds to one site, it changes the conformation of the protein in a fashion that alters the activity on the other side. Consequently, when the inducer (lactose) is added, the repressor undergoes an allosteric conformational change to a form that prevents it from interacting with the operator DNA, which leaves the operator so that the RNA polymerase is able to initiate transcription. This system of control of regulation is called **negative control** because transcription occurs only in the absence of repressor-operator interaction (i.e., failure of the repressor to bind to the operator). However, experimentation involving glucose showed that the control of the *lac* operon can also be positive. It was discovered that polymerase binding is made more efficient through a facilitator or ancillary protein called **catabolic activating protein (CAP),** which has a binding site within the promoter. In the presence of lactose, transcription may still be inhibited if glucose, a catabolic by-product of lactose metabolism, is present (called **catabolic repression**). Furthermore, for CAP to bind, it must be linked to **cyclic adenosine monophosphate (cAMP).** The operon model is supported by genetic evidence and crystallographic studies.

◼ REGULATION OF GENE EXPRESSION IN EUKARYOTES

DNA is universal, but are the mechanisms of gene expression regulation universal? Prokaryotic cells are not compartmentalized. Their genomic organization is much simpler. Prokaryotes are unicellular organisms. Eukaroytic cells are compartmentalized and contain more genomic information than prokaryotes. The genomic organization is more complex in eukaryotes. First, the genetic information is carried on many chromosomes instead of one as in prokaryotes. Then, the chromosomal DNA is packaged tightly in the form of chromatin. Some eukaryotic cells have a large genome comprised of DNA segments that are repeated from hundreds to millions of times. Such repetitive sequences are absent in prokaryotes (except a few repeats in rRNA and tRNA).

Operons in prokaryotes are typically polycistronic (involve translational units that code for multiple proteins that are involved in the same regulatory pathway). However, eukaryotic genes are typically **monocistronic,** with one transcriptional unit encoding one translational unit. Furthermore, eukaryotic genes are typically split, being interspersed with introns that need to be spliced out before a mature mRNA is produced.

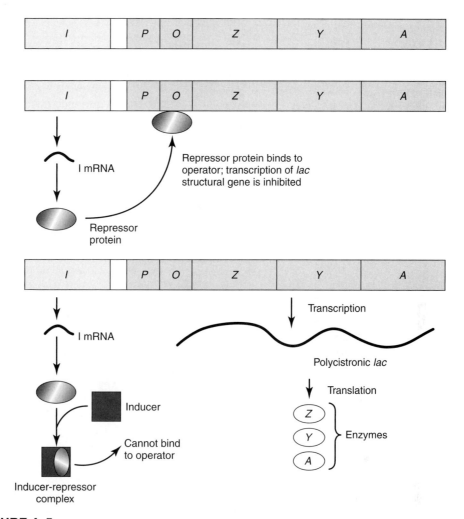

FIGURE 4–5
A sketch of the *lac* operon in its repressed state and then its induced state with the product of the expression of the *lac* operon.

Another difference is that transcription is spatially and temporally separated from translation in eukaryotes. After the complex process of RNA processing, the mRNA must be transported out of the nucleus into the cytoplasm for translation. In multicellular eukaryotes, there is differential regulation of gene expression. Cells often synthesize only a subset of possible gene products even though they contain the complete set of genetic information. These differences between prokaryotes and eukaryotes suggest that gene regulation in eukaryotes would be more complex.

POTENTIAL CONTROL LEVELS OF EUKARYOTIC GENE EXPRESSION

Potential control levels for the regulation of gene expression in eukaryotes occur in the nucleus as well as outside the nucleus (see Figure 4–6). These include (1) regulation of transcription, (2) regulation of RNA processing, (3) regulation of mRNA transport, (4) regulation of mRNA stability, (5) translational regulation, and (6) regulation of protein activity.

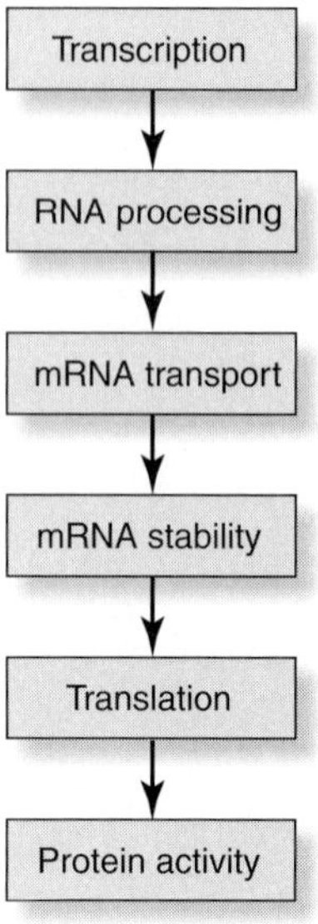

FIGURE 4–6

A summary of the key stages from gene transcription to post-translation, showing the opportunities for scientists to manipulate gene activity.

Transcription

↓

RNA processing

↓

mRNA transport

↓

mRNA stability

↓

Translation

↓

Protein activity

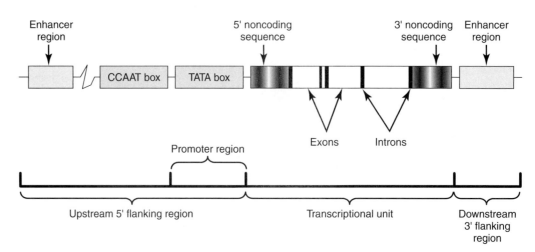

FIGURE 4–7

A sketch of the eukaryotic gene, showing its three regions and their constitution. The factors that control gene expression are located upstream of the transcriptional unit.

■ REGULATION OF TRANSCRIPTION

Gene expression begins with transcription. The level of gene transcription determines whether or not a gene is expressed. The organization of a typical eukaryotic gene and its **transcriptional control** regions is described in Figure 4–7. Transcription by RNA polymerase II is controlled by two types of regulatory sequences—promoters and enhancers. These regulatory elements may be found on either side of a gene, or at some distance from the gene.

In order for transcription to start, certain proteins called transcription factors, which are not part of the RNA polymerase II molecule itself, are needed to help RNA polymerase identify the transcription start site. This step is crucial because the polymerase does not recognize "naked" regulatory sequences (i.e., cannot bind directly to the DNA promoter). Transcriptional factors have several functional domains (clusters of amino acids that carry out a specific function). One domain, called the **DNA binding domain,** binds to DNA sequences present in promoters and enhancers. A second, **trans-activating domain,** activates transcription via protein-protein interaction.

Some transcription factors occur only in one or a few cell types (e.g., *MyoD* is found only in muscle cells). The DNA binding domains have distinctive three-dimensional structural proteins or **motifs.**

In summary, transcriptional regulation of gene expression in eukaryotes is significantly more complex than in prokaryotes. The primary level of regulation occurs at the stage where alterations in chromatin structure occur to allow the binding of transcription factors. In order for transcription to start, one or more factors must bind at the promoter region. Both promoter and enhancer sequences are recognized and bound by transcriptional factors.

■ *REGULATION OF RNA PROCESSING*

Because split genes occur in eukaryotes, splicing of the initial mRNA transcript occurs to remove introns and bring together the exons. Through alternative splicing, many different forms of a protein can be produced from a single gene, depending on the mode of splicing (see Figure 4–8). The diversity in gene products may impact developmental processes. This is found to be the case in *Drosophila,* in which the outcome of splicing a single exon determines the sex of an individual.

■ *REGULATION OF mRNA TRANSPORT*

Transcription and translation are spatially and temporally separated. The mRNA transcripts must be transported through **nuclear pore complexes** into the cytoplasm. Experiments

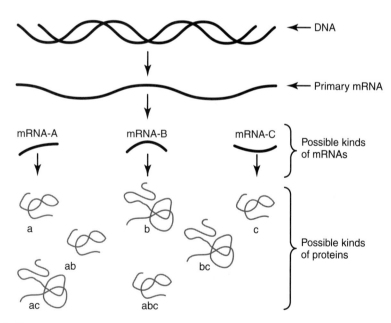

FIGURE 4–8

The introns in the eukaryotic gene must be removed to produce the mature mRNA that is translated into protein. The removal of introns and joining of exons may occur such that different mRNA may result. Consequently, different protein products may be produced from the alternate splicing of the gene. Furthermore, the primary proteins produced from translation may combine in various ways to produce secondary protein products.

with yeast indicate a relationship between splicing and the transport of mRNA. When primary transcripts lacked introns, they were transported immediately after synthesis. However, transcripts with introns were retained until after splicing and capping.

■ REGULATION OF mRNA STABILITY

Mature mRNA has a short life span (called half-life) in prokaryotes, usually degrading within a few minutes. The half-life (the time for 50 percent of the starting amount of mRNA to disappear) of mRNA in the cytoplasmic pool of eukaryotes is variable and characteristic of the mRNA molecule. However, eukaryotic mRNA is generally more stable than that of prokaryotes. The α-globin mRNA has been known to last for more than 10 hours, whereas mRNA stored in oocytes lasts for years. The longer an mRNA remains stable, the more protein it can produce. However, a short half-life allows the cell to respond rapidly to changes in the environment by synthesizing a new set of proteins needed under the new conditions.

Steroid (e.g., estrogen) receptors are known to increase the stability of certain mRNAs. For example, the addition of estrogen increases the half-life of the mRNA for vitellogen (the major proteins in the frog *Xenopus laevis*), by about 30 times. The stability of mRNAs that encode histones vary according to the stage in the cell cycle. Histone proteins are in great demand during the S-phase (DNA replication phase). The mRNAs in the S-phase have a half-life of about one hour. However, upon normal completion or abrupt interruption of replication by some chemical agent, the histone mRNA degrades within minutes.

■ REGULATION OF TRANSLATION

Whereas the exact mechanism is not clear, the available evidence suggests a regulatory mechanism that acts on mRNA in the act of translation. In prokaryotes, where polycistronic transcripts commonly occur, it is important that translation be regulated in order to produce the appropriate amount of each gene product for the specific metabolic pathway. **Embryogenesis** in higher plants is known to be subject to **negative translational control.** Many of the mRNAs synthesized during the development of the oocyte are deposited in the eggs. These so-called maternal mRNAs remain inactive in storage until activated by fertilization.

■ REGULATION OF PROTEIN ACTIVITY

Newly synthesized polypeptides may be taken up by organelles, namely the nucleus, mitochondria, peroxisomes, or chloroplasts. Alternatively, polypeptides may be transported via the endoplasmic reticulum. These constitute the principal pathways for intracellular protein translocation. The regulation of protein activity involves two processes: translocation of proteins and enzymatic modifications. Some modifications are permanent, such as the attachment of prosthetic groups to certain enzymes. Reversible modifications include protein phosphorylation. Phosphorylation can increase or decrease the activity of a specific protein. Other protein modifications include methylation, acetylation, and glycosylation. Experiments have shown that proteins in which the N-terminal amino acid is one of the following amino acids—methionine, serine, threonine, alanine, glycine, or valine—tend to be less vulnerable to protease activity and hence more stable.

KEY CONCEPTS

1. Genetic recombination is the principal phenomenon by which the genes or DNA sequences in the chromosomes are reshuffled.

2. Some traits are conditioned by one or a few genes (qualitative traits, or simply inherited traits), whereas others are conditioned by many-to-numerous genes (quantitative traits, polygenic traits).

3. Mutation is the ultimate source of biological variation without which natural selection would be hampered.

4. On the basis of how mutations arise, they may be classified as either spontaneous or induced mutations. Mutagenesis is the process of inducing mutations.

5. Agents that induce mutations are called mutagens. Mutagens may be physical agents (e.g., X-rays) or chemical agents (e.g., ethylmethane sulfonate (EMS)).

6. Genes can spontaneously relocate themselves in the genome.

7. These mobile genetic units relocate within the genome by the process called transposition. Transposable genetic elements (transposable elements, transposons, or "jumping genes") are widespread in nature.

8. Examples of transposable elements are the *Alu* family in humans and the P elements in *Drosophila.*

9. Gene expression can be controlled at different stages that may be grouped into three general categories: transcription, processing, and translation.

10. Generally, there are two basic types of regulation or control: negative regulation or positive regulation. In negative regulation, the genes are transcribed unless turned off by the regulator protein or the repressor protein. Positive regulation of transcription occurs when a regulator protein, called an activator, directly stimulates RNA production.

11. The operon is a unit of gene expression including structural genes and the regulatory genes.

12. Operons in prokaryotes are typically polycistronic, involving translational units that code for multiple proteins that are involved in the same regulatory pathway. However, eukaryotic genes are typically monocistronic, with one transcriptional unit encoding one translational unit.

OUTCOMES ASSESSMENT

Practice Problems

1. Problems on gene expression: *http://web.mit.edu/esgbio/www/chapters.html*
2. Components of the lac-operon: *http://www.biology.arizona.edu/molecular_bio/problem_sets/mol_genetics_of_prokaryotes/01Q.html*
3. Role of inducer in operon: *http://www.biology.arizona.edu/molecular_bio/problem_sets/mol_genetics_of_prokaryotes/02Q.html*
4. Problems associated with eukaryotic gene expression: *http://www.biology.arizona.edu/molecular_bio/problem_sets/mol_genetics_of_eukaryotes/eukaryotes.html*

ADDITIONAL QUESTIONS AND ACTIVITIES

1. Describe a typical structural gene.
2. How do prokaryotic and eukaryotic structural genes differ?
3. Discuss the concept of the operon.
4. What is the role of promoters in gene expression?

5. Discuss specific ways in which scientists can modify the expression of genes.
6. Distinguish between qualitative and quantitative traits.
7. Describe the *Ac–Ds* transposition system.

INTERNET RESOURCES

Mutations
1. Mutations: *http://www.ultranet.com/~jkimball/BiologyPages/M/Mutations.html*
2. Mutations: *http://www.accessexcellence.org/AB/GG/mutation.html*
3. Mutations: *http://www.accessexcellence.org/AB/GG/mutation2.html*
4. Radiations—physical mutagens: *http://www.ultranet.com/~jkimball/BiologyPages/R/Radiation.html*
5. Ames test: *http://www.ultranet.com/~jkimball/BiologyPages/A/AmesTest.html*
6. Using mice to test for chemical mutagens: *http://www.ultranet.com/~jkimball/BiologyPages/B/BigBlue.html*
7. Repair of mutations: *http://vector.cshl.org/dnaftb/28/concept/index.html*

Transposable Elements
1. Transposable genetic elements: *http://vector.cshl.org/dnaftb/32/concept/index.html*
2. Transposons: *http://www.ultranet.com/~jkimball/BiologyPages/T/Transposons.html*

Gene Regulation
1. Turning genes on and off: *http://vector.cshl.org/dnaftb/33/concept/index.html*
2. Regulation of genes in prokaryotes: *http://web.mit.edu/esgbio/www/chapters.html*
3. Operons: *http://www.ultranet.com/~jkimball/BiologyPages/L/LacOperon.html*
4. The lac operon: *http://web.mit.edu/esgbio/www/chapters.html*
5. Operons: *http://www.ultranet.com/~jkimball/BiologyPages/L/LacOperon.html*
6. The lac operon: *http://web.mit.edu/esgbio/www/chapters.html*
7. Regulation of gene expression in eukaryotes: *http://www.ultranet.com/~jkimball/BiologyPages/P/Promoter.html*
8. DNA response to signals outside the cell: *http://vector.cshl.org/dnaftb/36/concept/index.html*

Cell Division
1. Haploid sex cells: *http://vector.cshl.org/dnaftb/8/concept/index.html*
2. Cell division, growth, and death: *http://vector.cshl.org/dnaftb/38/concept/index.html*
3. Mitosis: *http://www.accessexcellence.org/AB/GG/mitosis2.html*
4. Meiosis: *http://www.accessexcellence.org/AB/GG/meiosis.html*
5. Comparison of mitosis and meiosis: *http://www.accessexcellence.org/AB/GG/comparison.html*
6. Crossing over: *http://www.accessexcellence.org/AB/GG/comeiosis.html*
7. Crossing over: *http://www.ultranet.com/~jkimball/BiologyPages/C/CrossingOver.html*
8. Chromosome concept: *http://www.accessexcellence.org/AB/GG/chromosome.html*
9. Details of mitosis and meiosis: *http://www.ultranet.com/~jkimball/BiologyPages/M/Mitosis.html*
10. Details of the cell cycle: *http://www.ultranet.com/~jkimball/BiologyPages/C/CellCycle.html*
11. Mitosis: *http://vector.cshl.org/dnaftb/7/concept/index.html*
12. Cell structure, division, and other cellular activities: *http://www.cellsalive.com/*

REFERENCES AND SUGGESTED READING

Ames, B. N., J. McCann, and E. Yamasaki. 1975. Method for detecting carcinogens and mutagens with the Salmonella/mammalian microsome mutagenicity test. *Mut. Res., 31:*347–364.

Auerbach, C., and B. J. Kilbey. 1971. Mutations in eukaryotes. *Annu. Rev. Genet., 5:*163–218.

Beckwith, J., and P. Rossow. 1974. Analysis of genetic regulatory mechanisms. *Annu. Rev. Genet., 8:*1–13.

Bertrand, K. 1975. New features of the regulation of the tryptophan operon. *Science, 189:*22–26.

Bragdo, M. 1955. Production of polyploids by colchicines. *Euphytica, 4:*76–82.

Buratowski, S. 1995. Mechanisms of gene activation. *Science, 270:*1773–1774.

Carter, P. 1986. Site-directed mutagenesis. *Biochem. J., 237:*1–7.

Cohen, S. N., and J. A. Shapiro. 1980. Transposable genetic elements. *Scientific American, 242* (Feb):40–49.

Drake, J. W. 1970. *Molecular basis of mutation.* San Franciso: Holden-Day.

Englesberg, E., and G. Wilcox. 1974. Regulation: Positive control. *Annu. Rev. Genet., 8:*219–224.

Glick, B. R., and J. J. Pasternak. 1994. *Molecular biotechnology, principles and applications of recombinant DNA.* Washington, DC: ASM Press.

Klug, W. S., and M. R. Cummings. 1997. *Concepts of genetics,* 5th ed. Upper Saddle River, NJ: Prentice Hall.

Knudson, A. G. 1979. Our load of mutations and its burden of disease. *Am. J. Hum. Genet., 31:*401–413.

McClintock, B. 1956. Controlling elements and the gene. *Cold Spring Harbor Symp. Quant. Biol., 21:*197–216.

Newcombe, H. B. 1971. The genetic effects of ionizing radiation. *Adv. Genet., 16:*239–303.

Platt, T. 1981. Termination of transcription and its regulation in the tryptophan operon of *E. coli. Cell, 24:*10–23.

Stoskopf, N. C. 1993. *Plant breeding.* San Francisco, CA: Westview Press.

Umbarger, H. E. 1978. Amino acid biosynthesis and its regulation. *Annu. Rev. Biochem., 47:*533–606.

5

Principles of Genetic Manipulation of Organisms: Conventional Approach

You may wonder why we need to manipulate organisms. Why can't we leave them alone like they were created? Humans, by nature, are creative beings. Society has evolved from the Stone Age to the Computer Age because of the creative genius of humans. The quality of life is what it is today because humans continue to exercise creativity, seeking new ways of doing things, and making good things better. This is not to say that everything created by man has been good. As they say, to err is human.

Science is all about discovery and the application of knowledge. Much of what scientists do, they learn by observing nature! Then, after understanding how things are done (the underlying laws and principles), scientists take the bold step of nudging nature to the advantage of humans. Instead of waiting for millions of years for evolution to produce its effect, breeders can intervene to bring about a desirable product in our lifetime.

Numerous reasons can be cited for manipulating organisms, depending on the organism and purpose of manipulation. Some of these reasons are discussed throughout the book, especially in sections on the applications of biotechnology. Organisms are genetically manipulated to improve their performance to satisfy the growing needs of modern society that cannot be met by organisms in their native state. Methods of manipulation can be grouped into two broad categories: conventional and genetic engineering. The two strategies are important and relevant to genetic manipulation of organisms. However, emphasis in this section will be on genetic engineering, using the plant system as the primary example. The principles are essentially the same for animal systems.

PURPOSE AND EXPECTED OUTCOMES

Breeders, plant or animal, are scientists who genetically manipulate organisms to fulfill specific purposes. Organisms can be manipulated in nongenetic ways to perform differently by modifying their production environment. This is what agronomists and animal producers do. The principles of genetic manipulation of plants and animals are essentially the same. The specific methodologies are variable, even within either group. Animals reproduce sexually. Plants can be propagated either sexually or asexually, depending on the species. The emphasis in this module is on the sexual propagation of organisms. Furthermore, the discussion excludes humans, since humans cannot be manipulated through breeding, for ethical and other social reasons.

In this chapter, you will learn:

1. The fundamentals of plant and animal improvement.
2. The limitations of conventional methods of plant and animal improvement.

THE BASIC PRINCIPLE OF GENETIC MANIPULATION

Breeders are not able to genetically manipulate every trait. Therefore, it is important for them to determine whether or not a trait can be genetically manipulated before embarking on an improvement program. An underlying concept in making this decision is embodied in the following equation:

$$P = G + E$$

where P = phenotype, G = genotype, and E = environment. Simply stated, what is seen is a product of the interaction of the genotype with its environment. To change the phenotype, the genes that code for the trait may be changed (e.g., through crossing to bring about genetic recombination, or mutagenesis), the environment may be changed, or both factors may be changed. Changing the environment is done through agronomic practices (e.g., irrigation, fertilization, and pest control). Changing the genotype is permanent (heritable); changing the environment is only temporary, meaning that the conditions must always be reintroduced for the trait to be expressed. It should be emphasized that a variety of a crop or breed of an animal is only as good as the environment in which it grows, since genes are not expressed in a vacuum. Investing in a high-yielding hybrid is useless, unless the proper conditions are provided for the optimal expression of the genes for attainment of high **heterosis** (hybrid vigor). Consequently, both the genotype and the environment are important in crop or animal improvement.

Another concept of importance to breeders is **heritability,** which is especially important when considering the improvement of a quantitative trait. Heritability is the degree of phenotypic expression of a trait that is under genetic control. Mathematically, it is expressed by the following equation:

$$H = V_g/V_p$$

where V_g = genetic variance and V_p = phenotypic variance. In this form, the formula estimates **heritability in the broad sense.** The values of this estimate range between 0.0 and 1.0 (or 0.0 and 100 percent). If the estimate of heritability is high, it indicates that the success of manipulating to improve a trait through breeding is likely. Otherwise, the trait may be enhanced through improving the cultural environment during crop or animal production.

GENERAL STEPS IN BREEDING

The general steps in a breeding program are as follows:

a. Determine the breeding objective(s).
b. Assemble genetic variability (or heritable variation).
c. Recombine the variation (cross or hybridize).
d. Select desirable recombinants.
e. Evaluate the selections.

■ BREEDING OBJECTIVES

A breeding program is initiated for a specific purpose or objective. This may be yield increase, disease resistance, improved quality (e.g., high protein content), herbicide tolerance, and others as determined by the breeder. The method used for breeding depends on the objective (the trait to be manipulated and the direction of manipulation). Some objectives are hence producer-oriented, facilitating the production process or improving the yield of the commercial product. These improvements usually bring about increased income to the producer. For example, pest resistance reduces the need to use expensive pesticides, whereas high-yielding cultivars produce higher returns per acre of crop. Higher milk production likewise brings additional income to the farmer. It should be pointed out that breeding environmentally responsive crops (crops that respond to production inputs like fertilizers) also benefits the agrochemical industry through increased purchases of their products by producers. Consumer-oriented breeding goals target the consumer first and bring profits to the producer as a result of increased purchases of products that meet customer needs. Examples include increased nutritional quality (e.g., high protein) and increased quality of industrial product (e.g., high oil content).

■ HERITABLE VARIATION

All variation is not heritable. Some variations are caused by differences in the environment. Without heritable variation it is not possible to conduct a breeding program, that is, to genetically manipulate plants by conventional methods. If a breeder desires to increase the protein content of an existing crop variety, there must exist somewhere a plant with high protein. Otherwise, such a trait must be induced, if possible, by mutation. Variability can be assembled from various sources including introductions (importation) from outside the breeding area.

■ RECOMBINATION

The conventional way of creating variation is through **hybridization** or the crossing of two different parents. In animals, mating is effected by introducing the desired sperm donor to the female at the right time. In plants, pollen grains from the desired source are deposited on the stigma of a receptive female plant. Pollination or mating is followed by fertilization and subsequently development into an embryo. The effect of this action is the reorganization of the genomes of the two parents into a new genetic matrix to create new individuals expressing traits from both parents. The ease of crossing or mating varies from one species to another.

Animals

The conventional way of producing hybrids in animals is to mate selected parents. This requires the physical presence of both parents and a physical act of mating. However, another procedure that has become routine in breeding animals such as cows, horses, and goats is called **artificial insemination (AI).** This procedure entails the artificial deposition of semen obtained from a male donor into the reproductive tract of a female. The semen may be fresh or frozen (if transported from a distant location). The main advantage of AI is that a male donor can be used in breeding irrespective of location. Semen from overseas may be imported for local breeding efforts. Other advantages include the elimination of the cost of maintaining a bull, buck, or other male sperm donor; prevention of transmission of infection; reduction in risk of injury to both parents during mating; ability to use desirable males that are unable to physically mate; and an increase in the rate of genetic improvement. However, it should be cautioned that artificial insemination is not a cheap option to traditional breeding by mating animals.

Plants

In plants, pollen from one parent is transferred to the stigma of the other parent by agents of pollination (e.g., wind, insect). In self-pollinated species where plants utilize pollen from the same plant, the breeder usually has a more difficult time with crossing. The flower of one plant must be designated as male and the other as female. The female flower is usually rid of all male organs by the often-tedious process of **emasculation.** The breeder then physically deposits the desired pollen on the stigma of the female parent. However, plants like corn that are naturally cross-pollinated, and thus can utilize pollen from other sources, may not need emasculation. The breeder simply plants the parents next to each other for pollen transfer to occur naturally by agents of pollination. Hybrid production is described in Figure 5–1, which illustrates crop breeding. The example shown in the figure is a more modern method in which a **male sterility gene** has been used to make one parent male infertile (or female), thereby eliminating the need for emasculation.

It is critical that a cross be authenticated (certified as a successful cross) before it is used to continue a breeding program. This is especially critical when crossing self-pollinated species. At the very least, a tag is used to identify the emasculated flower that becomes artificially pollinated. The seed from the putative cross can be further evaluated provided a genetic marker is incorporated in the breeding program (see Figure 5–2). Genetic markers are discussed elsewhere in detail.

There are different methods of breeding, depending on the breeding objective, the number of genes involved in the trait being improved, the species, and the resources available, among other factors.

■ SELECTION

After crossing, the breeding program proceeds with a series of selections (genetic discrimination) of desirable recombinants (individuals that show the desirable combination of traits for which the breeding program was initiated). The way selection is conducted depends on the method of breeding. Plant breeding programs are generally long duration, requiring 7 to 10 years and even longer in some cases. The breeding method illustrated in Figure 5–3 shows why. After crossing, the breeder needs to produce a large enough segregating population of plants from which to search for the recombinant individual that has the combination of the desired attributes of the two parents. The F_2 generation is the most heterogeneous, with heterozygosity decreasing by 50 percent in each subsequent generation (see Figure 5–4). Selection for the desired recombinant starts in the F_2. The minimum number of F_2 plants to produce for a chance of the presence of each possible genotype is given by the formula 4^n, where n = the number of heterozygous loci. The number of different genotypes in the F_2 is given by 3^n. Plant breeders typically plant 2,000 to 10,000 plants (and sometimes more) in the F_2.

It is clear from the foregoing information that breeding quantitative traits requires the planting of a large number of plants in the segregating population. Consequently, the search for the desirable recombinant is more challenging. The use of markers facilitates the breeding program by helping breeders readily identify recombinants by association with these markers. Conventional markers are detected by their gross phenotypic expression in the adult stage of the plant. The marker concept is discussed later, where the superiority of molecular markers over conventional markers will be made clear.

LIMITATIONS OF CONVENTIONAL BREEDING

Conventional breeding is beset by the following weaknesses:

a. **Long duration.**
   The breeding program lasts for several to many years in some cases.
b. **Limited to crossing within species.**
   To hybridize, the parents must be compatible and belong to the same species (occasionally, crosses between different species are possible, though problematic).
c. **Lower selection efficiency.**
   The methods used to sort among the enormous variation generated from a cross in the case of plants is not precise. This is the reason that markers are used to improve breeding efficiency.

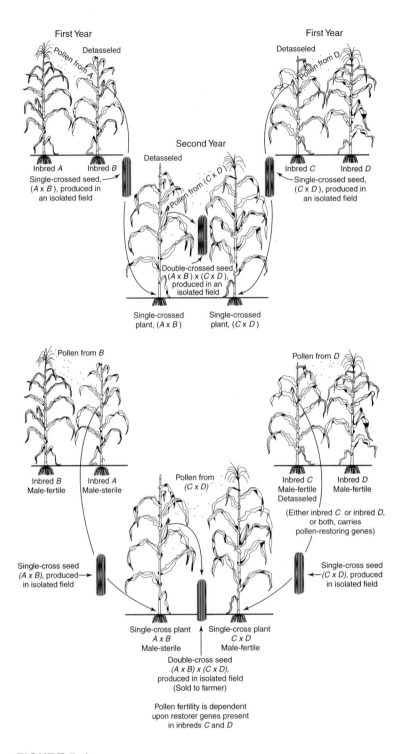

FIGURE 5–1

An illustration of the process of hybrid production in corn. One parent is designated as male (pollen source), and another the female. Each parent is selfed repeatedly to produce an inbred line that is highly homozygous. A cross between inbred lines A × B produces AB, called a single cross. A cross of two products of single crosses AB × CD produces ABCD, called a double cross. The use of male sterility removes the need of the tedious task of detasselling or emasculating (making a plant female) in breeding.

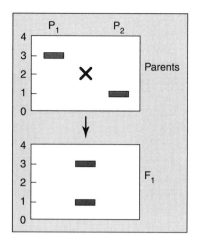

FIGURE 5–2
Genetic markers are used in plant breeding to facilitate the process of selection. Molecular markers can be assayed at any stage in the plant growth cycle. Isozymes are codominant markers. A cross between P_1 and P_2 should produce two bands following electrophoresis and the visualization process. The appearance of the two bands authenticates the hybrid.

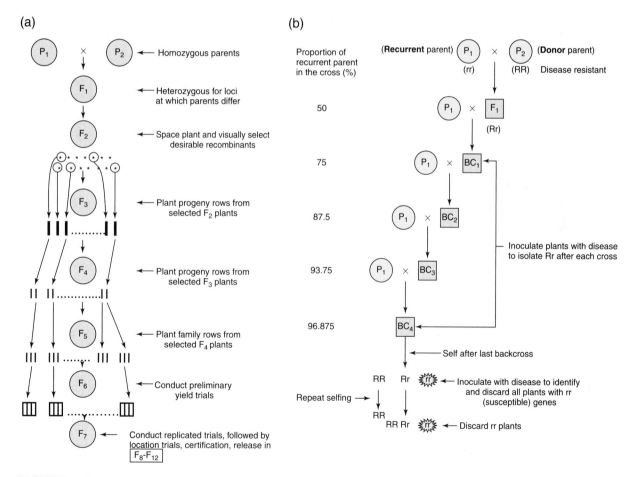

FIGURE 5–3
There are many types of plant breeding methods. (a) The pedigree method starts with a cross that is followed throughout the breeding program by maintaining records of lineage. The method is "exclusive" in that desirable crosses are retained while all others are discarded from one stage to the next. (b) In backcross breeding, the strategy is to conserve the genotype of one of the original parents (the recurrent parent) but incorporate only a specific gene or two from the donor parent. To achieve this, the product of the cross at each stage is crossed back to the recurrent parent. The donor parent on the other is used only once in the breeding program.

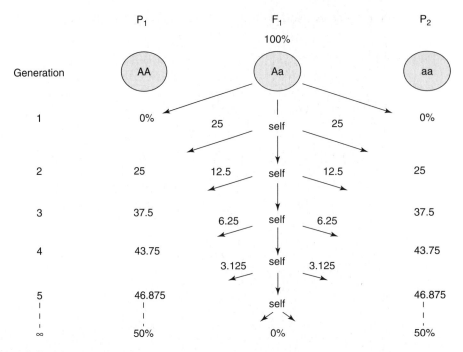

FIGURE 5–4
The effect of repeated selfing. Heterozygosity is reduced by 50 percent with each round of selfing but never completely exhausted in practice in a breeding program. The breeder decides the acceptable level of homozygosity and quits the breeding program at the appropriate generation.

  d. **Large segregating population.**
    In order to have a high chance of identifying the recombinant of interest, plant breeders usually plant large numbers of plants in the segregating population. This requires large amounts of space, and thus increases breeding expense.

  e. **Only genes for traits expressed within a restricted gene pool are accessible.**
    For example, you cannot transfer genes for traits from a dog to a cow, or from corn to pepper.

WHAT ABOUT PLANTS THAT DO NOT REPRODUCE SEXUALLY?

An advantage of asexual propagation of crops is the elimination of the gene mixing characteristic of sexual reproduction. Heterosis (hybrid vigor) is preserved if a plant is asexually propagated. Many plants that are agriculturally produced by asexual methods have the capacity to flower. However, because they do not breed true (i.e., significant variability exists from one generation to the next), sexual reproduction is used primarily to introduce the variability needed for breeding. In species that do not produce flowers, the conventional method of breeding such plants is through induced mutations. Plant tissue is exposed to an appropriate mutagen to induce mutations in somatic cells. These altered cells are nurtured to full plants. In some sophisticated procedures, cells from different parents may be fused *in vitro* to create new hybrids.

THE IMPORTANCE OF CONVENTIONAL BREEDING TO BIOTECHNOLOGY

The advantages of recombinant DNA technology will be discussed in Chapter 6. Whereas biotechnology has certain distinct advantages over conventional methods of plant and

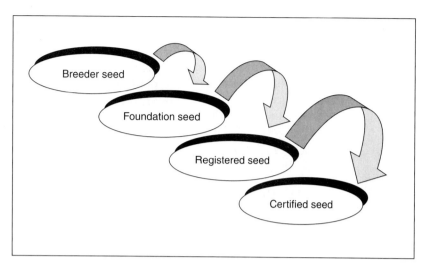

FIGURE 5–5
A summary of the seed certification process. After a plant breeder develops a new cultivar, it goes through several steps of seed increase before the commercial seed is produced for farmers. The seed that is marketed for farmers is called the certified seed.
Specific agencies are charged with certifying newly developed crop seed.

animal improvement, it should be emphasized that both approaches work together to
produce the finished product for the consumer. The customary process for the release of
a new breed of animal or crop cultivar (cultivated variety) is via the procedures of conventional breeding. The new gene introduced by biotechnology will have to be transferred into ecotypes for use in the different production systems. This will entail crossing
the biotechnology-produced plant with adapted varieties from the various regions of intended use. This will be followed by a number of field testings over years and locations.
Finally, before a farmer can obtain genetically improved seed for planting, it will have to
be subjected to a cultivar release process as outlined in Figure 5–5. Producers have access
to certified seed only.

KEY CONCEPTS

1. The conventional genetic manipulation of plants and animals is generally called
   plant or animal breeding. The basic strategy employed is to use the sexual mechanism to reorganize the genomes of two individuals in a new genetic matrix, and
   select for individuals in the progeny with the desirable combination of the parental
   characteristics.
2. The underlying principle in breeding is embodied in the equation P = G + E (phenotype = genotype + environment). Should you desire to change the phenotype,
   you may change the genotype (e.g., introduce genes by crossing; permanent or heritable change), change the environment (as done by agronomists; temporary
   change), or both.
3. A hybrid is the cross between two unrelated individuals.
4. Heritability is the degree of phenotypic expression of a trait that is under genetic
   control.

5. A breeding program is initiated for a specific purpose or objective. This may be an increase in yield of grain or carcass, disease resistance, improved nutritional quality (e.g., high oil or protein content, or low fat in meat animals), and others as determined by the breeder.

6. Without heritable variation it is not possible to conduct a classical breeding program. For example, if a breeder desires to increase the protein content of the seed of an existing cultivar or the milk of a cow, there must exist somewhere a genotype or breed with high protein in these parts or products that can be used as a parent in the cross.

7. Conventional breeding methodologies are limited by their long duration, need for sexual compatibility, low selection efficiency, and restricted gene pool.

8. Plants that do not flower are genetically improved through mutagenesis.

OUTCOMES ASSESSMENT

1. Without heritable variability, you cannot conduct a classical breeding program. Explain.
2. Breeders cannot successfully improve all traits by breeding methods. Explain.
3. How are breeding objectives decided?
4. Discuss how breeders and agronomists or animal producers work together to develop new crop cultivars or animal breeds.
5. Describe how hybrid corn is produced.
6. Explain how conventional breeding programs are typically long in duration.
7. Discuss the advantages of using artificial insemination in animal breeding.
8. Discuss the concept of heritability and its importance in the breeding of plants or animals.
9. P = G + E. Explain.
10. Explain why plant breeding usually takes a long time to complete.
11. Give the limitations of conventional breeding.
12. Discuss the role of genetic recombination in breeding plants or animals.

INTERNET RESOURCES

1. Genetic recombination: *http://vector.cshl.org/dnaftb/11/concept/index.html*
2. Corn breeding: *http://www.plant.uoguelph.ca/research/corn_breeding/*
3. Corn breeding: *http://corn.agronomy.wisc.edu/FISC/Corn/Breeding/ CornBreedingBiotechnology.htm*
4. Soybean breeding: *http://www.agron.iastate.edu/soybean/soybreed.html*
5. Bull semen and embryo: *http://www.bull-semen.com/*
6. Diverse information about breeding animals: *http://www.genaust.com.au/technologies.htm*

REFERENCES AND SUGGESTED READING

Hallauer, A. R. 1990. Methods used in the development of maize inbreds. *Maydica,* 35:1–16.

Richey, F. D. 1950. Corn breeding. *Advances in Genetics, 3:*159–192.

Shands, H. L., and L. E. Wiesner. 1991. *Cultivar development part I. Crop Science Society of America Special Publication No. 17.* Madison, WI: American Society of Agronomy.

Sprague, G. F., and S. A. Eberhart. 1977. Corn breeding. In *Corn and Corn Improvement.* Sprague, G. F. (ed). Madison, WI: American Society of Agronomy.

Stoskopf, N. C. 1993. *Plant breeding: Theory and practice.* San Francisco: Westview Press.

6 Principles of Genetic Manipulation of Organisms: Recombinant DNA (rDNA) Technology

PURPOSE AND EXPECTED OUTCOMES

In Chapter 5, you learned how plant and animal breeders conduct their business, and their limitations in doing so. A key limitation is the restriction on gene transfer. Recombinant DNA (rDNA) technology allows scientists to transfer genes from one organism to any other, circumventing the sexual process. For example, a gene from a bacterium can be transferred to corn. Consequently, rDNA technology allows scientists to treat all living things as belonging to one giant breeding gene pool.

In this chapter, you will learn:

1. The basic steps in genetic engineering or rDNA technology.
2. The enabling technologies of genetic engineering.
3. The importance of microorganisms in genetic engineering.
4. The fundamental difference between conventional breeding and genetic engineering.

rDNA technology is often referred to as **genetic engineering.** Unlike other natural genome rearrangement phenomena, rDNA introduces alien DNA sequences into the genome. Even though crossing of two sexually compatible individuals produces recombinant progeny, the term recombinant DNA is restricted to the product of the union of DNA segments of different biological origins. The product of recombinant DNA manipulation is called a **transgenic organism.** Because it is the core technology of biotechnology, rDNA will be discussed in detail.

GENERAL STEPS IN rDNA PROCEDURE

Certain basic steps are common to all rDNA experiments (see Figure 6–1):

a. The DNA of interest that is to be transferred (also called foreign DNA, insert DNA, cloned DNA, or **transgene**) is obtained by first extracting the DNA from the organism and then cutting out the specific DNA sequence using special enzymes.
b. The transgene is inserted into a special DNA molecule called a **cloning vector** and joined (by ligation) to produce a new recombinant DNA molecule (also called cloning vector-insert DNA construct, or simply DNA construct).
c. The DNA construct is transferred into, and maintained in, a host cell (bacterium) by the process of transformation. The vector replicates, producing identical copies (called clones) of the insert DNA.

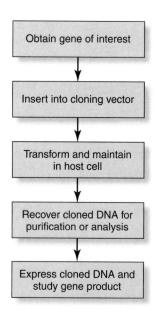

FIGURE 6–1
A summary of the general steps in recombinant DNA research. A gene of interest is first identified, isolated, and then cloned. Once cloned, a gene can be amplified (increased copies) to provide sufficient quantities for further research. The cloned gene may be transferred into another organism for expression.

d. The host cells that have incorporated the foreign DNA are identified and isolated from untransformed cells.

e. The cloned DNA can be manipulated such that the protein product it encodes can be expressed by the host cell.

RESTRICTION ENDONUCLEASES: CUTTING DNA

In order to transfer a gene (piece of DNA), it must be excised from the chromosome. DNA may be fragmented by a variety of methods, with the most common being enzymatic. Enzymes used for this purpose are found in bacteria where they play a defensive role against invading bacteriophage (a virus that attacks bacteria) by digesting the foreign DNA. Some endonucleases are base specific, cleaving only between specific bases. These base-specific endonucleases are called **restriction endonucleases** and have certain characteristics. Hundreds of these enzymes have been isolated. They are named according to a certain protocol following the binomial nomenclature (the scientific system for naming organisms by assigning a two-part name: genus and species). The genus is represented by the first letter (uppercase) followed by the first two letters of the species (lowercase), then the order of characterization (roman numerals). For example *HpaI* and *HpaII* represent the first and second type II restriction enzymes isolated from *Haemophilus parainfluenzae.*

   Restriction endonucleases cut DNA at specific sites called **recognition sites** (see Figure 6–2). The nucleotide sequence at these sites is palindromic (read the same from either end) and usually consists of four, five, six, or eight nucleotide pairs, and sometimes more. The four- and six-base cutters are most commonly used in molecular cloning research. Furthermore, a restriction enzyme may produce a blunt or flush cut (e.g., *HpaI*), or a staggered cut (sticky ends) in which there is a 5'-phosphate extension (e.g., *EcoRI*) or 3'-phosphate extension (e.g., *San 3AI*). Some enzymes leave a 3'-hydroxy extension (e.g., *PstI*). Four-base cutters cut more frequently than six- or eight-base cutters. When appropriate conditions are provided after cleavage, the sticky ends of DNA fragments can re-anneal through complementary bonding to create recombinant DNA molecules, with the enzyme T4 DNA ligase (from bacteriophage T4) producing the ligation.

Recognition sequence	End characteristics	Source
–G͜A–A–T–T–C– –C–T–T–A–A͜G–	Staggered	E. coli
–A͜A–G–C–T–T– –T–T–C–G–A͜A–	Staggered	Hemophilus influenza
–T͜C–G–A– –A–G–C͜T–	Staggered	Thermus aquaticus
–A–G͜C–T– –T–C͜G–A–	Blunt	Hemophilus aegypticus
–T–G–G͜C–C–A– –A–C–C͜G–G–T–	Blunt	Brevibacterium albidum

FIGURE 6–2
Selected restriction endonucleases and their specific recognition sequences.

There are three categories of restriction endonuclease:

1. **Type I enzymes**
   have different subunits for recognition, modification, and restriction or cleavage. The cleavage site is located more than 1,000 base pairs away from the recognition site. As a consequence, cleavage does not occur at a specific sequence even though certain regions are preferentially cleaved. It is not possible to define the recognition sites by characterizing the broken ends of the DNA.
2. **Type II enzymes**
   occur in about one in three bacterial strains. The enzymes are highly specific in action, as they are involved in only one act of restriction. The recognition sites or sequences are usually short (4 to 6 bp) and often palindromic. They cleave at or close to the target site and require no ATP for restriction. Some enzymes produce blunt ends while others produce staggered cuts (or sticky ends). The type II enzymes are the workhorses of recombinant DNA technology.
3. **Type III enzymes**
   have two subunits, one for recognition and methylation and the other for restriction. Like type I, type III restriction sites consist of assymetrical sequences that may be 5 to 7 bp long. Cleavage occurs some 24 to 26 bp downstream from the recognition site.

Whereas all three types of enzymes previously described are proteins with a catalytic effect, there is a unique class of non-protein enzymes called **ribozymes.** These are RNA enzymes with the capacity for cleaving specific phosphodiester bonds.

GENE ISOLATION

Gene isolation is one of the major activities of biotechnology. Before a gene can be genetically engineered, it must first be identified, isolated, and characterized (e.g., number and position of introns, the promoter and its elements). Isolating a gene enables researchers to determine its nucleotide sequence. From the DNA sequence several things can be deduced, including the amino acid sequence and the protein structure and function of the gene's product. In order to transfer a gene from one individual to another, it must first be identified and isolated. Isolation of a gene permits it to be amplified to obtain large quantities for studies.

A number of strategies may be used to isolate or clone a gene.

1. **Activation tagging**

   This strategy requires the availability of a well-characterized transposon system, something that is lacking in many species, except species like corn. The gene to be isolated is first inactivated by transposon insertion, resulting in the formation of a mutant. The DNA sequence of the transposon is used to identify the clones that contain the gene of interest.

2. **cDNA screening**

   A cDNA library is first created. A probe is then designed and used to screen the library to hybridize to the sequence of interest.

3. **Map-based gene cloning**

   Map-based cloning or positional cloning is an rDNA-based method for identifying a gene without first knowing its product. The first step in this method is to produce a high-resolution genetic map (average distance of less than 5 centi Morgans). This is followed by the production of a physical map (a map of the location of identifiable landmarks on DNA regardless of their inheritance). The principal procedures include physical mapping by contig construction using BACs, YACs, STS-content mapping, DNA fingerprinting, and pulse-field gel electrophoresis. Once a physical map is in place, the target gene may be identified by chromosome walking, using RFLP or other molecular markers. This entails starting with a closely linked RFLP probe and isolating genomic clones that it corresponds with, and then walking from these clones to the target genes. Alternatively, molecular markers that are tightly linked with the gene of interest are first identified. The DNA markers are used to screen a genomic library to isolate clones that contain the target gene (called chromosome landing). Genetic complementation through transformation is also part of this process of gene identification.

4. **Transformation-associated recombination**

   This method of gene isolation capitalizes on the natural ability of yeast cells to find and combine similar DNAs, regardless of their origin. Yeast cells are transformed with pieces of DNA along with a small fragment of the target DNA. As the yeast cells reproduce, only DNA that complements the small piece of DNA introduced into the cell are maintained (cloned).

CLONING VECTORS

How is a specific gene that has been extracted from a source transferred to a recipient? Vectors are entities for carrying the target DNA into a host cell for multiplication or cloning. Several kinds of vectors are used in genetic engineering research to accomplish various purposes. Some of them are suitable for cloning a small piece of DNA whereas others are used for large pieces of DNA. Regardless of type, all vectors consist of certain essential features as summarized in Table 6–1. Beyond these essentials, vectors may be designed for special applications by being equipped with additional characteristics such as the capacity to screen for inserts. Commonly used vectors include plasmids, bacteriophages, BACs, YACs, fosmids, P1, and PACs.

■ PLASMID CLONING VECTORS

Plasmids are double-stranded, circular DNA molecules that occur in bacteria. These extrachromosomal structures replicate autonomously, ranging in size from less than 1 to more than 500 kb (a kilobase equals 1,000 base pairs). Plasmids vary in certain other characteristics. One critical property of plasmids is the presence of a sequence that functions

TABLE 6–1
Features of cloning vectors. All cloning vectors have certain basic properties in common (e.g., replication origin, cloning site). Vectors are designed for specific purposes; therefore, some have additional unique features.

Basic Features
1. Selectable marker
2. Replication origin
3. Cloning sites (restriction sites)

Other Features for Specific Purposes Include the Following:
1. Capacity for screening for inserts (e.g., insertional activation)
2. Capacity for selection for inserts (e.g., selection against red-gram stuffers in lambda vectors)
3. Phage promoters (for high-level transcription of inserts)
4. Expression vectors
5. M13/fl origin (for replication of single-stranded DNA)

as an origin of replication (ori), without which autonomous replication within a host cell is impossible.

Plasmids are not used in their natural states for rDNA research, but are genetically engineered to have certain desirable features that facilitate research. Plasmids should have a desirable size to carry the insert DNA. They must contain a recognition site for a single (unique) restriction endonuclease to cleave to allow the cloning of the insert DNA. Finally, there has to be a system by which host cells that have incorporated the recombinant plasmid can be isolated from among the others that have not. A vector, therefore, has at least one selectable genetic marker; this may be a gene for antibiotic resistance.

A number of genetically engineered plasmid cloning vectors are available, including pUC18 (see Figure 6–3). Plasmid cloning vectors are designated by a lower case "p." The pUC18 is a sophisticated cloning vector. By being smaller in size, it can be used to clone longer DNA inserts. Also, it can produce about 500 copies per cell, which is 5 to 10 times more than pBR322, one of the earlier vectors. The pUC18 vector also has a **polylinker site** (a cluster of many restriction endonuclease recognition sites). Furthermore, the polylinker (or multiple cloning) site is located within an *E. coli* lacZ gene. This design facilitates the selection process to isolate transformed host cells. Bacterial cells carrying the intact plasmid produce blue colonies when cultured on a medium containing a compound called *X-gal* (5-bromo-4-chloroindolyl-galactosidase). To use, the researcher selects one of the restriction enzymes with a recognition site in the polylinker. Both the insert DNA source and vector are cleaved with the same restriction enzyme. Cleavage and insertion of a target DNA at the polylinker site inactivates the *lacZ* gene, causing host cells carrying the insert DNA to produce white colonies.

■ *VIRAL VECTORS*

Viruses that infect bacteria (bacteriophage, phage) have been engineered as vectors for cloning longer pieces of DNA. The *E. coli* virus lambda (λ) phage is a widely used cloning vehicle. Numerous vectors based on lambda phage have been developed. The DNA sequence of the phage lambda is known. It has been determined that the middle one-third of the lambda DNA consists of a 20-kb sequence (out of the 50-kb total length of the phage) that is required for the integration-excision events. This segment of DNA can be dispensed with and replaced with a large insert DNA. Scientists have genetically engi-

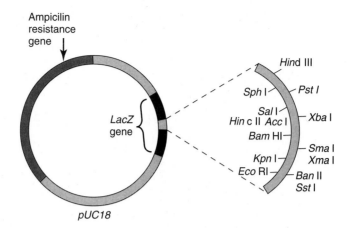

FIGURE 6–3
A sketch of the genetic map of a plasmid vector. The pUC18 vector is small in size and can hence be used for cloning large fragments for DNA. Also, it has a large number of restriction sites within the polylinker which are located in the *lacZ* gene.

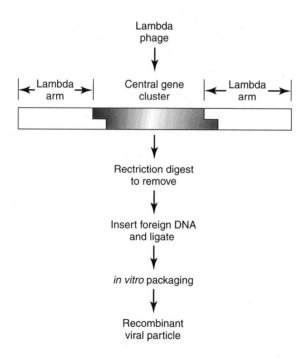

FIGURE 6–4
Summary of the cloning steps in phage lambda as a vector.

neered the phage such that the 20-kb section is bracketed by the recognition sequence for a restriction enzyme (see Figure 6–4). Both phage and insert DNA (which lacks an internal restriction enzyme site but is bracketed by restriction sites) are cleaved with the same enzyme (*BamHI* in this case) and incubated in the presence of T4 DNA ligase to create a recombinant bacteriophage λ.

Because they can carry large pieces of DNA, viral vectors are useful in the creation of DNA libraries. Phage libraries are screened by means of DNA probes or immunological assays. Instead of colonies as in plasmid cloning, plaques representing individual zones of lysis (disintegration of cells as a result of rupture of its membrane) and that contain phage are lifted onto a matrix for further analysis.

The M13 bacteriophage vector is a single-stranded phage. When it infects a bacterium, the single strand (+ strand) replicates to produce a double-stranded molecule called replicative form (RF). This RF is essentially similar to plasmids, and can be used like plasmid vectors. Once reinserted into bacteria, RF molecules replicate to produce single strands in addition to one strand of the inserted DNA.

■ VECTORS FOR CLONING VERY LARGE DNA FRAGMENTS

One of the most commonly used vectors for cloning large fragments of DNA is the bacterial artificial chromosome (BAC). Others are cosmids, yeast artificial chromosome (YAC), and P1.

Cosmids

A **cosmid** is a "hybrid" between a plasmid and a phage. It consists of the *cos* sequence of phage lambda (required for packing the phage DNA into the phage protein coat), the plasmid sequence for replication, and an antibiotic resistance gene to identify the host cell carrying the cosmids. Viral vectors can accommodate inserts of between 15 and 20 kb; cosmids can handle about 40 kb of cloned DNA. Furthermore, they can be maintained as either plasmids or bacteriophage λ vectors because they have an *E. coli* origin of replication. They also have cohesive ends (cos) sites found in phage (see Figure 6–5). An example of a cosmid is the pJB8-5.

Bacterial Artificial Chromosome

Bacteria consist of independently replicating plasmids that are involved in the transfer of genetic information during bacterial conjugation (temporary fusion of cells for transfer of genetic information). The F factor can carry up to 1 Mb (10^6 bp)-long fragments. Consequently, scientists have engineered such molecules into multipurpose vectors for mapping and analysis of complex eukaryotic genomes. These bacterial vectors are called **bacterial artificial chromosomes** (BACs) (see Figure 6–6). BAC vectors carry the F factor genes for replication and copy number, as well as antibiotic resistance marker and restriction enzyme sites.

Yeast Artificial Chromosome

Yeast, a eukaryote, can be manipulated and cultured like bacteria and used as a vector for the cloning of DNA. The **yeast artificial chromosome** (YAC), in its linear form, con-

FIGURE 6–5
A sketch of the genetic map of a cosmid vector, pJB8. The origin of replication (*ori*) from a bacterium allows this vector to operate as a bacterial plasmid. The *cos* site allows cosmids carrying large inserts to be packaged into lambda viral coat proteins as though they were viral chromosomes.

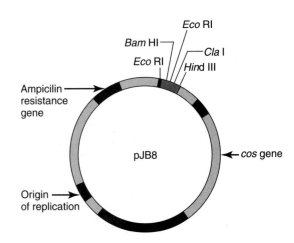

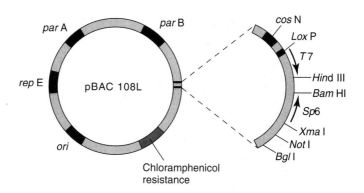

FIGURE 6–6

A sketch of the genetic map of a bacterial artificial chromosome (BAC), pBAC 108L. This vector also has a polylinker site with restriction sites for a large number of enzymes.

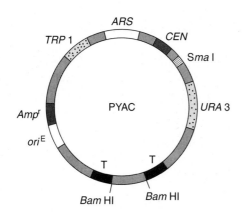

FIGURE 6–7

A sketch of the genetic map of a yeast artificial chromosome (YAC).

tains yeast telomeres for the distribution of replicated YACs to daughter cells at cell division (see Figure 6–7). In addition to these features, a YAC also contains a selectable marker on each arm (TRP1 and URA3) and a cluster of unique restriction sites for DNA inserts. YACs are capable of receiving DNA inserts longer than 1 Mb, making them useful in mapping eukaryotic genomes (e.g., the *Drosophila* and human genomes).

Shuttle Vectors

When hybrid vectors are constructed with origins of replication from different sources (e.g., plasmids and animal viruses such as SV40), they can replicate in more than one host cell. This capacity allows such vectors, called **shuttle vectors,** to be used by researchers to move DNA inserts back and forth between different host cells (e.g., bacteria and yeast). Shuttle vectors contain genetic markers that are selectable in different host systems. They are useful in studying gene expression.

BACTERIAL TRANSFORMATION

Transformation is the process of introducing free DNA into a host cell. After the foreign DNA has been successfully inserted into a vector, the next step is to place the recombinant vector into a host for the replication of the foreign DNA. A common host for vector replication is the *E. coli* strain called K12. Genetic transformation of prokaryotes can be accomplished by one of several methods.

1. **CaCl$_2$-heat transformation**

    Bacteria cells are treated with ice-cold CaCl$_2$ and then exposed to high temperature (42°C) for about 90 seconds. The exact mechanism of action of CaCl$_2$-heat transformation is not known. It is assumed that this treatment breaks down the cell wall of the bacterial cells, in localized regions, thereby allowing the free DNA to be taken up into the interior of the host cell. This method has a transformation frequency of about 1 transformed cell per 1,000 cells. Some bacteria are naturally competent (able to take up DNA) whereas others like *E. coli* require a chemical induction. The transformation efficiency (number of transformants per microgram of DNA added) is about 10^{-7} to 10^{-8} colonies.

2. **Electroporation**

    Instead of chemical induction, a strong electric field may be used to render the host cell wall permeable to the free DNA through the use of equipment called **electroporators** (see Figure 6–8). The bacterial cells and the free DNA are placed in a chamber fitted with electrodes. A single pulse (about 25 microfarads, 2 to 5 kilovolts, and 200 ohms) is administered for about 4.6 milliseconds. The transformation efficiency for small plasmids (about 3 kb) is 10^{-9} transformants per microgram of added DNA, while it is 10^{-6} for large (136 kb) plasmids. The protocol for electroporation varies among kinds of hosts. The mechanism of DNA uptake during electroporation is not exactly known. It is believed that transient pores are formed in the cell wall as a result of electrical shock. When the free DNA contacts the lipid bilayer, it is taken into the cell through these reversible pores.

3. **Conjugation**

    Conjugation is a natural process by which cell-to-cell genetic transfer occurs in prokaryotes. However, most of the plasmids used in rDNA research are incapable of conjugation.

TRANSGENE DELIVERY

If recombinant DNA technology allows scientists to nonsexually introduce specific alien genes that confer specific desirable traits into recipient plants or other organisms, how is

FIGURE 6–8

An electroporator is a device used to transfer DNA fragments into a host cell. There are different models and capabilities of electroporators.

Source: Photo taken at Oklahoma State University Nobel Research Center.

such a feat accomplished? The methods of transgene delivery may be grouped into two broad categories:

1. Direct gene transfer.
2. Mediated (indirect) gene transfer.

In plants, transgenes may be delivered to meristematic cells of an intact plant or seed (called *in planta* transformation). This eliminates the need for tissue culture induction and regeneration. DNA is delivered to meristems or developing floral organs. This technique has limited success so far, with transient gene expression reported in a few cases like *Arabidopsis*. Established transformation techniques for transgene delivery in all the major crops deliver DNA to cultured cells *in vitro*, followed by regeneration of plants.

■ DIRECT GENE TRANSFER

Although several methods of direct gene transfer have been developed, only about three or four are used routinely: protoplast transformation, tissue/cell electroporation, silicon carbide fiber vortexing, and microprojectile bombardment.

1. **Protoplast transformation**

    Protoplast transformation was the first method used to demonstrate that direct gene delivery to plants was feasible. However, this method poses some technical challenges and therefore is not a method of choice.

2. **Tissue/cell electroporation**

    Callus cultures or primary explants such as immature embryos or inflorescence may be used as target material. The transformation process occurs in an electroporator. Transformation efficiency levels by electroporation are sufficiently high.

3. **Silicon carbide fiber vortexing**

    Silicon carbide fibers (about 0.3 to 0.6 m diameter and 10 to 80 m long) are mixed with a suspension culture (cells replicating and growing in a liquid medium as opposed to a callus growing on a solid medium such as agar) as explant, plus plasmid DNA, and vortexed. The mixture is cultured on a medium with selectable markers. To use this technique, there must be a regeneration system in place for regenerating plants from single cells.

4. **Microprojectile bombardment**

    Microprojectile bombardment or **biolistics** is the transfer of target DNA into intact cells by literally shooting the DNA from a biolistic device (hence the nickname of **shotgun transformation**). The device is sometimes called a **gene gun.** To use the device, micron-sized (1 to 5 m diameter) carrier particles (tungsten or gold) are coated with the DNA of interest and accelerated in the barrel of the biolistic device at energies powerful enough to penetrate the cell. About 50 µg of tungsten is required for each DNA transfer event. The rate of acceleration may be up to 430 m/s in a partial vacuum. The carrier particles are placed on a support film and the film mounted in the particle acceleration device. The support film is accelerated (usually by gas pressure; the original device utilized gun powder) and then stopped by a protective mesh. The carrier particles pass through the mesh, hitting the target tissue mounted in a petri dish below the biolistic device (see Figure 6–9). The survival of bombarded cells is highest with a low penetration number of projectiles (1 to 5 per cell). More than 80 percent of bombarded cells may die if particle penetration reaches 21 per cell.

Different kinds of propelling forces are used to deliver the particles to the target tissue. The commercially available particle delivery system is marketed by DuPont and is called the **DuPont Biolistic® PDS1000/He** device. Helium is the propellant gas. Electric discharge may also be used as the propelling force. A high-voltage discharge is delivered to a small water droplet, which is vaporized rapidly to release energy to propel the carrier particles. Its advantage is that the electrical voltage can be manipulated such that

FIGURE 6–9
A gene gun. This table-top model uses compressed helium gas to propel the DNA-coated particles into the target material. Hand-held models are also available.

Source: Photo taken at Langston University Center for Biotechnology Research and Education.

the ideal particle acceleration velocity is attained. Other noncommercial delivery devices have been developed and used successfully to transform plants. These include air-gun and gas-stream devices. Lately, a hand-held particle gun has been developed, thus allowing transgene delivery to tissues of large intact plants.

Many variables impact the success of biolistic transformation, the key ones being chamber vacuum level, particle size range, and shot distance. Others include the amount of DNA per particle, explant type and physiological conditions, and gas type and pressure. Biolistics has been used to transform both dicots (e.g., soybean, peanuts, and tobacco) and monocots (e.g., corn, wheat, and rice). Organelle transformations have also been reported with this technique.

■ *MICROINJECTION*

Microinjection is a technically demanding technique for gene transfer. It is labor intensive, requiring great skill to be successful, not to mention the sophisticated equipment required (microcapillaries and microscopic devices). The equipment is used to deliver selectable DNA solution directly into the nucleus of a single protoplast at any given time. If the cells are cultured individually, it eliminates the need for having drug resistance or marker genes to discriminate between transformed and untransformed cells. The method has been used successfully in a number of species including tobacco and rapeseed, in which stable transformation was first demonstrated. Transformation frequencies of 15 to 25 percent have been reported.

The technique has several drawbacks apart from equipment and operator skill. To facilitate physical coordination of the injection pipette and the protoplasts, the latter may be immobilized (e.g., by embedding in agarose or agar or applying suction through a holding pipette). Because plants have large vacuoles, locating the nucleus is often problematic.

■ *MEDIATED (INDIRECT) GENE TRANSFER*

Target genes may be transferred into a plant or animal through the mediation of various agents, the common ones being vectors (microorganisms), chemicals, and electrical current.

Biological Vectors

The biological vectors in use for transformation of plants are plant DNA viruses, *Agrobacterium tumifaciens* and *Agrobacterium rhizogenes*.

Viral Vectors

The most commonly used viral vector is the **cauliflower mosaic virus (CaMV).** Because this virus is able to infect plants and move systematically through the host, there is no need for cell culture when this technique is used. Viral vectors are limited to the host species they naturally infect. Consequently, their use as DNA transfer vehicles is limited.

Agrobacterium-mediated Transformation

Agrobacterium-mediated transformation is the most common technique for transforming plants. Two species of this soil-borne bacterium are used in biotechnology. *A. tumifaciens*, also called the crown gall bacterium, causes tumors or galls on plants (dicots) it infects. *A. rhizogenes* also infects plants through wounds but causes hairy root disease.

The oncogenic (tumor-inducing) properties of *A. tumifaciens* resides on a large tumor-inducing **Ti plasmid** (called **Ri plasmid** in *A. rhizogenes)*. The Ti plasmid has two regions. The T-region (called T-DNA or transferred DNA) is what is transferred into the plant cell and integrated into the host chromosome. A second region on the Ti/Ri plasmid, called the virulence (*vir*) region, carries the genes for tumor induction and is also involved in the transfer of T-DNA. T-DNA also carries genes for synthesizing metabolic substrates for bacteria called **opines.** Further, the T-DNA is flanked or bordered by short (25 bp) direct repeats of DNA. The deletion of the tumor-inducing segment of the DNA does not prevent the transfer of T-DNA.

Systems

Two classes of *Agrobacterium* vectors have been engineered: **co-integrate** and **binary.** In both systems, the gene to be introduced into the plant must be inserted between the borders or adjacent to one of the borders of the T-DNA. The *Agrobacterium* used is first disarmed (unable to induce tumors). In the co-integrate vector system, the gene to be transferred is inserted at the integration site by a one-step recombinational event within the bacterium. The binary system requires two plasmids: Ti plasmid containing the *vir* gene (which may be disarmed or with oncogenes) and a genetically engineered T-DNA plasmid, which acts in the *trans* configuration to transfer the T-DNA. The use of binary vectors eliminates the recombinational step. A binary T-DNA plasmid vector has a broad host range origin of replication and a marker gene for selection and maintenance of the cloned gene in both *Agrobacterium* and *E. coli*. It also has a T-DNA segment containing a plant selectable marker and at least one restriction site for cloning DNA inserts. Even though binary vectors simplify the cloning task, they are more difficult to use, among other drawbacks.

Requirements for Successful Transformation by *Agrobacterium* Mediation

The following are some key considerations for successful transformation research:

1. **Efficient plant regeneration system**

    An efficient regeneration system is the first critical step in any transformation undertaking. In effect, it may be said that "thou shall not transform unless thou can regenerate!" For best results, the regeneration tissue should be at or close to the surface of the explant to make it readily accessible to the *Agrobacterium*. The regeneration system should involve a minimum of callus growth so as to reduce the incidence of somaclonal variation (spontaneous mutations that arise among cells in tissue culture).

2. **Determination that the cells are susceptible to *Agrobacterium* transformation**
    It is important that the *Agrobacterium* be able to transform most of the cells at the target site.
3. **An efficient and sensitive selection method**
    Transformed cells should be readily identified and selected from among untransformed cells.
4. **Stable transformation**
    The transformation material should not only be regenerated, but the introduced gene should be expressed in subsequent generations. The fewer the number of cells producing the regenerated plant the better; otherwise, a chimeric product will result.

Methodology of *Agrobacterium* Transformation

The general steps involved are:

1. Incubate bacteria with plant cells (a few hours to about two days).
2. Wash cells and treat with antibiotic (to remove bacteria).
3. Culture in the presence of selectable agents.
4. Regenerate transformed cells.

The incubation of plant cells with bacterium is called **co-cultivation.** There are two *Agrobacterium* co-cultivation systems: explant and protoplast.

Explant Co-cultivation System

Even though stem segments are commonly used, the explant may consist of other plant parts, depending on the ease of regeneration from somatic cells. Depending on the explant, certain treatments (e.g., puncturing with sterile needles) may be required for effective transformation. Explant transformation produces more rapid regeneration. Suspension culture may also be used for co-cultivation. This is used effectively with woody species.

Co-cultivation success is influenced by certain factors including the duration of co-cultivation, presence of inducing (induce activity of the *vir* genes) compounds (e.g., acetosyringone), the types of bacteria strains, and selection environment (or medium). Co-cultivation with *Agrobacterium* is less successful with cereals (or monocots) in general. Wound response is critical to *Agrobacterium* transformation and, consequently, cereals with appropriate wound response (as well as dicots) are responsive to this technique.

Use of Protoplasts

The use of isolated protoplasts has several advantages over explants. Each protoplast is in a state of wound repair and hence highly susceptible to *Agrobacterium* infection. Large numbers of protoplasts can be treated to produce large numbers of transformants. Furthermore, cells are more uniformly exposed to agents during the selection stage, leading to a reduced incidence of non-transformed cells (escapes) after selection (low-frequency transformation events are easy to detect). However, the protoplast system is more difficult to use and regeneration from protoplasts is slower.

Chemically Mediated Transformation

Certain chemicals that make the cell membrane permeable can facilitate DNA uptake. The common ones are polyethylene glycol (PEG) and polyvinyl alcohol (PVA). Together with Ca^{2+} and high pH, these chemicals permeabilize the cell membrane (like electroporation) to allow the uptake of foreign DNA. Compared to *Agrobacterium* mediation, this system has low transformation frequency. Furthermore, it requires the use of protoplasts. PEG-mediated DNA uptake has been successfully used to transform grass species including rice, Italian ryegrass (*Lolium multiforum*), and diploid wheat (*Triticum monococcum*). However, such success in monocots is not widely documented. Chemically mediated transformation also enables scientists to utilize other vectors besides *Agrobacterium*. *E. coli* vectors yield a higher copy number than Ti plasmids.

Electroporation-mediated DNA Uptake

Originally developed for transformation of mammalian cell lines, electroporation-mediated DNA uptake is similar to chemically mediated systems: they both render the membrane reversibly permeable. Electroporation accomplishes this by using short electrical pulses. This technique has been adapted for use in plants with great success. Species that do not respond to other techniques or have no established standard transformation protocols benefit from this technique. The exact mechanism by which the membrane becomes permeable is not completely understood.

Certain factors affect the efficiency of electroporation. Strong DC power reduces cell viability, whereas a protracted exposure of cells to weak current promotes regeneration from callus. AC current has been used in electroporation with success (e.g., in sugar beet). Generally, the electric field strength must exceed a certain critical value to be effective. Optimum field strength for barley (*Hordeum vulgare*) was found to be fine pulses of 100 to 400s at 120 V/mm. The issue of an upper size limit of plasmids that can be uptaken during electroporation is not settled. Plasmids that have been studied widely are less than 20 kb in size.

The concentration and type (linear or supercoiled) of exogenous DNA affect transformation frequency. Linearized plasmid DNA gives a higher transformation frequency than supercoiled DNA. Similarly, increasing the plasmid DNA concentration of the electroporation medium increases its transformation frequency. The composition of the electroporation medium and the incubation temperature affect membrane permeability. Increasing the conductivity of the medium decreases the pulse strength. Furthermore, Ca^{2+} ions in the medium stabilize the plasma membrane. It has also been determined that electroporating at a lower temperature increases DNA uptake. Cultivars of the same species differ in their responses to transformation.

Transient Versus Stable Transformation

The goal in developing transgenic plants is to express the transgene in adequate amount, and also to have it propagated through the germline (heritable). Such an expression is said to be **stable.** Stable expression may take at least several months before materials are available for this evaluation. Furthermore, certain species are difficult to regenerate from the explant and hence cannot be evaluated for stable expression of a transgene.

Consequently, researchers may evaluate the expression of the transgene at an intermediate stage in the transformation process. For example, the target DNA may be added to protoplasts by electroporation (or chemical mediation) and the protoplasts isolated only after one to two hours for evaluation of the expression of encoded genes. This preliminary expression of the transgene is described as **transient expression.**

Gene Delivery into Mammalian Cells

Most of the discussion on transgene delivery so far has emphasized methodologies for plant cells. Plant and animal cells can take up DNA from their environment by the general process called **transfection.** Gene transfer can also be accomplished by using vectors. The common methods for DNA uptake in mammalian cells include electroporation, coprecipitation with calcium phosphate, endocytosis, direct microinjection, and direct encapsulation of DNA into artificial membranes (liposomes) followed by fusion with cell membranes. In addition to these methods, DNA may be transferred into mammalian cells by using YACs and vectors based on retroviruses (RNA viruses). The application of retroviruses is further discussed in the section on gene therapy.

TISSUE CULTURE AND SELECTION

Transgene delivery is commonly conducted into cells and tissues rather than whole organisms, as is the case in conventional breeding. To this end, there is a need for an environment in which cells and tissues can be maintained and nurtured as desired. Furthermore, because the transgene may not be successfully delivered into a cell (in the case of microinjection) or into all cells in a mass of callus (as in biolistics), it is important to have a selection system to discriminate among cells to identify and isolate only genuine transformants.

■ TISSUE CULTURE

Tissue culture is an integral part of most transformation systems (the methodology is discussed in more detail elsewhere in the text). It is critical to establish a reliable regeneration system prior to embarking on transformation in plants. When somatic explants (the part of the plants used to initiate tissue culture) are used as targets for transformation, the transgene are usually delivered into cells before the process of differentiation (morphogenesis) begins. A variety of explants are used, including protoplasts, immature embryo, cell suspension, shoot meristem, and immature inflorescence. The genotype is as important as the choice of explant. The transgene is usually delivered into undifferentiated cells of primary explants before callus initiation or into proliferating embryogenic tissue (see Figure 6–10). The embryogenic cells then divide to produce one or several somatic embryos and subsequently develop into full plants. The culture medium is manipulated for optimal recovery of transformants (see Figure 6–11).

■ SELECTION SYSTEMS

The purpose of selection is to identify transformants (cells that have incorporated the transgene) or transgenic cells, following gene delivery, by establishing their preferential growth over non-transgenic cells. To accomplish this, cloning vectors have selectable markers built into their design (see Figure 6–12). Markers are used in biotechnology for a variety of purposes and will be discussed further elsewhere. The present discussion pertains to markers used for selection in transformation research. Selectable marker systems used may be grouped as follows: antibiotic selection, herbicide selection, scorable gene-mediated selection, and positive selection. Selection systems differ in their ease of use and efficiency. The selection agent is included in the tissue culture medium.

Antibiotic Selection

Antibiotic marker genes may be described as the first generation selectable markers for biotechnology. Vector designs incorporate markers that confer resistance to antibiotics such as aminoglycosides, kanamycin, neomycin, and paromomycin (see Figure 6–13). One of the most common antibiotic genes is *nptII* (which encodes the enzyme neomycin phosphotransferase). It confers kanamycin or neomycin resistance.

Herbicide Selection

Herbicide resistance genes generally encode modified target proteins that are insensitive to the herbicide, or an enzyme that degrades or detoxifies the herbicide in the plant. The bacterial phosphinothricin acetyltransferase gene (*bar*) is one of the most commonly used herbicide resistance genes. This gene encodes enzymes that detoxify common herbicides like glufosinate, bialaphos, Basta, and Liberty. Bacterial and plant EPSP synthase (an enzyme that is the primary target of the herbicide glyphosate) genes and acetolactate synthase genes are also used in herbicide selection (for glyphosate or sulfonylurea herbicides).

Scorable Gene-mediated Selection

Scorable marker genes (**reporter genes**) are typically used as markers for rapid visual confirmation for transient expression following DNA delivery. They are not usually used for establishing preferential growth of transgenic cells like the other marker systems. Reporter genes are expressed in cells without integration into the genome and are assayed in a variety

Genetic transformation in cassava: Induction of embryogenic target tissue

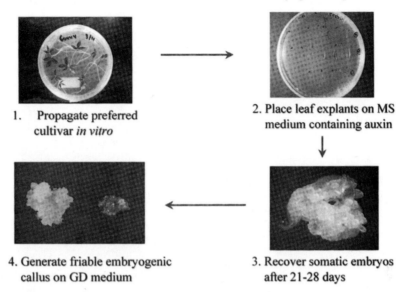

1. Propagate preferred cultivar *in vitro*

2. Place leaf explants on MS medium containing auxin

4. Generate friable embryogenic callus on GD medium

3. Recover somatic embryos after 21-28 days

FIGURE 6–10

Induction of the embryogenic target tissue for genetic transformation. The appropriate tissue is always critical to initiation of genetic transformation. In cassava, a responsive friable callus is obtained by producing embryos from explants on MS medium and finishing on GD medium.

Source: Photos are courtesy of Nigel Taylor, Danforth Plant Science Center, St. Louis, Missouri. Used with permission.

Genetic transformation of cassava: Recovery of transgene plants

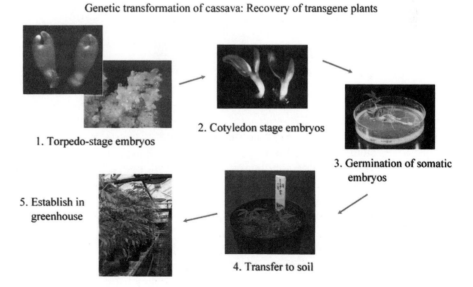

1. Torpedo-stage embryos

2. Cotyledon stage embryos

3. Germination of somatic embryos

5. Establish in greenhouse

4. Transfer to soil

FIGURE 6–11

Recovery of transgenic plants. Transgenic callus lines develop into somatic embryos and then progressively move into plantlets, which are eventually established in the soil in the greenhouse.

Source: Photos are courtesy of Nigel Taylor, Danforth Plant Science Center, St. Louis, Missouri. Used with permission.

FIGURE 6–12
One of the required
characteristics of a vector
is at least one selectable
marker gene. A selectable
marker is needed to be
able to distinguish
between a cell that has
incorporated a foreign
gene and one that has not.

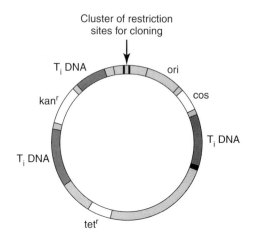

Genetic transformation in cassava: Gene transfer and selection of transgenic tissue

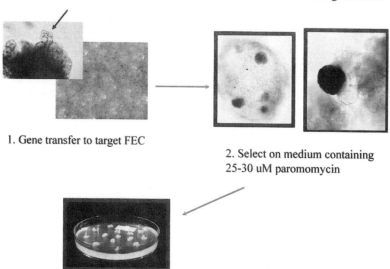

1. Gene transfer to target FEC

2. Select on medium containing
25-30 uM paromomycin

3. Recover transgenic callus lines

FIGURE 6–13
Gene transfer and selection of transgenic tissue. In this figure, friable embryogenic
callus (FEC) is the target of transgene delivery. Photo 2 shows the presence of multiple
transformation events appearing as dark spots. The transformed cells are selected and
cultured to produce transgenic callus lines.

Source: Photos are courtesy of Nigel Taylor, Danforth Plant Science Center, St. Louis, Missouri. Used with
permission.

of ways, depending upon the type of gene. Reporter gene assay is a transient assay that can
be conducted within 24 hours after transformation of the cells. The common reporter genes
used often encode enzymes that have distinct substrate specificities. Consequently, they can
be monitored and visualized readily through a variety of visualization protocols. Examples of
reporter genes are **CAT (chloramphenicol acetyltransferase),** which is visualized by a ra-
diochemical protocol; **GUS (β-glucuronidase),** which is visualized by a histochemical or flu-
orometric protocol; and **LUC (luciferase),** which is visualized by a luminescence system (see
Figure 6–14). A more recent addition to scorable markers is the **green fluorescent protein
(GFP)** system, which allows the nondestructive, visual identification of transgenic cells by
standard fluorescence microscopy. Another reporter gene in use was developed by using the

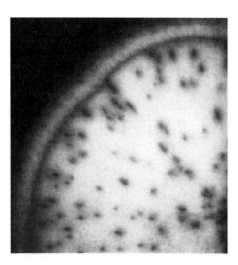

FIGURE 6–14
GUS expression in cassava tissue. The event is manifested as a bright blue stain.
Source: Photo is courtesy of Nigel Taylor, Danforth Plant Science Center, St. Louis, Missouri. Used with permission.

anthocyanin genes of maize. Upon bombardment into cells, a purple pigmentation is observable in cells that have incorporated the target DNA. This assay is nondestructive and consequently the fate of the target DNA can be studied over a period of time.

Positive Selection

Because of the persistent protests by activists against the use of environmentally risky materials in biotechnology, scientists are searching for less controversial marker systems, the so-called **benign markers.** Positive selection systems are based on metabolic pathways. Transgenic cells are given a metabolic (and consequently, growth) advantage over non-transgenic cells. In one example, the mannose selection system, phosphomannose isomerase gene (*pmi*) is used as a selectable marker, with mannose as the selecting agent. The gene confirms the ability to convert mannose-6-phosphate, produced by endogenous plant hexokinase from mannose, to fructose-6-phosphate. However, nontransgenic cells accumulate the mannose-6-phosphate to a cytotoxic level, thereby eliminating them from the culture. Several other positive selection systems are available.

■ TRANSGENE INTEGRATION

The goal of transgene delivery is that the transgene becomes stably introduced in the cell. The transgene that is delivered, either by direct or mediated transfer, first enters the nucleus and then may become integrated into the genome. Research indicates that transgenes integrate into the genome at sites located randomly throughout the genome, but predominantly in transcriptionally active regions. The number of insertions at a locus is highly variable, and may range from 1 to 5. *Agrobacterium*-mediated transformation generally produces fewer insertions per locus. This variability in transgene insertion number is affected by the methodology, including factors like the preparation, physical status of the plasmid DNA (coiled, linear), and concentration and amount of DNA delivered.

Once inside the nucleus, the foreign DNA sequence may be transcribed (producing transient expression) or become integrated into the chromosome (for stable transformation). The plant DNA of necessity must be broken at the point of insertion. After insertion, there is DNA repair and re-ligation.

TRANSGENE EXPRESSION

The ultimate goal of transformation is for the transgene to be stably expressed in the desired amounts in the desired tissue. Promoters are the "engines" that drive DNA expression by determining the level of transcription of a selectable coding sequence. Promoters are derived from a variety of sources including viruses, bacteria, and plants. There are three major classes of promoters that are used.

■ CONSTITUTIVE PROMOTERS

Structural genes are transcribed either continuously (constitutive expression) or periodically as the gene product is required (regulated expression). **Constitutive promoters** (which have high affinity for RNA polymerase and consequently promote frequent transcription of the adjacent region) are used for a variety of purposes, but mainly to drive the expression of the selectable marker gene for identification of transgenic tissues *in vitro*. Using a "strong" promoter with ubiquitous activity increases the chance of recovering stable transformants. Expressing the selectable marker gene to a high degree in all cell types protects the transformed cell from the action of the selection agent. In applications like the development of herbicide resistance in plants, it is desirable to express the genes highly in all tissues and at all stages of the plant's development. In this case, a constitutive promoter may be used to drive the expression of the herbicide resistance gene.

Specific examples of constitutive promoters are the 5′ regulatory regions derived from T-DNA genes of opine or nopaline synthase, and the 35S promoter from cauliflower mosaic virus (CaMV). CaMV 35S is one of the most widely used promoters in plant transformation research, providing constitutive and high levels of expression of an array of genes in plant cells (see Figure 6–15). It is effective in monocots, though not the optimum for all monocots. The ADH 1 (alcohol dehydrogenase 1) promoter and the 2′ promoter of the octopine T-DNA have produced better results in certain cases. The only setback with ADH 1 is that it is anaerobically induced and not present at all times.

Marker gene expression in cassava cv TMS 60444

35S promoter

Cassava vein mosaic
virus (CsVMV) promoter

  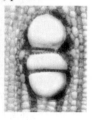

840 bp cassava PAL promoter

FIGURE 6–15

Marker gene expression in cassava cv TMS 60444. The 35S promoter is a constitutive promoter and is hence expressed throughout the plant. The CsVMV promoter, however, is expressed only in the veins of the leaf. Similarly, the cassava PAL promoter is tissue specific, expressed in specific tissues as seen in dark spots, streaks, or borders around cells, depending on the tissue.

Source: Photos are courtesy of Nigel Taylor, Danforth Plant Science Center, St. Louis, Missouri. Used with permission.

Eukaryotic promoters have certain features that enable temporal and spatial regulation of gene expression. This occurs because of the interactions between units called *cis*-acting elements (also called enhancers). These enhancers can increase the level of transcription of a gene from a promoter and thereby increase its level of expression. Promoter effect can be drastically increased by what is described as a mix-and-match design in which several copies of an enhancer are combined with several copies of the core promoter sequence. This was demonstrated by a promoter construct consisting of six ARE (anaerobic response element) and four OCS (octopine synthase), which gave 10 to 50 times the increase in gene expression over CaMV 35S.

■ TISSUE-SPECIFIC AND DEVELOPMENTALLY REGULATED PROMOTERS

Unlike herbicide resistance, in which it is desirable for the gene to be expressed throughout the plant, there are some occasions where targeted gene expression is needed. For example, in animals, you would want milk to be expressed only in the mammary system, not in the muscles. In plants, genes for grain quality should be targeted for expression (e.g., in the endosperm). Promoters for targeted expression are available for plant tissues like endosperm (e.g., for storage protein), anther tapetum (e.g., for engineering male sterility), embryo (engineering grain quality traits), and phloem (e.g., for engineering pest resistance to sucking insects). There are some promoters that are responsive to environmental stimuli (e.g., light). Some genes need to be expressed only at certain developmental stages, and hence need to be regulated for such specific roles.

■ INDUCIBLE PROMOTERS

Sometimes, genes need to be expressed only under certain conditions where the organism needs the product. For example, a transgene may be developed to trigger the expression of a gene in response to pathogenic invasion or wounding. Wound-induced promoters have been isolated in species like rice (e.g., the rice basic chitinase—RC24, which drives *uidA* gene expression in wounded roots and stems).

STABILITY OF TRANSGENE EXPRESSION

Whereas it is hoped that transgenes will be stably expressed indefinitely, the reality is that some level of instability may occur over generations due to a variety of causes. Often, transgenic breakdown is caused by structural defects in the gene construct inserted (it may not be intact). However, the phenomenon of **progressive transgenic silencing** or **failure** is not completely understood. It is suspected that ancillaries of an expression system (e.g., terminal sequence promoters) may have a role in the event, especially systems using heterologous sequences from nonrelated sources (e.g., plants having bacterial, viral sequences, instead of native sequences or those from closely related species). These systems are more susceptible to malfunction. In addition, they may be recognized as foreign DNA and consequently excised or methylated. It is also suspected that the structural integrity of the introduced foreign sequence may be responsible. This relates to the presence or absence of introns and the nature of the 5′ and 3′ sequences. Transgenics produced from cDNA clones instead of genomic sequences lack introns and possibly other 3′ gene sequences, which may reduce the accumulation of mRNA (as some researchers have reported). Researchers tend to gravitate toward cDNA sequences from ESTs (expressed sequence tags—unique short DNA segments within genes that are used as markers) and gene isolation programs because they are more readily available. Furthermore, there is more flexibility with their use (e.g., they are amenable to the use of different promoters).

The pattern of transgenic integration into the genome is also a suspect in transgene instability. Complex patterns of integration (e.g., involving inverted repeats, tandem arrays) are more likely to be destabilized than simple patterns. Some promoters (e.g., CaMV 35S) have certain sequences in their structural genes that make them prone to structural changes that could trigger transgene silencing.

Several strategies may be employed to reduce the chance of transgene instability. One strategy is to ensure that the introduced foreign sequence matches the isochore of the host genome. That is, the coding regions should have the right codon usage and GC (guanine-cytosine) content. For example, monocot genes have 44 to 70 percent GC value while dicots have 40 to 56 percent. The *Bt* toxin gene used for corn was modified to increase the GC value for a fiftyfold expression of the encoded protein in dicots. Another strategy is to avoid the use of duplicated sequences in transformation vectors. For example, vectors like pAHC25, which is widely used in cereals, contain two copies of ubiquitin promoter and two copies of the *nos* terminator. Such duplication provides opportunity for recombination to form secondary structures. Designers of transformation vectors should also use genomic clones instead of cDNA, since there is ample evidence that introns, 5′ unsaturated regions, and specific sequences downstream of the polyadenylation site all impact the expression levels and stability of mRNA.

MARKER-INDEPENDENT TRANSGENIC PRODUCTION

To avoid distraction of the development and application of biotechnology, scientists are addressing the thorny issue of the use of molecular markers, both herbicides and antibiotics. Two approaches are being employed—development of new markers (benign markers) or post-transformational removal of marker genes in transgenic products. Removing the marker gene after transformation could be accomplished naturally in cases where the marker gene integrates at a locus different from that of the trait gene. After one generation of seed production, it should be possible to recover from the segregating population individuals containing only the "trait gene" locus. Unfortunately, direct gene transformation systems rarely integrate at more than one locus. The use of two T-DNA vectors in *Agrobacterium*-mediated transformation has resulted in successful isolation of marker-free transgenic lines in rice. Post-transformational removal of markers may be accomplished artificially by the techniques of transposition and site-specific recombination.

KEY CONCEPTS

1. Recombinant DNA (rDNA) technology, also called gene cloning, molecular cloning, or genetic engineering, refers to the experimental methodologies utilized to transfer genetic information from one organism to any other disregarding natural breeding barriers.
2. Restriction endonucleases are bacterial enzymes used for cleaving DNA.
3. The type II class of restriction endonucleases is the workhorse of biotechnology.
4. Vectors are entities for carrying the target DNA into a host cell for replication or cloning. There are bacterial, viral, and cosmid vectors. A vector may have one or more reporter genes and one or more selectable markers.
5. Plasmids are double-stranded, circular DNA molecules found in bacteria.
6. Viruses that infect bacteria are called bacteriophages.
7. One of the most common viruses used in plants is called the CaMV 35S.

8. Bacterial artificial chromosomes (BACs) and yeast artificial chromosomes (YACs) are widely used for cloning in eukaryotes.
9. Shuttle vectors can be used to clone in multiple hosts because they have ori (origin of replication) from different sources.
10. Promoters are the "engines" that drive DNA expression by determining the level of transcription of a selectable coding sequence.
11. Promoters are classified as either constitutive (continuously expressed) or regulated (periodic expression).
12. Selectable markers are depended upon for recovering transformed cell lines. These may be anitibiotic or herbicide resistance genes, among other types.
13. Transformation is the process of introducing free DNA into a host cell.
14. Gene transfer can be accomplished by direct methods (e.g., electroporation or particle bombardment) or mediated methods (e.g., *Agrobacterium* or chemical).
15. Transgenes may be targeted for expression only in specific tissues or throughout the organism.
16. Sometimes, transgenes may be unstably expressed (fail).
17. There are methodologies for conducting transformation research without using markers.

OUTCOMES ASSESSMENT

Internet-based
1. Problems on recombinant DNA: *http://web.mit.edu/esgbio/www/chapters.html*

ADDITIONAL QUESTIONS AND ACTIVITIES

1. What is genetic engineering?
2. Compare and contrast genetic improvement of an organism by conventional and genetic engineering approaches.
3. Describe the typical structure of a cloning vector.
4. Describe gene transfer by *Agrobacterium* mediation.
5. Discuss the concept of a universal gene pool.
6. Give the general characteristics of restriction endonucleases.
7. Discuss the plasmid vectors used in rDNA.
8. Distinguish between YACs and BACs.
9. What are shuttle vectors?
10. Describe the method of gene transfer by biolistics.
11. Describe scorable markers and how they are used in biotechnology.
12. Discuss the phenomenon of progressive transgenic failure.

INTERNET RESOURCES

1. Moving gene across species boundary: *http://vector.cshl.org/dnaftb/34/concept/index.html*
2. Overview of recombinant DNA technology: *http://www.ultranet.com/~jkimball/ BiologyPages/R/RecombinantDNA.html*

3. PowerPoint presentations on the variety of laboratory protocols: *http://www.biotech.iastate.edu/publications/ppt_presentations/default.html*

4. Restriction enzymes: *http://www.accessexcellence.org/AB/GG/restriction.html*

5. Restriction endonucleases: *http://www.ultranet.com/~jkimball/BiologyPages/R/RestrictionEnzymes.html*

6. Inserting gene into a plasmid: *http://www.accessexcellence.org/AB/GG/inserting.html*

7. Cloning into a plasmid: *http://www.accessexcellence.org/AB/GG/plasmid.html*

8. Cloning into a plasmid: *http://web.mit.edu/esgbio/www/chapters.html*

9. Promoters in gene regulation: *http://www.ultranet.com/~jkimball/BiologyPages/P/Promoter.html*

10. Cloning into YAC: *http://www.accessexcellence.org/AB/GG/YAC.html*

REFERENCES AND SUGGESTED READING

Asano, Y., Y. Otsuki, and M. Ugaki. 1991. Electroporation-mediated and silicon carbon fiber-mediated DNA delivery in *Agrostis alba* L. (Redtop). *Plant Science, 79:*249–252.

Barcelo, P., S. Rasco-Gaunt, C. Thorpe, and P. A. Lazzeri. 2001. Transformation and gene expression. In *Advances in botanical research incorporating advances in plant pathology, biotechnology of cereals, Vol 34.* P. R. Shewry, P. A. Lazzeri, K. J. Edwards, and J. A. Callow (eds), New York: Academic Press.

Birch, R. G. 1997. Plant transformation: Problems and strategies for practical application. *Annual Review of Plant Physiology and Plant Molecular Biology, 48:*297–326.

Crystal, R. G. 1995. Transfer of genes to humans: Early lessons and obstacles to success. *Science, 270:*404–410.

Geneve, R. L., J. E. Preece, and S. A. Merkle (eds). 1997. *Biotechnology of ornamental plants.* New York: CAB International.

Glick, B. R., and J. J. Pasternak. 1994. *Molecular biotechnology, principles and applications of recombinant DNA.* Washington, DC: ASM Press.

Gray, D. J., and J. J. Finer (eds). Development and operation of five particle guns for introduction of DNA into plant cells. *Special section on particle bombardment: Plant Cell Tissue and Organ Cult., 33:*219–257.

Hamilton, A. J., and D. C. Baulcombe. 1999. A species of small antisense RNA in post-transcriptional gene silencing in plants. *Science, 286:*950–952.

Hansen, G., and M. S. Wright. 1999. Recent advances in the transformation of plants. *Trends in Plant Science, 4:*226–231.

Hooykass, P. J. J., and R. A. Schilperoort. 1992. *Agrobacterium* and plant genetic engineering. *Plant Molecular Biology, 19:*15–18.

Jefferson, R. A., T. A. Kavanagh, and M. W. Bevan. 1987. GUS fusions: b-glucuronidase as a sensitive and versatile gene fusion marker in higher plants. *EMBO J., 6:*3901–3907.

Kikkert, J. R. 1993. The biolistic PDS-1000/He device. *Plant Cell Tissue Organ Cult., 33:*221–226.

Klug, W. S., and M. R. Cummings. 1997. *Concepts of genetics,* 5th ed. Upper Saddle River, NJ: Prentice Hall.

Lal, R., and S. Lal. 1993a. Electroporation. *Genetic engineering of plants for crop improvement* (pp. 14–23). Boca Raton, FL: CRC Press, Inc.

Lal, R., and S. Lal. 1993b. *Genetic engineering of plants for crop improvement.* Boca Raton, FL: CRC Press, Inc.

Potrykus, I. 1990. Gene transfer to plants: Assessment and perspectives. *Physiologia Plantarium, 79:*125–134.

Zambryski, P. C. 1988. Basic processes underlying *Agrobacterium*-mediated DNA transfer to plant cells. *Annu. Rev. Genet., 22:*1–30.

Enabling Technologies of Biotechnology

There are certain key technologies that are critical to biotechnology. Some of these were mentioned in passing during the discussion of rDNA technology, the core technology of biotechnology. The purpose of this section is to discuss, in some detail, selected technologies that are fundamental or indispensable to biotechnology, or help scientists to perform certain unique tasks. Some of these key technologies include tissue and cell cultures, markers, electrophoresis, PCR, DNA sequencing, cDNA, blotting, gene transfer, restriction enzymes, and informatics. Another key technology that enables microbes to be used as bioreactors is the fermentation technology. This technology is described in detail in the discussion covering industrial applications.

7 Cell and Tissue Culture

PURPOSE AND EXPECTED OUTCOMES

Manipulating organisms in biotechnology frequently involves manipulating individual cells or tissues. Target genes are inserted into cells, the fundamental units of organization in higher organisms. In plants, the single cells are then nurtured back into full-fledged plants. Tissue and cell culture techniques are used to nurture and sustain single cells or tissues to make this possible. Tissue and cell cultures are used in microbial and general animal biotechnology. Plant tissue and cell culture will be described to introduce this technology. It offers a much wider scope of application than animal tissue culture. Specifically, plant tissue culture may be used not only to help manipulate an organism at the molecular level but also to nurture a cell to complete the adult stage in the laboratory.

In this chapter, you will learn:

1. The general properties of a tissue culture medium.
2. How cells and tissues can be regenerated into full plants.
3. The importance of cell and tissue culture in biotechnology.

CONCEPT OF TOTIPOTENCY

Raising plants from seed is seen as a normal part of a plant's life cycle. In nonflowering species, plants can be raised from cuttings (obtained from the leaf, stem, root, or other parts). However, is it possible to raise full plants starting from portions of plants such as tissues or even single cells? Tissue and cell culture is based on the concept that higher organisms can be separated into component parts that can, in turn, each be cultured *in vitro* back into the full-fledged plants. Each cell in a complex organism is potentially **totipotent** (i.e., endowed with the full complement of genes to direct the development of the cell into a full organism). In theory, a cell can be taken from a root, leaf, or stem, and manipulated *in vitro* to grow back into a complete plant. Some exceptions do exist in which cells are unable to differentiate into all the various kinds of cells present in an adult organism (**multipotent** or **pluripotent**).

ENVIRONMENTAL REQUIREMENTS FOR TISSUE CULTURE—OVERVIEW

Nurturing a cell or tissue into a plant requires that the appropriate environment (biotic and abiotic) be provided.

1. **General environment**

    One of the most critical requirements for tissue culture is a sterile environment. A basic piece of equipment in a tissue culture lab is hence an autoclave. All glassware and other tools and materials are routinely sterilized before use. In addition to heat sterilization in an autoclave, ethanol and Clorox (household bleach) are standard chemicals in a tissue culture lab. Another key piece of tissue culture laboratory equipment for maintaining a sterile environment is the laminar flow hood (see Figure 7–1). Instead of the conventional upward draft of air current associated with fume cupboards in chemistry laboratories, filtered air is blown over the work area toward the worker. This horizontal draft of air reduces the chance of contaminants dropping under the force of gravity into open containers.

2. **Cultural environment**

    The growth environment for growing plants in the soil under natural conditions should provide adequate moisture, nutrients, light, temperature, and air. Similarly, in tissue and cell culture, plant materials are grown in totally artificial environments in which these growth factors are provided. The cultural environment may be manipulated by the researcher to control the growth and development of the cultured material. For example, in plant tissue culture, the researcher may cause the callus to develop only roots or shoots by manipulating the hormonal balance in the culture medium.

FIGURE 7–1

A laminar flow hood is a critical part of a biotechnology laboratory. It is used to provide a sterile environment for tissue culture and other *in vitro* manipulations.

Source: Photo taken at the Center for Biotechnology Research and Education at Langston University.

CULTURE MEDIUM

What kind of medium is used to culture cells and tissues? A tissue culture medium contains components that can be categorized into four groups: mineral elements, organic compounds, plant growth regulators, and support systems.

1. **Mineral elements**

    Mineral elements consist of both macro- and micronutrients like those obtained in a mineral soil. Several recipes for mineral elements have been developed over the years, of which the most commonly used is Murashige and Skoog medium (popularly called the **MS salts**) developed in 1962. The mineral elements and their functions are summarized in Table 7–1. It is informative to know that the ions as well as the form in which they are provided are both critical.

2. **Organic compounds**

    The organic compounds provide a carbon source and other growth-enhancing factors. The common ones are sugars, vitamins, and myo-inositol (Table 7–2). The most common sugar is sucrose, the major sugar that is transported in most plant species.

3. **Plant growth regulators**

    Plant growth regulators are the equivalent of growth hormones in animals. They have morphogenic effects on cultured tissues. Scientists, as previously indicated, manipulate the development and growth of cultured cells and tissues by varying the concentrations of growth regulators. The common growth regulators in tissue culture are auxins and cytokinins. The ratio of auxin to cytokinin in the medium determines whether shoots or roots would be enhanced. A balance in favor of auxins promotes rooting, whereas the reverse promotes shoot formation. Other classes of plant growth are presented in Table 7–3.

TABLE 7–1

Inorganic nutrients used in the preparation of tissue culture media. The nutrients together provide all the essential micro- and macronutrients for plant growth.

1. Macronutrients	Nitrogen—NO_3
	NH_4
	Phosphorus—P
	Potassium—K
2. Micronutrients	Ca, Mg, Cl, Fe, S, Na, B, Mn, Zn, Cu, Mo, Co, I

TABLE 7–2

Common organic compounds used in the preparation of tissue culture media. The specific combinations of compounds used depends on the purpose of the medium.

Compound	Functions
Sugars	Usually sucrose (but sometimes fructose, glucose, sorbitol); energy or carbon source; contributes to osmotic potential (osmoticum).
Vitamins	Usually thiamine (B_1), but also nicotinic acid (niacin), pyridoxine (B_6), vitamin C, E; required for carbohydrate metabolism.
Myo-inositol	Sugar alcohol; has a role in membrane and cell wall development.
Complex organics	Includes coconut milk, yeast extract, fruit juices; promotes general growth; difficult to quantify.
Activated charcoal	Absorbs toxic compounds released by the plant tissues; aids in root induction.

TABLE 7–3

Plant growth regulators commonly used in tissue culture media. Auxins generally promote rooting, while cytokinins promote shoot growth.

Compound	Function
Auxins	Affects cell elongation, adventitious rooting, and apical dominance. Used also for callus induction from explants and for somatic embryogenesis. Examples of natural auxins are indole-3-acetic acid (IAA), indole-3-butyric acid (IBA); synthetic auxins include 1-naphthalene acetic acid (NAA), 2,4-dichlorophenoxyacetic acid (2,4-D).
Cytokinins	Antagonistic in effect to auxins regarding apical dominance. Often inhibits embryogenesis and root induction. May be natural (e.g., zeatin) or synthetic (e.g., benzyladenine or BA, kinetin).
Gibberellins	Main role is to promote stem elongation and flowering. Only two forms used in tissue culture (GA_3, and GA_{4+7}).

4. **Support systems**

The medium components are delivered either in a liquid or solid medium. An aspect of media preparation is pH buffering. Improper pH has a variety of adverse consequences including failure of gel to solidify. In liquid media culture (or suspension culture, as it is also called), tissues or cells are suspended in water fortified with nutrients. To improve aeration, the medium has to be agitated frequently. Solid media are prepared by using gelling agents, the most common of which are being agar and agarose. Agar, the most widely used gelling agent, is a mixture of polysaccharides derived from red algae. Agar is easy to prepare. It melts at about 100° C and solidifies at about 45° C. However, when pH is acidic (especially less than 4.5) agar does not gel properly. It is important to prepare agar to the right concentration, usually between 0.5 and 1.0 percent. Agar is resistant to enzymes and does not react with media components. However, it has impurities.

Agarose is a purer support material that is extracted from agar. It is without the impurities (agaropectin and sulfate groups) found in agar. It also has higher gel strength and thus requires smaller amounts for preparing a solid medium. There are other gelling agents such as gellan gums (e.g., Phytagel®), which provide clear gels (rather than translucent gels). They are derived from bacterium *Pseudomonas elodea*. However, these gels have certain limitations. The choice of a support system is based on experimentation.

Standard media are available for various purposes. The ingredients of the popular MS medium are presented in Table 7–4. Normally, tissue culture researchers prepare stock solutions of these minerals and vitamins. Media may be sterilized by autoclave (at 121° C and pressure of 1.05 kg/cm^2 for 15 mins). Heat-labile materials may be sterilized by filter sterilization. The media may be stored at 4° C.

MICROPROPAGATION

Micropropagation is the *in vitro* clonal propagation of plants. The term is sometimes used synonymously with tissue culture. Micropropagation can utilize preexisting meristems or utilize nonmeristematic tissue. It is a method that is used in commercial production of certain high value horticultural species. There are four basic *in vitro* micropropagation methods: shoot culture, node culture, shoot organogenesis, and nonzygotic embryogenesis. Of these methods, shoot culture (or enhanced axillary shoot proliferation) is the most commonly used for micropropagation (see Figure 7–2).

TABLE 7–4
The sources of compounds used in the preparation of MS salts. Frequently, researchers prepare stock solutions (e.g., 10x, 20x) from these salts, and use them over a long period of time.

Nutrient	Source
Nitrate	NH_4NO_3
	KNO_3
Sulfate	$MgSO_4.7H_2O$
	$MnSO_4.H_2O$
	$ZnSO_4.7H_2O$
	$CuSO_4.5H_2O$
Halide	$CaCl_2.2H_2O$
	KI
	$CoCl_2.6H_2O$
P, B, Mo	KH_2PO_4
	H_3BO_3
	Na_2MoO_4
NaFeEDTA	$FeSO_4.7H_2O$
	NaEDTA

FIGURE 7–2
Micropropagation in cassava. Small cuttings are rooted in a sterile environment *in vitro* until they attain a certain size. Then, they are transferred into pots.
Source: Photo is courtesy of Nigel Taylor, Danforth Plant Science Center, St. Louis, Missouri. Used with permission.

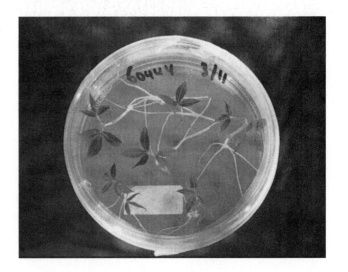

■ *SHOOT AND NODE CULTURE*

These methods entail the propagation from preexisting meristems. In shoot culture (or **shoot-tip culture,** as it is sometimes called), axillary shoot proliferation is stimulated by disrupting apical dominance through the supplementation of the culture medium with cytokinin. The shoot may be divided into shoot tips and nodal segments that may be used as secondary explants for another cycle of axillary shoot proliferation, or as microcuttings that are induced to root to produce plantlets.

Nodes (rather than shoots) are preferred in species (e.g., sweet potato) that do not respond well to cytokinin stimulation of axillary shoot proliferation. In this case, shoots consisting of single or multiple nodes per segment are placed horizontally on the medium. From these materials single unbranched shoots arise that may be induced to root to produce plantlets.

Micropropagation occurs in five stages:

1. **Selection of explant source and preparation**

    The plant part used to initiate tissue culture is called the **explant,** which must be in good physiological condition and disease free. The size of the explant, its location on the plant, its age, or its developmental phase are among factors that affect the success of micropropagation.

2. **Initiation and aseptic establishment**

    The explant is surface sterilized before placing it on the medium. Exogenous plant growth regulators may be added to the medium according to the source of the explant. Dependence on supplemental growth regulators is determined in part by the size and location of the explant.

3. **Proliferation of axillary shoots**

    Cytokinin-enriched medium enhances axillary shoot proliferation. The shoots may be subcultured at an interval of about four weeks.

4. **Rooting**

    The purpose of this stage is to prepare shoots for transfer into the soil. This requires application of auxin in the medium. Some commercial producers bypass this stage and plant directly into the soil.

5. **Transfer to natural environment**

    Plants are put through the process of hardening off to ready them for the natural environment. They are gradually moved from ideal lab conditions to more natural conditions by reducing the relative humidity and increasing light intensity.

■ ORGANOGENESIS

Organogenesis is the ability of nonmeristematic plant tissues to form various organs *de novo* (e.g., embryos, flowers, leaves, shoots, and roots). This is possible because plant cells have the capacity to dedifferentiate from their current structural and functional state, and then embark upon a new developmental path to produce new characteristics.

There are two basic pathways by which organogenesis occurs: indirect and direct.

1. **Indirect organogenesis**

    This pathway includes a callus stage (see Figure 7–3). **Callus** is a mass of dedifferentiated cells like meristematic cells. The callus forms meristemoids or an aggregation of meristem-like cells that are developmentally plastic (can be manipulated to redirect a morphogenic end point). The callus phase provides an opportunity for the introduction of variation (somaclonal variation).

2. **Direct organogenesis**

    Direct organogenesis bypasses a callus stage. The cells in the explant act as direct precursors of a new primordium.

■ NONZYGOTIC EMBRYOGENESIS

In sexual reproduction, embryos form after fertilization to produce a zygote (zygotic embryo). Nonzygotic embryos can be produced in tissue culture from somatic tissue. Somatic embryos arise from a single cell rather than budding from a cell mass as in zygotic embryos. This event is very important in biotechnology since genetic engineering of plants may involve the manipulation of single somatic cells. Transformation that targets somatic explants is put through morphogenesis whereby regeneration occurs via somatic

a.   b.   c.

d.   e.

f.

FIGURE 7–3

An illustration of organogenesis in wheat. By manipulating the cultural environment, a mass of callus (b) differentiates into organs and eventually into a full plant.

Source: Photo is courtesy of Bryan Kindiger, USDA-ARS Grazinglands Research Station, El Reno, Oklahoma. Used with permission.

embryo formation. It is clear that the capacity for embryogenesis has a genetic basis. However, the exact nature of the triggering mechanism(s) is unknown.

Induction of embryogenesis requires the use of plant growth regulators, the common ones being the synthetic auxin 2,4-D (2,4-dichlorophenoxyacetic acid). Other auxins that are used include IBA (indole butyric acid), dicamba, and picloram. Cytokinins are also used, the most common being BA (benzyladenine). Others are TDZ (thidiazuron) and kinetin.

Embryo development, zygotic or somatic, goes through certain stages: globular, scutellar, and coleoptilar (in monocots), and globular, heart, torpedo, and cotyledonary stages (in dicots). It is generally difficult to obtain plants from somatic embryos. However, without successful regeneration, plant transformation cannot be undertaken. Somatic embryogenesis has potential commercial applications, one of which is in the synthetic seed technology (production of artificial seeds).

PROTOPLAST CULTURE

A **protoplast** consists of all the cellular components of a cell, excluding the cell wall. Protoplasts may be isolated by either mechanical or enzymatic procedures. Mechanical isolation involves slicing or chopping of the plant tissue to allow the protoplast to slip out through a cut in the cell wall. This method yields low numbers of protoplasts. The preferred method is the use of hydrolytic enzymes to degrade the cell wall. A combination of three enzymes—cellulase, hemicellulase, and pectinase—is used in the hydrolysis. The tissue used should be from a source that would provide stable and metabolically active protoplasts. This calls for monitoring plant nutrition, humidity, daylength, and other growth factors. Often, protoplasts are extracted from leaf mesophyll or plants grown in cell culture. The isolated protoplast is then purified, usually by the method of flotation.

This method entails first centrifuging the mixture from hydrolysis at about $50 \times g$, and then resuspending the protoplasts in a high concentration of fructose. Clean, intact protoplasts float and can be retrieved by pipetting.

Protoplasts are used in biotechnology in various ways. For example, as was previously described, transgene delivery via protoplasts is one of the methods used in plant transformation. Protoplasts can also be used to create hybrids *in vitro* (as opposed to crossing mature plants in conventional plant breeding).

SOMATIC HYBRIDIZATION

Protoplasts from two different plants can be fused to create a hybrid. Fusion can occur spontaneously in certain cell lines called **fusogenic protoplasts.** The most common methods of fusion are by chemical agents or electrical manipulation. Fusogenic agents include salt solutions (e.g., KCl, NaCl). However, the most commonly used agent is polyethylene glycol (PEG). The protoplasts are agglutinated (brought together) by the application of PEG to facilitate the fusion. Addition of the compound called concanavalin A to PEG enhances the fusion. Cell fusion by **electrofusion** is also used. Protoplasts are agglutinated by the technique of dielectrophoresis in which they are subjected to a nonuniform AC field of low intensity. This is followed by an application of high-voltage AC pulse to destabilize the cell membrane at specific sites to facilitate the fusion.

The products of fusion are cultured on an appropriate medium (e.g., MS). Fusion of cells does not necessarily guarantee fusion of nuclei. For a stable hybrid to form, the two nuclei must fuse within a single cell, followed by mitosis involving the two genomes. There is the need to have a selection system to verify hybridity since fusion is nonspecific and therefore allows the formation of various products (multiple fusions, homokaryons, heterokaryons, and unfused protoplasts). Various methods are used to verify hybridity, including genetic complementation of non-allelic mutants, use of selective media, isozyme analysis, and microisolation.

Somatic hybridization provides another opportunity for circumventing barriers to sexual hybridization. Sexual hybridization involves fusing two haploid nuclei and one maternal cytoplasm; somatic hybridization combines diploid nuclear genomes and two maternal (cytoplasmic) genomes. Whereas sexual hybrids are uniform, somatic hybrids produce significant variability in the population, resulting from genetic instability, mitotic recombination, somaclonal variation, and cytoplasmic segregation. Products of somatic hybridization may be true hybrids or parasexual hybrids with the complete genomes of two parents, partial genomes of the parents (called **assymetrical hybrids**), or **cybrids** (combination of the nuclear genome of one parent and the cytoplasm of another). When closely related species are used, somatic hybrids that are polyploids are formed. For example, a somatic hybrid of potato was produced that combined the genomes of *Solanum tuberosum* ($2n = 48$) and *S. brevidens* Phil (a wild non-tuberous variety; $2n = 24$) to produce a somatic hybrid with 72 chromosomes.

ANIMAL TISSUE AND CULTURE

Animal tissues are cultured for a variety of purposes. Unlike plants, the goal is not to produce a full-grown organism from a cell, but rather to produce organs (e.g., skin), antibodies (as is undertaken by tissue culture and hybridoma facilities), or even chemical products. The media used are different, but the principles for producing them are the same as for plant tissue culture.

One of the controversial areas of biotech research is stem cell research, which is discussed under applications in medicine. Cell or tissue culture is required in animal cloning and the development of transgenic animals. Animal cells in culture are injected with transgenes, then placed in females for nurturing into adult animals.

KEY CONCEPTS

1. The cell is the fundamental unit of organization of all living organisms.
2. Micropropagation is the *in vitro* clonal propagation of plants.
3. A tissue culture medium consists of four primary components: mineral elements (e.g., NPK and micronutrients), organic compounds (sugars, vitamins, and myoinositols), growth regulators, and support systems (e.g., agar, agarose).
4. Each cell is potentially totipotent, capable of being manipulated *in vitro* to grow back into a complete plant.
5. Callus is a mass of undifferentiated cells.
6. There are four basic *in vitro* micropropagation methods: shoot culture, node culture, shoot organogenesis, and non-zygotic embryogenesis.
7. The plant part used to initiate tissue culture is called the explant.
8. Organogenesis is the ability of nonmeristematic plant tissue to form various organs *de novo* (e.g., embryos, flowers, leaves, shoots, and roots). It can occur by direct or indirect means.
9. Somatic embryogenesis goes through a callus phase.
10. Growth regulators are important in tissue culture where they are used to influence events like organogenesis.
11. A protoplast consists of all the cellular components of a cell, excluding the cell wall. Protoplasts from two different plants can be fused to create a hybrid.
12. Whole plants can be regenerated from single plant cells.
13. Tissue culture is a part of most transformation projects in plants.

OUTCOMES ASSESSMENT

1. Why is it possible (at least theoretically) to raise a full plant or for that matter any organism from just one of its own cells?
2. All cells are not totipotent. Explain.
3. What is the importance of a callus phase in plant tissue culture research?
4. Discuss the rationale for the composition of a tissue culture medium.
5. What is clonal propagation?
6. Describe the *in vitro* production of hybrids.

Internet-based
1. *http://www.biology.arizona.edu/human_bio/problem_sets/human_genetics/12t.html*

INTERNET RESOURCES

1. Journal of plant tissue and organ culture: *http://www.kluweronline.com/issn/0167-6857*
2. Soybean tissue culture: *http://mars.cropsoil.uga.edu/homesoybean/*

3. Links to numerous aspects of plant micropropagation: *http://aggie-horticulture.tamu.edu/tisscult/microprop/microprop.html*
4. Repositories for somatic cell hybrids: *http://locus.umdnj.edu/nigms/hybrids/hybrids.html*

REFERENCES AND SUGGESTED READING

Fosket, D. E. 1994. *Plant growth and development—a molecular approach.* New York: Academic Press.

Frey, L., Y. Saranga, and J. Janick. 1992. Somatic embryogenesis in carnation. *HortScience, 27:*63–65.

Gamborg, O., R. Miller, and K. Ojima. 1968. Nutrient requirements of suspension cultures of soybean root cells. *Exp. Cell Res., 50:*473–497.

Geneve, R. L., J. E. Preece, and S. A. Merkle (eds). 1997. *Biotechnology of ornamental plants.* New York: CAB International.

Glick, B. R., and J. J. Pasternak. 1994. *Molecular biotechnology, principles and applications of recombinant DNA.* Washington, DC: ASM Press.

Jones, P. G., and J. M. Sutton. 1997. *Plant molecular biology: Essential techniques.* New York: John Wiley and Sons, Ltd.

Kamo, K. 1995. A cultivar comparison of plant regeneration from suspension cells, callus, and cormel slices of *Gladiolus. In Vitro Cellular and Developmental Biology, 31:* 113–115.

Kelly, A., and S. R. Bowley. 1992. Genetic control of somatic embryogenesis in alfalfa. *Genome, 35:*474–477.

Murashige, T, and T. Skoog. 1962. A revised medium for rapid growth and bioassays with tobacco tissue culture. *Physiologia Plantarum, 15:*473–497.

Stefaniak, B. 1994. Somatic embryogenesis and plant regeneration of gladiolus. *Plant Cell Reports, 13:*386–389.

Trigiano, R. N., and D. J. Gray (eds). 1996. *Plant tissue culture concepts and laboratory exercises.* New York: CRC Press.

8 Electrophoresis and Blotting

PURPOSE AND EXPECTED OUTCOMES

One of the specific objectives of molecular biotechnology research is gene isolation. However, the process of isolating a single gene from a genome is like searching for a needle in a haystack. As a typical eukaryotic genome is about 10^9 bp, and a single gene may average about 5,000 bp, the proportion of a genome represented by a gene is about 1.0×10^{-5} percent.

In this chapter, you will learn:

1. The basic techniques of electrophoresis.
2. The identification of specific biomolecules from a mixture.
3. The isolation of specific biomolecules from a mixture.

The Southern blot, a technique for transferring DNA fragments from an agarose gel to a matrix (nitrocellulose filter), facilitates gene isolation. To do this, DNA fragments are produced from a restriction digest, which must then be separated or spread out to expose individual DNA fragments so the desired one can be identified and isolated.

WHAT IS ELECTROPHORESIS?

Electrophoresis is the biochemical technique used for separating compounds in an electrical gradient. The separation of compounds is based on variations in molecular or physical structure and chemical properties (e.g., size, shape, and natural charges). If significant natural differences do not occur, the experimental environment (e.g., pH) may be manipulated to effect separation.

■ ROLE OF ELECTRICAL CURRENT AND CHARGE

Biomolecules such as amino acids may be neutral, positively charged, or negatively charged. At a pH of 7 (neutral), many amino acids are **amphoteric**; that is, they carry both positive and negative charges. However, it is not the charge *per se* that is important in electrophoresis, but the charge density (charge per unit mass or net charge per surface area) that determines the rate of movement in the electrical field. The charge per mass ratio is constant for nucleic acids because the negative charges are evenly spaced along the molecule. However, this property is variable for proteins. Consequently, these two key biomolecules (nucleic acids and proteins) are separated by methods that take into account this fundamental difference.

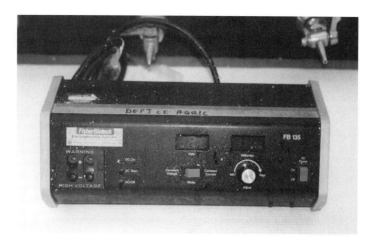

FIGURE 8–1

Direct current (DC) is used in electrophoresis to separate charged molecules. The sources of DC vary in output as well as capacity.

Source: Photo taken at the Center for Biotechnology Research and Education at Langston University.

Electrophoresis uses DC (direct current) only (see Figure 8–1). The goal of electrophoresis is to accomplish differential separation of samples. Each molecule has an electrophoretic mobility which is a standard reference value measured as cm^2/V. Migration is affected by charge density, molecular size and shape, medium characteristics, and buffer properties. Support media may have sieving properties, retarding the mobility of large molecules while allowing small ones of identical charge density to move with relative ease. Molecules with large charge density will migrate faster than those with smaller charge densities, and some electrophoretic procedures (SDS-PAGE) are able to cancel out the effect of charge so that molecules are separated on the basis of their molecular weight.

■ SUPPORT SYSTEMS

Electrophoresis is commonly conducted in a matrix or support system, the common ones being gels prepared from starch, agar, agarose, or polyacrylamide. Gels are prepared to give desirable physical properties to effect separation. Starch gel is relatively inexpensive; however, its weaknesses include low strength (fragile) and variability among sources that prevent the reproducibility of experimental results. Agar is widely used in microbial research because of its versatility, strength, and clarity, among other properties. Agarose is clearer, stronger, and even more versatile. Its large pores suit the separation of large biomolecules like nucleic acids. Polyacrylamide gels have more strength than agarose and can be prepared to provide thin gel thickness and a wide range of pore sizes. Preparation of polyacrylamide is more complex, not to mention the fact that unpolymerized polyacrylamide is toxic. Agarose is routinely used in molecular biotechnology research.

Gels may be cast in one of three basic ways—**rod (cylindrical), vertical slab,** or **horizontal slab.** Slabs are more commonly used in biotechnology because they allow multiple samples to be run in parallel on the same gel for easy comparison. Slab gel sizes are also variable and can be classified into four groups: micro, mini, standard, and long (see Figure 8–2). These gels are cast to about 1 mm thickness or even less.

VISUALIZATION

Once electrophoresis is completed, the researcher needs to visualize the relative locations of the separated molecules, which usually occur as either bands or spots. Depending on

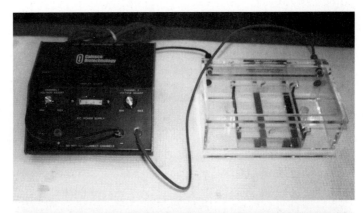

(a)

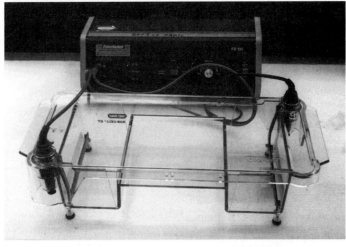

(b)

FIGURE 8–2
Electrophoresis units are designed to run gels in either a horizontal or vertical position. Gel units differ in size. The smaller units (a) are used for short-duration runs of less than one hour. Larger units (b) may be used for overnight runs.
Source: Photos taken at the Center for Biotechnology Research and Education at Langston University.

the purpose of the investigation, the migration of unknown molecules may be compared with those of standards included in the process. There is sometimes interest in quantifying the amount of a specific component by determining the intensity of the signal. Because biomolecules (especially nucleic acids) are often noncolored, scientists need additional techniques to make these bands or spots visible.

Protein and nucleic acids may be visualized under ultraviolet (uv) illumination. However, direct and immediate evaluation may not be convenient or suitable in all situations. There are general steps in gel visualization. The gel is initially removed from the electrophoresis chamber. This is followed by some treatment to fix the bands or spots (by denaturation, cross-linking, or blotting). The fixation step is desirable for minimizing distortion of bands. However, some studies require that the biomolecules remain biologically active; hence, they cannot be denatured by fixation.

Gel visualization techniques may be grouped into two categories: direct treatment and use of probes.

1. **Direct treatment**

In protein electrophoresis, gels are often exposed to dyes (including color dyes) to stain these biomolecules. Dye staining is a quick and inexpensive technique. However, it has significant setbacks, including significant background noise. Furthermore, dyes tend to fade with time. Among the dyes commonly used are Amido Black 10B, Ponceau S Fast Green, and Coomassie Blue dyes (especially type R 250). In addition, nucleic acids can be stained with fluorescent dyes and visualized under uv light. This technique requires special equipment. Radioisotopes are more challenging to use and require autoradiography, but are applicable to proteins and nucleic acids.

2. **Use of probes**

Probes utilize matching sequences to select specific bands, thereby reducing the number of bands visualized to a minimum. Stains can be labeled in various ways: by antibodies, nucleic acids, radioisotopes, or those that are enzyme-linked. Antibodies are best used on blots but are prone to cross-reactions. Nucleic acids and proteins require denaturing. Labeling with antibodies utilizes the natural biological activities of proteins. Antibody labeling is based on immunological binding to specific proteins (antigens). Enzyme-linked probes require an appropriate substrate that is converted into a visible product. Some modern methods utilize chemiluminescent products that can be visualized under ultraviolet light. The product in both cases must be photographed in order to obtain a permanent record of the electrophoretic run.

TWO-DIMENSIONAL (2-D) ELECTROPHORESIS

Two-dimensional (2-D) electrophoresis is a technique for separating complex compounds in two directions, one after the other (i.e., two consecutive runs at right angles to each other) (see Figure 8–3). This technique is able to separate molecules that co-migrated to the same location in the first run, but are not identical, to be separated in the second run. The separation in the first direction is accomplished on the basis of charge, usually by **isoelectric focusing** (IEF). IEF is a high-resolution technique for separating proteins and other amphoteric molecules. Conventional electrophoresis is conducted at a constant buffer pH. IEF effects separation in a uniform pH gradient, which is produced by using carrier ampholytes (synthetic aliphatic compounds with different charges) in the support medium. When placed in an electric field, the current causes the ampholytes to be arranged in the medium according to their isoelectric points (points of zero net charge). This creates a pH gradient in which the anodal region of the medium is occupied by the most anodic ampholytes while the most basic ones migrate to the cathodal region. In the second stage, protein samples migrate along the pH gradient in the support matrix until they reach their respective isoelectric point (pI). At these locations, the proteins completely lose their electrophoretic mobility due to their zero net charge. As a result, proteins of identical pI become concentrated in a very narrow band. Should a molecule stray away from its pI location, it immediately acquires a charge and becomes electrophoretically mobile once again, whereupon it must then migrate back to its pI. It is this anti-diffusion property of IEF that is responsible for the high resolution of bands in focused gels.

IEF is followed by separation in the second direction on the basis of molecular weight. Polyacrylamide gels can be manipulated such that pore size varies systematically across the gel, with the gel density increasing (while pore size is decreasing) the farther away one is from the origin of electrophoresis. This technique improves the resolution of bands on the gel. For protein analysis, chemicals such as SDS (sodium diodecylsulfate), a detergent, may be added to polyacrylamide gels to denature proteins by dissociation (called **SDS-PAGE** or **polyacrylamide gel electrophoresis**).

After electrophoresis, the proteins may be visualized by staining them with Coomassie Brilliant Blue 250 (silver stain) fluorescent dyes, or through radioisotope detection after labeling the proteins with 3H-, 14C-, or 35S-labeled amino acids. Two-dimensional gels may also be electroblotted to PVDF or nitrocellulose membranes for further analysis. The analysis of 2-D gels is complicated, requiring the assignment of

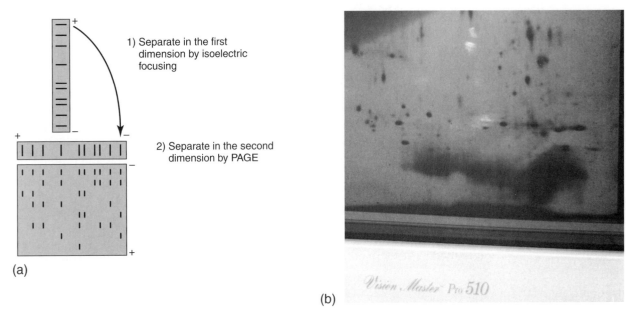

1) Separate in the first dimension by isoelectric focusing

2) Separate in the second dimension by PAGE

(a)

(b)

FIGURE 8–3

Two-dimensional (2-D) electrophoresis is used to separate molecules in two ways. The first separation is conducted using the technique of isoelectric focusing (IEF) to separate the material along a pH gradient. The second separation is conducted by the method of polyacrylamide gel electrophoresis (PAGE) in which separation is based on molecular size. Because the results of 2-D electrophoresis can be very complex, (see the photo labeled b), computer-aided interpretation of data, using x-y coordinates, may be helpful.

x-y coordinates to the bands. A key limitation to 2-D electrophoresis is that only one sample can be electrophoresed at a time, preventing the use of standards for comparison. Furthermore, reproducibility of the separation is not good. 2-D electrophoresis is critical to protein engineering and some other biotechnology strategies for studying the genome that will be discussed later.

CAPILLARY ELECTROPHORESIS

Another powerful separation technology, the **capillary electrophoresis,** can be applied to proteins, peptides, amino acids, and nucleic acids, organic acids, and inorganic ions, among other molecules. The separation is effected in tiny tubes of about 0.010 to 0.075 mm inner diameter (hence the name capillary electrophoresis). These tubes are made of fused silica (silica that has been treated to get rid of the defects of crystal structure). Fused silica allows uninhibited penetration of uv light from a detector, resulting in a more accurate reading. The capillary tube is filled with buffer while each end is immersed in a vial of the same buffer. A voltage of about 100 to 700 volts/cm is applied across the capillary. This induces endoosmotic flow of H+ ions causing both anions and cations to migrate to the cathode, separating into components along the way (see Figure 8–4). At the other end of the capillary, each separated analyte is detected and quantified. Modes of detection include mass spectrometry, fluorescence, radioactive detection, and electrochemical detection. The types of data collected are migration time (mobility), peak width (diffusion coefficient), mobility (charge around molecule), and diffusion coefficient (size of molecule).

Capillary electrophoresis is widely used in genomic and proteomic research. Its advantages include high separation efficiency, small sample required (1 to 10μl), fast separation (1 to 45 mins), automation, reproducibility, linear quantitation, coupling to mass

FIGURE 8–4
A schematic presentation of the components of a capillary electrophoresis system.

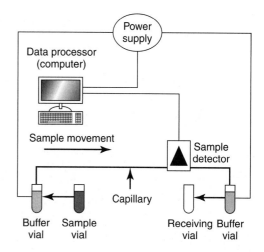

spectrophotometer, and fundamental data collection. Because proteins are large and amphoteric molecules, they tend to stick on the walls of the tube. Using high salts (0.5 to 01 M) may help prevent this sticking, but it generates heat in the process.

BLOTTING

Blotting is the technique of transferring electrophoretic products onto other materials prior to visualization. Proteins and nucleic acids may be blotted after electrophoresis. Blotting allows the separated materials to be more accessible to the dyes and reagents used in visualization. It also facilitates the staining and destaining processes, since the membranes are thinner. The blotted copy can be dried and stored for analysis at a later time.

Materials used for blotting are variable. Two of the most common materials are nitrocellulose and cellulose. Nitrocellulose is not stable in alkali, which is a significant limitation to its use, since blotting methods tend to use high pH buffers to minimize the tendency of single-stranded DNA molecules from reassociating and thereby increasing their transfer to the blotting membrane. A single nitrocellulose blot can be subjected to several rounds of visualization probing, one after the other, with removal by elution between successive probing. Nitrocellulose derivatives with improved binding capacity are also available.

Nylon-66 derivates provide the other popular blotting membrane. Nylon is several times stronger than nitrocellulose and has a higher binding capacity, allowing it to bind to nucleic acids in low ionic strength buffers. It is also able to handle more rounds of reprobing.

Several blotting systems are in use (see Figure 8–5). They are, however, variations of the technique originally developed by Edward Southern in 1975 for DNA, called the **Southern blot.** The subsequent protocols were named in a geographic fashion following Southern: **Western blot** (for proteins) and **Northern blot** (for RNA). The basic concepts for all the blots are the same. The compound to be blotted is first immobilized on the membrane, and all the vacant sites are blocked. The hardcopy is then subjected to staining. The transfer process originally occurred by capillary action effected through a pile of dry papers and heavy weight, while later versions used air vacuum to reduce the transfer time from about 18 hours to only 1 to 3 hours.

■ SOUTHERN BLOT

The steps in Southern blot are as follows:

1. Cloned DNA is cut by one or more restriction enzymes.
2. Fragments are submitted to gel electrophoresis for separation, then stained with ethidium bromide for visualization and photographed under ultraviolet light.

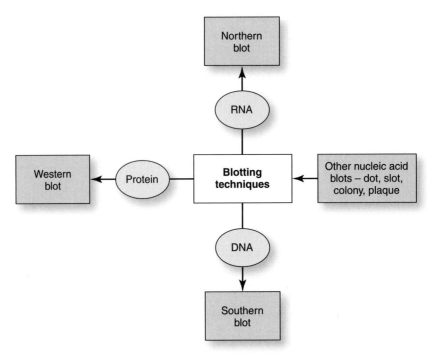

FIGURE 8–5
A summary of the types of blotting techniques and the materials for which they are used.

3. DNA in the gel is denatured into single-stranded fragments.
4. Denatured DNA is transferred to DNA-binding material (commonly, nitrocellulose or nylon). The transfer procedure gives the name to the technique—blot (see Figure 8–6). The gel is placed on a sponge wick that is partially immersed in a tray containing a buffer solution, which is covered with a DNA-binding filter (nitrocellulose or nylon). A stack of blotting paper is placed on the filter, followed by a weight. Capillary action draws buffer from the tray through the gel and transfers the pattern of DNA fragments from the gel to the filter.
5. DNA in the filter is immobilized by baking at 80°C or exposing to uv light to cross-link the fragments to the membrane.
6. The DNA is hybridized *in situ* with a radioactive probe. This may be done in a heat-sealed food bag containing probe solution.

Only the single-stranded DNA fragments that are complementary to the nucleotide sequence of the probe will form hybrids. The hybridization may be visualized with autoradiography (see Figure 8–7). The filter is washed to remove excess (unbound) probe and overlaid with an X-ray film for autoradiography.

Selected Uses of Southern Blot
Experimental uses of Southern blots include the following:

1. Characterizing cloned DNA.
2. Mapping restriction sites within or near a gene.
3. Identifying DNA fragments carrying a single gene from a mixture of other fragments.
4. Identifying related genes in different species.
5. Detecting rearrangements and duplications in genes associated with human genetic disorders and cancers.

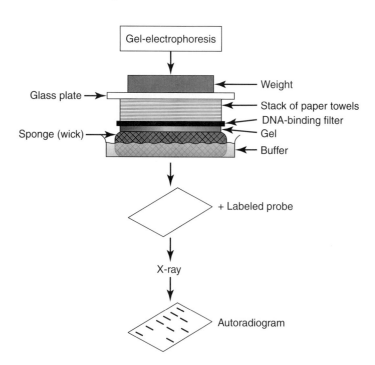

FIGURE 8–6

A sketch of a set for Southern blotting. Following electrophoresis of restriction digested DNA, the DNA in the cell is denatured, the fragments are transferred onto a DNA-binding filter, and the filter and a solution containing labeled probe are placed in a heat-sealed bag for a period. The filter is washed to remove excess probe and autoradiographed.

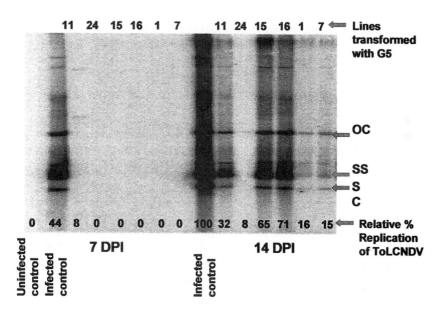

FIGURE 8–7

An example of a Southern blot analysis. This blot shows levels of ToLCNDV DNA accumulation in *N. benthamiana* plants transformed with G5 inoculated with ToLCNDV.

Source: Photo is courtesy of Nigel Taylor, Danforth Plant Science Center, St. Louis, Missouri. Used with permission.

■ *NORTHERN BLOT*

The Northern blot technique is a variation of the Southern blot technique. It is used to blot RNA (not DNA) from agarose gel. RNA is extracted from the experimental material (cells or tissues) and subjected to electrophoresis, after which the pattern of RNA bands is blotted onto an RNA-binding sheet (as in Southern blot). A labeled single-stranded

DNA probe (cDNA or cloned genomic DNA) is used for hybridization, followed by autoradiography.

Selected Uses of Northern Blot

Uses of Northern blot include:

1. Studying patterns of gene expression in embryonic and adult tissues.
2. Detecting alternative splicing of mRNA and multiple transcripts derived from a single gene.
3. Characterizing and quantifying transcriptional activity of a specific gene in different cells, tissues, or organisms (the density of RNA band on X-ray film relates to the amount of RNA transcribed).

■ *DOT BLOT*

Dot blot is a variant of the Northern blot technique where cloned RNAs are spotted adjacent to each other on a filter, followed by hybridization with a radioactive DNA probe representing the target sequence. The intensity of the dot corresponds with the amount of transcribed RNA in the cell or tissue being studied.

Selected Uses of Dot Blot

Uses of the dot blot include:

1. Detecting the presence and extent of a specific sequence in the RNA of various cell types.
2. Rapid testing of homology between a labeled probe and many cloned sequences.
3. Gene expression pattern analysis.

■ *ELECTROBLOTTING*

Electroblotting involves sandwiching the gel between supports connected to a power source with electrodes and subjected to a uniform electric field. Electroblotting is quick, requiring about one hour to complete, and is applicable to proteins and nucleic acids, as well as polyacrylamide gels and agarose.

■ *WESTERN BLOT*

Western blot is used for protein analysis. Protein identification is based on both antibody reactions and antigens. Proteins are separated by a denaturing SDS polyacrylamide gel. Following electrophoresis, the proteins are transferred (blotted) to a nylon membrane. The membrane is then exposed sequentially to solutions containing a primary antibody, followed by a secondary antibody to which an enzyme is coupled. All sites on the membrane that do not contain blotted protein from the gel are nonspecifically "blocked" so that antibody (serum) will not nonspecifically bind to them to produce a false positive result. When a large number of samples are being tested, the membrane may be cut into strips to facilitate testing for antibodies directed against the blotted protein (antigen). The membrane is then soaked in a substrate solution to develop the color reaction, which results in identifying the antigen as a band. Apparent molecular weights of the antigens are measured using protein markers of known molecular weight.

KEY CONCEPTS

1. Electrophoresis is the biochemical technique used for separation of compounds in an electrical gradient. The separation of compounds is based on variations in molec-

ular or physical structure and chemical properties (e.g., size, shape, and natural charges).

2. Electrophoresis is commonly conducted in a matrix or support system, the common ones being gels prepared from starch, agar, agarose, or polyacrylamide. Agarose is clearer, stronger, and more versatile. Its large pores suit the separation of large biomolecules like nucleic acids. Polyacrylamide gels have more strength than agarose and can be prepared to provide thin gel thickness and a wide range of pore sizes.

3. Gels may be cast in one of three basic ways—rod (cylindrical), vertical slab, or horizontal slab. Slabs are more commonly used in biotechnology because it allows multiple samples to be run in parallel on the same gel for easy comparison.

4. Because biomolecules (especially nucleic acids) are often noncolored, scientists need additional techniques to make these bands or spots visible. Gel visualization techniques may be grouped into two categories: direct treatment and use of probes.

5. Two-dimensional (2-D) electrophoresis is a technique for separating complex compounds in two directions, one after the other (i.e., two consecutive runs at right angles to each other).

6. Blotting is the technique of transferring electrophoretic products onto other materials prior to visualization. Proteins and nucleic acids may be blotted after electrophoresis.

7. The protocols are named in a geographic fashion: Southern blotting (for DNA), Western blotting (for proteins), and Northern blotting (for RNA).

OUTCOMES ASSESSMENT

Internet-based
1. Problems on blotting: *http://web.mit.edu/esgbio/www/chapters.html*
2. Applications with blotting: *http://www.biology.arizona.edu/molecular_bio/problem_sets/ Recombinant_DNA_Technology/06t.html*

ADDITIONAL QUESTIONS AND ACTIVITIES

1. What is the rationale of electrophoresis?
2. Why does electrophoresis utilize only direct current?
3. Why does 2-D electrophoresis separate only one sample at a time?
4. Describe the Southern blot technique.
5. Contrast the Southern blot and the Western blot techniques.
6. Discuss the technique of capillary electrophoresis.
7. Give three specific applications of the Southern blot technology.
8. Give specific advantages of capillary electrophoresis over conventional electrophoresis.

INTERNET RESOURCES

1. Application of electrophoresis to blood studies: *http://www.ultranet.com/~jkimball/ BiologyPages/B/Blood.html#serum*
2. Role of pH in electrophoresis: *http://www.ultranet.com/~jkimball/BiologyPages/P/pH.html*
3. Capillary electrophoresis; animation: *http://ntri.tamuk.edu/ce/ce.html*

4. Capillary electrophoresis: *http://hobbes.chem.ualberta.ca/~karl/titlepg.html*
5. Gel blotting: *http://www.ultranet.com/~jkimball/BiologyPages/G/GelBlotting.html*
6. Gel blotting principles: *http://web.mit.edu/esgbio/www/chapters.html*
7. Southern blotting: *http://www.accessexcellence.org/AB/GG/southBlotg.html*
8. Two-dimensional electrophoresis: *http://www.expasy.ch/ch2d/protocols/*

REFERENCES AND SUGGESTED READING

Acquaah, G. 1992. *Practical protein electrophoresis for genetic research.* Portland, OR: Dioscoredes Press.

Allen, R. C., and B. Budowle. 1994. *Gel electrophoresis of proteins and nucleic acids.* New York: Walter de Gruyter.

Glick, B. R., and J. J. Pasternak. 1994. *Molecular biotechnology, principles and applications of recombinant DNA.* Washington, DC: ASM Press.

O'Farrell, P. H. 1975. High-resolution two-dimensional electrophoresis of proteins. *Journal of Biological Chemistry, 250*:4007–4021.

Southern, E. M. 1975. Detection of specific sequences among DNA fragments separated by gel electrophoresis. *J. Mol. Biol., 98*:503–517.

9 Molecular Markers

PURPOSE AND EXPECTED OUTCOMES

Genetic markers, or simply **markers,** are depended upon in studying genomic organization, localizing genes of interest, and facilitating crop and animal breeding. Genetic markers are essentially landmarks on chromosomes, which are essential to finding out where genes are on a genetic map. Genes can be located by how close they are to known landmarks or markers. To be useful, markers should be heritable and readily assayable or detected. The concept of the application of markers is that certain traits or biological events are difficult to observe or detect. Therefore, by finding easy-to-observe or detect traits that are linked to these difficult-to-observe traits or events, you can assume that the latter is present or has occurred when you observe the former trait or see certain events. A goal of the Human Genome Project is to identify markers that are linked to genetic diseases.

In this chapter, you will learn:

1. The types of markers.
2. The types of molecular markers.
3. The properties of molecular markers.
4. The importance of markers in biotechnology.

TYPES OF MARKERS

There are two general categories of markers: **morphological markers** and **molecular markers.** Morphological markers are manifested on the outside of the organism as a product of the interaction of genes and the environment (i.e., an adult phenotype). On the other hand, molecular markers are detected at the subcellular level and can be assayed before the adult stage in the life cycle of the organism. Molecular markers of necessity are assayed by chemical procedures.

Molecular markers are useful for germplasm evaluation, map-based cloning of genes, and molecular breeding. Markers are used in choosing parents for crossing on the basis that more distant relatives combine to give favorable heterosis or hybrid vigor.

Plant and animal breeders use marker-assisted selection with or without phenotypic selection to increase gain from selection, and thereby speed up their breeding programs, bringing new breeds and cultivars to the market sooner. Markers are very useful for breeding quantitative traits and traits with low heritability. Markers linked to genomic regions containing quantitative traits loci (QTL) that influence the expression of these traits are identified and employed in selection. They provide a tool for selecting traits in environments that do not favor the expression of these traits.

MOLECULAR MARKERS

Isozymes were the first true molecular markers to be discovered. Apart from the insufficient number of assays available (only about several dozen protocols exist), isozyme markers are unevenly distributed on the genetic map. **Restriction fragment length polymorphisms (RFLPs)** arrived on the scene in the 1980s as superior to both morphological and isozyme markers. Since then, several other molecular markers have been developed. DNA markers that are currently widely used include RFLP, AFLP (amplified fragment length polymorphism), PCR (polymerase chain reaction)-based, SNPs (single-nucleotide polymorphisms), and microsatellites. The PCR technology is discussed in detail in Chapter 10.

Molecular markers may be grouped into two general categories:

1. **Single-locus, multi-allelic, codominant markers**

    This group of markers includes RFLPs and microsatellites (SSR). Microsatellites are capable of detecting higher levels of polymorphisms than RFLPs and AFLPs. The RFLPs marker system will be described to illustrate the key characteristics of this group of markers in breeding.

2. **Multi-locus, single-allelic, dominant markers**

    This group of markers includes AFLP and RAPD (random amplified polymorphic DNA) (which is less important these days).

■ *RESTRICTION FRAGMENT LENGTH POLYMORPHISMS (RFLPs)*

Restriction fragment length polymorphisms (RFLPs) were some of the first, and arguably the best, markers for plant genome mapping. They originate from natural variations in the genome that cause homologous chromosomes to produce different restriction products when digested with restriction enzymes. These variations, which are codominantly inherited, arise from insertion and deletion events that abolish recognition sites for certain restriction enzymes. RFLPs are randomly distributed throughout the genome of an organism and may occur in both exons and introns. A restriction enzyme is used to digest DNA, followed by electrophoresis, Southern blotting, and then probing to detect polymorphisms. The DNA profiles or fingerprints produced are specific to the combination of restriction enzyme and probe. Probes may be derived from a random genomic DNA library, cDNA library (from the species or heterologous species), or microsatellites from other organisms.

One of the advantages of using RFLPs is that the sequence (the cloned DNA fragments) need not be known. All that a researcher needs is a genomic clone that can be used to detect the polymorphism. Very few RFLPs have been sequenced to determine what sequence variation is responsible for the polymorphism. However, in the absence of sequence information, interpreting complex RFLP allelic systems may be problematic. There are different types of RFLP polymorphisms, the simplest being the two-allele system involving the presence or absence of a restriction site, which is detected by a single restriction enzyme (see Figure 9–1). Screening produces three different types of banding patterns: a large band (homozygous), two smaller bands (restriction site occurs on both homologs), and all three bands (heterozygote). It is assumed that a single base-pair change within the recognition site will result in a chromosome that either would or would not have the restriction site. In another allele system, one band corresponds to one allele. This system is also easy to score, as one variable band corresponds to a homozygote. Although a wide range of variable fragment sizes exist in the population, an individual does not have more than two of the variable bands representing the alleles of the homologs. This system is often displayed by polymorphisms produced by VNTRs (variable number tandem repeats) (see Chapter 17). There are other RFLP allele systems, some of them detected by more than one restriction system.

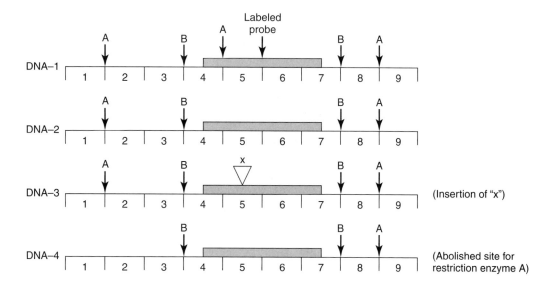

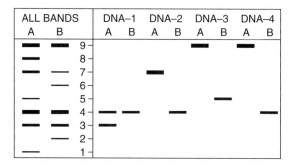

FIGURE 9–1

The concept of restriction fragment length polymorphism (RFLP). DNA is digested with two different enzymes, A and B. The arrows indicate restriction sites. The DNA may be cleaved into nine fragments. After electrophoresis, the gel is submitted to Southern blotting. The labeled probe recognizes sequences associated with fragments 4, 5, 6, and 7. DNA-3 has been extended by the length of an insertion ("x"), whereas DNA-4 has an abolished enzyme A restriction site. These variations will cause different fragments to be produced following restriction enzyme digestion of the four DNA strands. Following Southern blotting, only fragments that have portions containing a sequence that can be identified by the probe will appear on the autoradiogram. Because the separation of fragments is based on size, a band on the autoradiogram may comprise multiple fragments. For example, the lowest band produced by enzyme A for DNA-1 is made up of fragments 2, 3, and 4.

■ *RANDOM AMPLIFIED POLYMORPHIC DNA (RAPD)*

Random amplified polymorphic DNA (RAPD) is a PCR-based marker system. The total genomic DNA is amplified using a single short (about 10 bases) random primer (see Figure 9–2). The PCR product is submitted to gel electrophoresis. This method yields high levels of polymorphism and is simple and quick to conduct. PCR is readily automated. However, care must be taken to optimize the PCR environment for best results. Because of the sensitivity of PCR technology to contamination, it is common to observe a variety of bands that are not associated with the target genome but are artifacts of the PCR condition. Consequently, certain bands may not be reproducible. When using RAPD markers, using only the reproducible major bands for identification may minimize these shortcomings. Furthermore, one may include parental genomes where available to help determine bands of genetic origin.

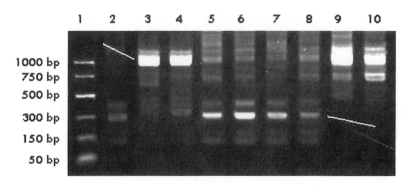

FIGURE 9–2
An example of electrophoresis results on the products of an RAPD analysis, a PCR-based marker.
Source: Photo is courtesy of Bryan Kindiger, USDA-ARS-Grazinglands Research Station, El Reno, Oklahoma. Used with permission.

■ *DNA Amplification Fingerprinting (DAF)*

DNA amplification fingerprinting (DAF) is a variation of the RAPD methodology which uses very short (five to eight bases) random primers and consequently produces more variation than RAPDs. DAF is hence more effective in distinguishing among closely related cultivars (e.g., bioengineered cultivars that differ only by the transgenes they carry). Digesting the template DNA with restriction enzymes prior to conducting PCR further enhances the sensitivity of the technique. Because of the great capacity for producing polymorphisms, DAF is best used in cases where plants are genetically closely related. Distinguishing among species of plants at a higher taxonomic level where genetic variation is already pronounced is problematic. Just like RAPD, it is imperative that the PCR environment be optimized for reproducible results.

■ *Simple Sequence Repeats (SSRs)*

Simple sequence repeats (SSRs) occur in microsatellites in eukaryotic genomes. These are random tandem repeats of two to five nucleotides (e.g., GT, GACA) that are present in copy numbers that vary among individuals and are sources of polymorphism in plants and animals. The SSR technique is also PCR-based. Because the DNA sequences that flank microsatellite regions are usually conserved, primers specific for these regions are designed for use in the PCR reaction. The SSR and RFLP techniques are more reliable than the RAPD and DAF techniques, but they are more tedious to conduct.

■ *AFLP*

Amplified fragment length polymorphism (AFLP) is a highly sensitive method for DNA fingerprinting, discovered by Keygene, which is applicable to a wide variety of fields—plant and animal breeding, medical diagnostics, forensic analysis, and others. AFLP is simply RFLPs visualized by selective PCR amplification of DNA restriction fragments. The steps to this random amplification technique are described in Figure 9–3. The technique has several advantages. It uses primers that are 17 to 21 nucleotides in length and are capable of annealing perfectly to their target sequences (the adapter and restriction sites) as well as a small number of nucleotides adjacent to the restriction sites. The effect of this capacity is that AFLP is very reliable and robust, immune to small variations in ampli-

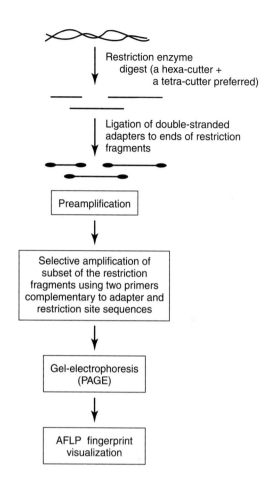

FIGURE 9–3
Steps in the application of the AFLP technique.

Restriction enzyme digest (a hexa-cutter + a tetra-cutter preferred)

Ligation of double-stranded adapters to ends of restriction fragments

Preamplification

Selective amplification of subset of the restriction fragments using two primers complementary to adapter and restriction site sequences

Gel-electrophoresis (PAGE)

AFLP fingerprint visualization

fication parameters (e.g., thermal cyclers, template concentration). The technique also produces a high marker density. A typical AFLP fingerprint (the restriction fragment patterns generated by the technique) contains between 50 to 100 amplified fragments, of which up to 80 percent may serve as genetic markers. Another advantage of the technology is that it does not require sequence information or probe collections prior to generating the fingerprints (useful in studies where DNA markers are scarce). Furthermore, markers generated are unique DNA fragments (which usually exhibit Mendelian inheritance). Most of the markers are mono-allelic (where the corresponding allele is not detected).

AFLP markers are used in a variety of ways, including biodiversity studies, analysis of germplasm collections, genotyping of individuals, identification of closely linked DNA markers, construction of genetic DNA marker maps, construction of physical maps, gene mapping, and transcript profiling.

■ SINGLE NUCLEOTIDE POLYMORPHISMS (SNPS)

A **single nucleotide polymorphism (SNP)** is simply a single base pair site in the genome that is different from one individual to another. SNPs (pronounced "snips") are the most frequently occurring genetic markers in humans, occurring at a rate of every 100 to 1,000 base pairs. They arise from deletions, insertions, or substitutions. The more common the marker, the more likely it is for scientists to discover a difference among individuals in the population.

SNPs are often linked to genes, making them very attractive subjects to study by scientists interested in locating, for example, disease genes. Sometimes, SNPs have no detectable phenotypic effect. However, in other cases, SNPs are responsible for dramatic changes. For example, the ABO blood group system is based on variations in only four nucleotide polymorphisms. A gene encoding type A blood may have the sequence CGTGGTGACCCCTT. An inheritance of four snips from a parent will result in the production of antigen B with the sequence CGTCGTCACCGCTA. To encode type O blood, the sequence has a deletion, producing the sequence CGTGGT-ACCCCTT (the dash represents a deletion). The consequence of this deletion is a lack of antigen produced.

To understand the genetic basis of disease, the SNP consortium was formed to create a database of these genetic markers for producing a SNP map of humans. The results of the Human Genome Project facilitate the discovery of SNPs. By comparing the SNP patterns in various patient and control populations, researchers hope to identify genetic differences that predispose some people but not others to diseases and to learn what underlies differences in response to therapy. In sickle cell disease, a point mutation in the sixth amino acid of the beta-globin gene occurs, whereby the adenine in the healthy person is substituted by thymine in the afflicted individual. The consequence of this SNP is that the amino acid valine is produced instead of glutamic acid.

MARKER-ASSISTED SELECTION

Markers are used in biotechnology in various ways. These include plant cultivar identification, linkage mapping, selection system in breeding, and selection in transformation research.

■ PLANT CULTIVAR IDENTIFICATION

Molecular markers are more versatile in cultivar identification for protecting proprietary rights as well as authenticating plant varieties, and can be assayed at the seedling stage. Isozyme markers have been used in cultivar identification since the 1960s. Isozymes are still used as quick and reliable methods for authenticating hybridity in hybrid breeding programs. However, isozymes have shortcomings—there are not many of them available for the purpose of identification in any cultivar. Furthermore, isozymes may be tissue specific, in which case sampling becomes critical in crop cultivar identification. Identical tissues (in terms of plant part, age, and other characteristics) are essential for reliable results.

DNA markers are superior to isozymes by being unlimited in number and detectable in all tissues and at all ages in the plant's life. They are stably expressed in all tissues and not subject to the environment. There are a number of molecular genetic methods used in cultivar identification, the most common being RFLPs, RAPD, DNA amplification finger printing (DFA), and SSR polymorphism.

■ CONVENTIONAL BREEDING (AND LINKAGE MAPPING)

Markers may be used to facilitate a breeding program, making identification and selection of desirable recombinants more effective and efficient. In using markers, breeders search for linkages between a trait of interest and a marker.

Lack of suitable genetic markers hampered the construction of linkage maps in many important plant species. RFLP markers have been identified for numerous quantitative and qualitative traits. In plants, a cross is made between two divergent parents. DNA is extracted from the parents, F_1, and a random sample of F_2 plants. A number of restriction enzymes are used to digest the DNA samples, followed by electrophoresis, Southern blotting, and then probing to detect polymorphism between the parents. Segregation of RFLP markers in F_2 plants is used in constructing the linkage map. The markers are arranged in linkage groups (linkage of markers to chromosomes is established). These analyses are made possible by using statistical packages and computer mapping software. Linkage relationships among markers with a recombination frequency of less than 50 percent LOD (logarithm of the odds, or lod) score of 3 or greater indicates that a gene and an RFLP marker are linked. LOD is the ratio of two probabilities: no linkage/a certain degree of linkage. The gene and RFLP marker are first assumed to be unlinked. Then, the probability of the observed pattern of inheritance if the gene and the marker have a certain degree of linkage is calculated.

In humans, three or more generations in which an RFLP marker and a gene (e.g., for a disease) are segregating are required for the analysis. It is best to select RFLP markers for which most members in these populations are heterozygous.

■ *MARKERS FOR SELECTION IN TRANSFORMATION STUDIES*

As already indicated, rDNA technology usually requires the use of a marker. Cloning vectors have built-in selection marker systems. The next activity after transformation is selection, the process by which host cells that have incorporated recombinant vectors are identified and isolated from untransformed cells. The strategy employed in selection depends on the properties of the cloning vector (i.e., the selectable marker genes present). For a vector like pBR322, and cloning in Bam HI, the transformation mixture is first plated onto a medium containing ampicillin. This step is designed to eliminate all nontransformed cells (those that lack the genetically engineered plasmid with *Amp* gene for ampicillin resistance). In the second step, an inoculum of cells from each colony growing on the ampicillin-fortified medium is transferred to an agar plate containing tetracycline. It is critical to maintain correspondence between the origin (site on ampicillin plate) and destination (site on tetracycline plate). The method of replica plating may also be used (see Figure 9–4). Because the Bam HI site occurs in the tetracycline gene, plasmids that contain the insert DNA will have the deactivated tetracycline gene

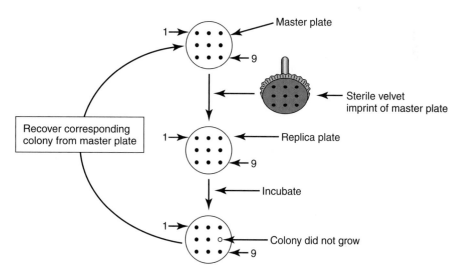

FIGURE 9–4

The procedure of replica plating. Transformed cells are cultured on a medium containing a selection agent, such as ampicillin. Only cells containing plasmids will survive and grow to produce colonies, because they contain the ampicillin resistance gene. Samples of the colonies on the master plate are transferred onto a fresh plate by making imprints of the master plate onto a velvet pad. Care is taken to keep track of the location of each colony (1 to 9). The fresh plate contains another selection agent (e.g, tetracycline). The replica plate is incubated. The colony that contains the recombinant insert will not grow because the tetracycline gene has been inactivated by the insertion of the transgene. The corresponding colony on the master plate is identified and selected for further analysis.

and hence be sensitive to the antibiotic. However, cleaved plasmids that recircularized without incorporating the insert DNA will still have active *Tet^r* genes and confer tetracycline resistance on the host cell. Researchers go back to the ampicillin plate to locate (by correspondence) the colonies with transformed cells. The colonies may be subcultured individually or pooled.

KEY CONCEPTS

1. There are two general categories of markers: morphological and molecular.
2. Molecular markers may be single locus or multiple loci, single allele or multiple allelic, and dominant or codominant.
3. Molecular markers usually have wider genomic coverage (except isozyme markers).
4. Morphological markers are manifested on the outside of the organism as a product of the interaction of genotype and environment (i.e., as adult phenotype).
5. Molecular markers are detected at the subcellular level and can be assayed before the adult stage by chemical analysis.
6. Commonly used molecular markers include RFLPs, AFLPs, SSR, and RAPD.
7. Some molecular markers are PCR-based.
8. PCR-based markers are rapid to generate and easy to use in research.
9. Markers are widely used in biotechnology for a variety of purposes including selection in transformation, cultivar identification, selection in breeding, and mapping.

OUTCOMES ASSESSMENT

1. Give the rationale of markers.
2. Describe how RFLPs arise.
3. Give four specific examples of molecular markers.
4. Describe a specific application of molecular markers in biotechnology.
5. Discuss the advantages of the AFLP technology.
6. Discuss how microsatellites are used as markers.
7. Discuss the importance of markers in plant and animal breeding.

INTERNET RESOURCES

1. Three-point test cross procedure: *http://www.ultranet.com/~jkimball/BiologyPages/T/ThreePointCross.html*
2. Genetic mapping problems: *http://web.mit.edu/esgbio/www/chapters.html*
3. AFLP technology: *http://www.keygene.com/html/aflp.htm*
4. Types of RFLP polymorphisms: *http://hdklab.wustl.edu/lab_manual/14/14_3.html*
5. SNPs: *http://www.artsci.wustl.edu/~jstader/Dickinson.htm*

REFERENCES AND SUGGESTED READING

Barrett, B. A., and K. K. Kidwell. 1998. AFLP-based genetic diversity assessment among wheat cultivars from the Pacific Northwest. *Crop Science, 38*:1261–1271.

Davila, J. A., Y. Loarce, and E. Ferrer, 1999. Molecular characterization and genetic mapping of random amplified microsatellite polymorphism in barley. *Theoretical and Applied Genetics, 98*:265–273.

Dawson, E. 1999. SNP maps: More markers needed? *Molecular Medicine Today, 10*:419–420.

Edwards, K. J., and D. Stevenson. 2001. Cereal genomics. In *Advances in botanical research incorporating advances in plant pathology, biotechnology of cereals.* Vol. 34. P. R. Shewry, P. A. Lazzeri, K. J. Edwards, and J. A. Callow (eds). New York: Academic Press.

Kwok, P. Y., and Z. Gu. 1999. SNP libraries: Why and how are we building them? *Molecular Medicine Today, 12*: 538–543.

Law, C. N. 1995. Genetic manipulation in plant breeding—prospects and limitations. *Euphytica, 85*:1–12.

Rosenberg, M., M. Przybylska, and D. Strauss. 1994. RFLP subtraction: A method for making libraries of polymorphic markers. *Proc. Nat. Acad. Sci. USA, 91*:6113–6117.

Smigielski, E. M., K. Sirotkin, M. Ward, and S. T. Sherry, 2000. dbSNP: A database of single nucleotide polymorphisms. *Nucleic Acid Research, 28(1)*:352–355.

Williams, J. G. K., A. R. Kubelik, Rafalski, J. A., and S. V. Tingey. 1990. DNA polymorphisms amplified by arbitrary primers are useful as genetic markers. *Nucl. Acids Res., 18*:6531–6535.

Xiao, J., J. Li, L. Yuan, S. R. McCouch, and S. D. Tanksley. 1996. Genetic diversity and its relationship to hybrid performance and heterosis in rice as revealed by PCR-based markers. *Theoretical and Applied Genetics, 92*:637–643.

10

The Polymerase Chain Reaction (PCR) and DNA Synthesis

PURPOSE AND EXPECTED OUTCOMES

What do scientists do when the sample of DNA of interest is not sufficient for their purpose? Can DNA be synthesized *in vitro*? Why would scientists like to synthesize DNA *in vitro*?

In this chapter, you will learn:

1. How the cutting edge technology of Polymerase Chain Reaction (PCR) works.
2. How DNA is synthesized in the laboratory.
3. The importance of DNA synthesis in biotechnology.
4. Some applications of PCR technology in biotechnology.

WHAT IS PCR?

The **Polymerase Chain Reaction (PCR)** technique, discovered in 1986, is a technique for directly amplifying a specific short segment of DNA without the use of a cloning method. Prior to PCR, scientists had to clone and reclone target DNA in order to obtain sufficient quantities for research. The PCR technique can utilize minuscule amounts of DNA that do not have to be purified. Sources of DNA may be genomic DNA, forensic samples (e.g., semen, hair, or dry blood), mummified remains, fossils (as in the movie *Jurassic Park*), and others. Two specific oligonucleotides (primers) are synthesized to flank the target DNA segment, with one being complementary to the 3' end on one DNA strand and the other one complementary on the 3' end on the opposite strand. For this to occur, the DNA sequence information about the segment to be amplified must be known for the design and production of the appropriate primers.

PCR is one of the most widely used techniques in molecular biology. It is a rapid, inexpensive, and simple means of producing relatively large numbers of copies of DNA molecules from minute quantities of source DNA material, even when the source DNA is of relatively poor quality.

CONDUCTING PCR

■ THE PCR SAMPLE

Even though PCR may use many types of samples for nucleic acid analysis, DNA rather than RNA is the most commonly used, because of the stability of the DNA molecule and the ease with which DNA can be isolated. The most basic and critical requirements for any DNA sample are that it contains at least one intact DNA strand encompassing the region to be amplified, and that any impurities are sufficiently diluted so as not to inhibit the polymerization step of the PCR reaction. Sample preparation may be undertaken by using one of the many protocols available. However, it is advisable to use a protocol with the fewest possible steps. This strategy would reduce the incidence of accidental contamination with undesirable DNA. As previously indicated, the PCR technology is very sensitive. Researchers usually dilute the sample (1:5 dilution) as a means of diluting out impurities that may originate from the purifying protocol.

Another precautionary measure routinely taken in PCR analysis is the use of duplicate DNA samples. This strategy provides a control for the relative quality and purity of the original sample. A small amount of DNA may be added to the control just after the reaction mixture is complete to allow the detection of anything that might inhibit the PCR reaction.

■ THE PCR MIXTURE

A PCR mixture (sometimes called the master mix) consists of four key components—two primers (about 20 nucleotides long), a target DNA sequence (about 100 to 5,000 bp), a thermostable DNA polymerase (AmpliTaq polymerase, which can remain stable at 95° C or higher), and four deoxynucleotides. These components are added to a water and buffer mixture. As little as 25 or 50 µl of this mixture may be used as a cost-cutting strategy, but 100 µl is a more common amount. Volume not withstanding, the critical requirement is to maintain constant final concentrations of the reagents.

The role of primers in the PCR mixture is as an initiation site for the elongation of the new DNA molecule. The primers used may be specific to a particular DNA nucleotide sequence or they may be what is classified as **universal primers** (which are complementary to nucleotide sequences that are very common in a particular set of DNA molecules). Universal primers are versatile; they are able to bind to a wide variety of DNA templates. Universal primers are created as sequences that are complementary to certain nucleotide sequences known to be common to a group of organisms. For example, commonly used bacterial universal primers have the following sequence: Forward 5′ GAT CCT GGC TCA GGA TGA AC 3′ (20-mer); Reverse 5′ GGA CTA CCA GGG TAT CTA ATC 3′ (21-mer). In animal cell lines, a commonly occurring sequence is called the "*alu* gene," of which about 900,000 copies are known to occur in the human genome. An *alu*-based universal primer may be used for PCR analysis in animal studies. It binds in both forward and reverse directions: 5′ GTG GAT CAC CTG AGG TCA GGA GTT TC 3′ (26-mer). When using universal primers, the annealing temperature on the thermal cycle is lowered to 40 to 55° C.

■ THE PCR CYCLE

The PCR process occurs in cycles, with each cycle entailing three steps (that constitute a PCR-cycle). The three steps are carried out in the same vial, but at different temperatures:

1. **Denaturation**

    The PCR reaction mixture is heated to 95° C to denature the DNA into single-stranded molecules. The DNA source does not have to be purified. The thermostable DNA polymerase may be **Taq polymerase** (obtained from the bacterium *Thermus aquaticus*). The denaturation temperature is maintained for about one minute.

2. **Renaturation**

    This step involves slowly cooling the mixture to about 55° C to allow the primers to anneal to their complementary sequences in single-stranded source DNA (primers cannot anneal to the DNA at the denaturation temperature).

3. **Extension**

The third step in the PCR cycle involves heating the mixture again to raise the temperature to about 75° C. At this optimum temperature, Taq DNA polymerase begins to extend the primers by adding nucleotides to the 3'-hydroxyl end. The replication of the original DNA target includes the flanking primer regions and beyond, producing new DNA strands (called "long strands") that will be used in the second cycle.

In practice, the PCR process is automated and occurs in a special machine called the **thermocycler** that can be programmed (temperatures, duration of process). The machine contains a heat block in which the reaction tubes are inserted (see Figure 10–1). The duration of a typical PCR cycle is about three to five minutes. The reaction is repeated in the second and subsequent cycles. A significant feature of PCR is that each product serves as a template in the next reaction cycle, thus setting in motion a chain reaction (see Figure 10–2). After about 20 to 25 cycles (lasting several hours), a several million-fold amplification of the discrete target DNA sequence would be produced (see Figure 10–3).

PCR is designed to amplify very small amounts of DNA sequences. Consequently, it is highly susceptible to contamination. Even body cells from the researcher that accidentally fall in the PCR reaction mixture can be problematic. Strict laboratory sanitation rules must be observed when conducting PCR research. Care must be observed when recycling PCR solutions or cleaning the reaction tubes. It is important to include adequate and appropriate controls in the PCR project.

■ *THE PCR DETECTION AND ANALYSIS*

At the end of the predetermined duration of the PCR process, DNA fragments of a specified length would be produced. A sample from the reaction product is subjected to electrophoresis along with appropriate molecular-weight markers, using an agarose gel that contains 0.8 to 4.0 percent ethidium bromide. The resulting bands may be visualized under ultraviolet transillumination. The markers provide a means of identifying the corresponding fragments.

FIGURE 10–1

A thermocycler is used for DNA amplification. The designs and capacities differ widely.

Source: Photo was taken at the Center for Biotechnology Research and Education at Langston University.

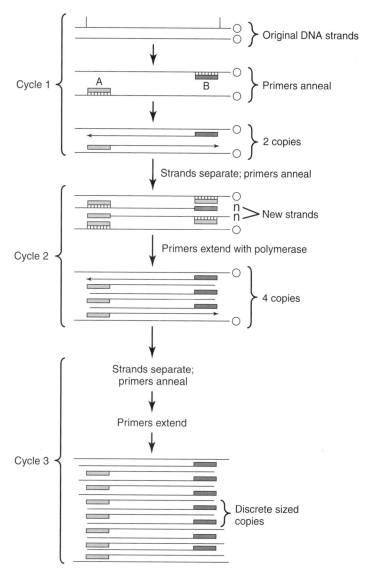

FIGURE 10–2

PCR amplification starts with the denaturing of the target DNA into single strands. Each strand is then annealed to a complementary primer such that the two primers flank the region to be amplified. The DNA polymerase and added nucleotides are used to extend the primers in the 3′ direction, resulting in the production of a double-stranded DNA molecule with the primers included in the newly synthesized strands. The products of the first cycle are used as templates in the next PCR cycle. The third cycle produces four discrete-sized copies of the target DNA. Cycles 3 to 25 yield a more than 10^6-fold increase in the target DNA.

FIGURE 10–3

Relative number of copies of PCR products corresponding to the number of PCR cycles.

Cycle	Relative number of copies of PCR product
1	2
5	32
10	1,024
15	32,768
20	1,048,576
25	33,554,432
30	1,073,741,824

ADVANCES IN PCR TECHNOLOGY

Since its invention by the Nobel laureate Kevin Mullins, PCR technology has become widely used because of several key breakthroughs. The first was the discovery and purification of a heat-stable DNA polymerase from the thermophilic bacterium, *Thermus aquaticus,* in Yellowstone National Park. This eliminated the need for multiple water baths maintained at the different temperatures needed for each PCR cycle, as well as the need to manually move vials frantically around from one bath to another. In addition, the need to replenish the polymerase for each repetition of the cycle was removed. The second breakthrough in PCR technology was automation with the invention of the thermocycler by Cetus Instrument Systems.

These two basic improvements have received further enhancements over the years. There are alternative polymerases (e.g., designer polymerase), new detection strategies (e.g., microwell, probe arrays), anticontaminants, quantitation, cycle sequencing, and multiplex PCR. The frontline of PCR technology is currently the **real-time PCR (RT-PCR).** In this modification, a fluorogenic probe is added to the PCR mixture. This probe emits fluorescence that accumulates in the mixture and can be detected and quantified in real time. This entails the use of Double-Dye oligonucleotide probes that anneal between the upstream and downstream primer. The 20 to 24-bases-long probes contain a 5' fluorophore and a 3' quencher. The 3' terminus may be blocked (e.g., with a PO_4 or NH_2). As the reaction proceeds, the 5'-3' exonuclease activity of the Taq polymerase cleaves the fluorophore from the probe. Because the fluorophore is no longer in close proximity to the quencher, it is free to fluoresce. The intensity of the fluorescence is directly proportional to the amount of target DNA that accumulates during the PCR cycles.

SELECTED USES

Uses of PCR technology include:

1. Amplifying DNA from even a single cell. This helps in identifying rare genes.
2. Supporting other studies by increasing the amount of existing DNA to sufficient quantities.
3. Identifying infectious diseases (including HIV, bacteria, and other viruses). To detect a viral infection, for example, primers are designed from a known sequence of the virus. Amplification will occur after PCR only if the viral DNA is present in the DNA sample.
4. Purifying a crude DNA sample.
5. Detecting mutations.
6. Forensic applications (discussed further later in the book).
7. Sequencing of cloned DNA.

PCR-BASED MARKERS

The PCR technology is used in marker development. Some of the previously discussed markers in use include RAPD, SSR, AFLP, and DAF.

CHEMICAL SYNTHESIS DNA

Can DNA be synthesized in the laboratory? Certainly. Why would such an effort be necessary? Chemical synthesis of DNA is an indispensable part of biotechnology, as there are many applications of this technology in biotechnology.

■ APPLICATIONS OF DNA SYNTHESIS

Single-stranded DNA oligonucleotides have many uses including:

1. **Hybridization probes**
   The codons from amino acids of a protein can be deduced and used to construct single-stranded oligonucleotide probes of 20- to 40-mers.
2. **Linkers**
   Double-stranded sequences that are used in the genetic cloning of DNA fragments are created by synthesizing an oligomer that is a palindromic single-stranded DNA sequence that base pairs to itself and contains a restriction endonuclease recognition site.
3. **Directed (site-specific) mutagenesis**
   These are used to create mutations *in vitro.*
4. **Primers**
   These are used in PCR studies.
5. **Facilitate/optimize gene cloning**
   Sometimes, a gene is poorly expressed in a host, in which case one with a codon optimization may be chemically synthesized (without changing the true amino acid sequence of the protein). Also, chemical DNA synthesis may be necessary when, even though an amino acid sequence is known, it still proves difficult to clone the gene. The nucleotide sequence of the gene of interest is deduced from the established amino acid sequence and chemically synthesized.

■ STEPS IN DNA SYNTHESIS

DNA synthesis is currently a routine activity largely because of the invention of DNA synthesizers (gene machines) that automate the complex process of chemical synthesis. Single-stranded oligonucleotides of up to 50 nucleotides can be readily synthesized in these machines following highly precise timing and dispensing of nucleotides and reagents to the reaction column. Even though several methods are available, the **phosphoramidite method** is considered the method of choice. To avoid side chains from forming during chain elongation, the amino groups of the nitrogenous bases that have them (adenine, guanine, and cytosine) are treated with chemicals (e.g., benzyol for adenine) so they are derivatized.

The process of chemical synthesis is summarized in Figure 10–4. The five general steps alternate with washings.

1. **Linking of first nucleotide to column**
   The starting complex consists of the initial nucleotide, a spacer, and a solid support (usually, controlled pore glass). A dimethoxytrityl (DMT) is attached to the 5'-hydroxyl group of the starting nucleotide to prevent it from reacting nonspecifically before the second nucleotide is added.
2. **Detritylation**
   The 5'-DMT group is removed by using trichloroacetic acid.

FIGURE 10–4
A summary of the steps of
DNA synthesis.

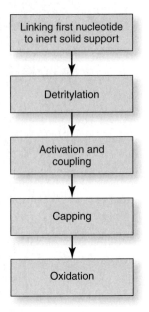

3. **Activation and coupling**

This is the addition of the next prescribed nitrogenous base and tetrazole (activates the phosphoramidite or base so that its 3'-phosphite forms a covalent bond with the 5'-hydroxyl group of the starting nucleotide that has been detritylated).

4. **Capping**

The purpose of this step is to acetylate the 5'-hydroxyl group of unreacted detritylated nucleotides to prevent them from participating in the coupling reaction of the next cycle.

5. **Oxidation**

This stage is designed to strengthen the internucleotide bonds. The phosphite triester bonds are unstable and susceptible to breakage when exposed to acid or base. The addition of an iodine mixture oxidizes these bonds into more stable pentavalent phosphate triester bonds.

To add the next set of residues, these five steps are repeated for as many cycles as there are residues to be added. At the end of the reaction, the oligonucleotide is removed from the column and purified.

KEY CONCEPTS

1. Polymerase chain reaction (PCR) technique is a technique for directly amplifying a specific short segment of DNA without the use of cloning.
2. A PCR cycle entails three steps: denaturation, renaturation, and extension.
3. Various modifications of the basic PCR method exist.
4. A minuscule amount of DNA can be amplified a million-fold using PCR.
5. DNA can be synthesized *in vitro*.
6. PCR uses primers that have to be synthesized *in vitro*.
7. Synthesis of biomolecules is critical to biotechnology research and applications.

OUTCOMES ASSESSMENT

1. Describe how PCR works.
2. How important is PCR technology to the development and application of biotechnology?
3. How important is DNA synthesis to the development and application of biotechnology?
4. In what way is PCR itself dependent on biotechnology?
5. Describe two specific uses of PCR technology.
6. Describe how DNA is synthesized.

INTERNET RESOURCES

1. Polymerase chain reaction procedure: *http://www.ultranet.com/~jkimball/BiologyPages/P/PCR.html*
2. PCR: *http://www.accessexcellence.org/AB/GG/polymerase.html*
3. PCR principles: *http://web.mit.edu/esgbio/www/chapters/html*
4. PCR lectures: *http://www.idahotech.com/lightcycler_u/lectures/applicationsofrc.htm*

REFERENCES AND SUGGESTED READING

Climie, S., and D. V. Santi. 1990. Chemical synthesis of the thymidylate synthase gene. *Proc. Natl. Acad. Sci. USA.*, 87:633–637.

Erlich, H. A., D. H. Gelfand, and J. J. Sninsky. 1991. Recent advances in the polymerase chain reaction. *Science*, 252:1643–1651.

Innis, M. A., D. H. Gelfand, and J. J. Sninsky (eds). 1999. *PCR applications: Protocols for functional genomics.* London, UK: Academic Press.

Itakura, K., J. J. Rossi, and R. B. Wallace. 1984. Synthesis and use of synthetic oligonucleotides. *Annu. Rev. Biochem.*, 53:323–356.

Mollis, K. B. 1990. The unusual origin of the polymerase chain reaction. *Sci. Am.*, (April) 262:56–65.

Saiki, R. K., D. H. Gelfand, S. Stoffel, S. Scharf, R. Higuchi, G. T. Horn, K. B. Mullis, and H. A. Erlich. 1988. Primer-directed enzymatic amplification of DNA with a thermostable DNA polymerase. *Science*, 239:487–491.

11 Genome Mapping and DNA Sequencing

Why do we need to map the genome? Maps are indispensable in everyday life. They help us navigate our way about a locality and travel from one point to another. They provide an overview of a geographic area (e.g., a neighborhood, a city, a state, country, and so on). Naturally, the larger the coverage of the map, the fewer details that can be provided. To navigate around the area described by a map, there are street signs, names of streets, landmarks, north–south orientation, and other things that help us find our way about. Similarly, in biology, genomic maps are created by scientists to help us know where genes are located in the genome. You may consider the chromosomes as streets. From one tip to the other, there are genes and other structures that are identified by a variety of names and symbols called markers. These symbols depend upon how the markers were developed. A significant difference between a geographic map and a genomic map is that, whereas the former can be multidimensional (having length, width, latitude, and longitude), genomic maps are linear (one-dimensional).

In this chapter, you will learn:

1. The types of genome maps and how they differ.
2. The strategies for mapping the genome.
3. The methods of DNA sequencing.

COMPARATIVE SCALE OF MAPPING

Genomic maps depend on markers, which have been discussed previously. Generally, the more markers on a map, the better (a **saturated map** refers to a map with a large number of markers). Creating a map by markers is like drawing a street with dashes (- - - - -); there are gaps in-between the markers. Currently, scientists are pursuing a more precise way of mapping genes, called **genome sequencing,** where the map produced is a solid line (_____).

Even though the genomic sequence gives the most complete picture of the genome, the genome map is still useful in constructing genome sequences utilizing the clone-by-clone method. The map helps to determine which clones belong to the genome being sequenced and helps in assembling the clones to produce a complete map. Furthermore, after producing a genomic sequence, all scientists have is an endless chain of letters that represent both the coding and noncoding regions of the chromosome. Markers are needed to help them better understand and use the sequence. A marker linked to a disease helps the scientists locate the approximate region of the chromosome where the disease gene resides. They can then begin to zero in to isolate the precise sequence and clone it. Animal and plant breeders use gene maps to help in identifying parents for mating in a breeding program and to facilitate breeding by increasing selection efficiency to shorten breeding time.

KARYOTYPE

The **karyotype** of an individual represents the chromosome complement of a somatic cell. In practice, a karyotype is used to refer to the descriptive diagrams produced by photographing mitotic chromosomes at metaphase, printing the photo, matching homologous chromosomes, and arranging the chromosome pairs according to banding patterns, length or size, and centromere placements. A karyotype does not show locations of genes and is not considered when mapping strategies are discussed.

GENOME MAPPING

Genome mapping determines the order of genes (or other genetic markers) and the spacing between them on each chromosome. Two categories of strategies are employed in genome mapping: (1) genetic linkage mapping, and (2) physical mapping. How do these basic kinds of maps differ?

■ GENETIC LINKAGE MAPS

Genetic linkage mapping is based on the concept that genes that are spatially located close together on the same chromosome tend to be transmitted from the parent to the offspring together (i.e., gene block). However, meiotic recombination provides opportunities for homologous chromosomes to exchange parts. This phenomenon may cause genes on the same chromosomes to not be inherited together. The frequency of crossovers is used to calculate the distance between two markers on a genetic map. A linkage map does not tell where on a chromosome (or even necessarily on which specific chromosome) a gene of interest is located. It only gives the probability that a gene of interest would be co-inherited with a marker.

Genetic linkage mapping uses information about recombination frequency between genes or markers. Only markers that are polymorphic (alternative forms of the gene exist among individuals in the population) can be used in genetic linkage mapping. Meiotic mapping requires that two copies of a chromosome be distinguishable by polymorphic markers. Common polymorphic traits in the human population are blood types (A, B, AB, and O) and eye color (brown, green, blue, and so on). Whereas some polymorphisms are observable because they are products of gene expression (i.e., originate within exons), most variations originate within introns (unexpressed DNA sequences) and are detectable at the DNA level (molecular markers).

Genetic maps consist of diagrams in which distances between markers are determined by meiotic recombinational frequencies between the markers (or genes). A unit of recombination is equivalent to 1 percent of recombination between two markers or genes. The unit of measurement on a genetic map is centimorgans (cM); 1 cM is equivalent to 1 percent of recombination. Further, 1 cM is equal to approximately 1 million bp (1 Mb) of physical distance on the chromosome.

Classical genetic mapping uses the three-point mapping strategy to map three or more linked genes in a single cross. The three criteria for three-point mapping are:

1. The genotype of the organism producing the crossover gametes must be heterozygous at all loci of interest.

2. The cross should allow the investigator to correctly determine the genotypes of all gametes from their phenotypes (since gametes and their genotypes cannot be observed directly).

3. The cross should generate a large enough offspring to ensure that all crossover classes are represented.

The location of genes for certain traits in plants (e.g., corn) and humans (e.g., sickle cell disease, cystic fibrosis, or Tay-Sachs disease) have been identified through the study of the transmission pattern of markers associated with these traits (i.e., markers found in individuals with the trait but absent in those without). Genetic linkage maps are often of low resolution because of insufficient markers. However, the use of molecular biological techniques (e.g., RFLP) can help produce higher resolution maps.

■ PHYSICAL MAPS

Physical maps are constructed from information obtained from the chemical characteristics of the DNA itself. This may involve DNA/chromosome fragmentation, histochemical staining and visualization, and DNA synthesis. Physical maps may be grouped into two categories on the basis of resolution.

Low-resolution Maps

Two mapping techniques are commonly used to produce low-resolution maps:

1. **Chromosomal map**

    Genetic linkage maps are also referred to as **chromosome maps** because they provide the relative location of genes on the chromosomes. The distances between genes or markers are measured in centimorgans. In chromosome mapping, the genes or genetic markers are identified on specific chromosomes through cytogenetic staining. Furthermore, the distances between them are estimated in terms of base pairs. Conventional techniques allow DNA sequences that are separated by about 10 Mb to be located on chromosomes. Improvements in the technique of **fluorescent *in situ* hybridization** (FISH) has narrowed the distance between genes to only about 2 to 5 Mb (see Figure 11–1). Increased resolution is achievable when interphase chromosomes are targeted. FISH

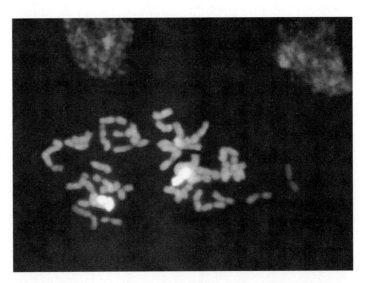

FIGURE 11–1

An example of fluorescent *in situ* hybridization (FISH) showing wheat with a *Leymus multicaulis* disomic addition appearing as the two light spots. The disomic addition was labeled with digoxigenin-11-dUTP (DIG-Nick Translation Mix for *in situ* probes, Boehringer Mannheim).

Source: Photo courtesy of Zhao Mao-Lin, Beijing Agricultural Biotechnological Research Center, Beijing Academy of Agricultural and Forestry Science, Beijing, China.

technology involves tagging of the DNA marker with a chemiluminiscent compound that fluoresces, or a radioactive material that can be visualized by autoradiography.

2. DNA map

Complementary DNA (cDNA) is produced from mRNA transcribed from coding parts (exons) of the genome. The synthetic DNA is then mapped to corresponding regions of the chromosomes. Genes with known functions can be mapped in this fashion.

High-resolution Maps

High-resolution maps are possible because of rDNA technology. Classical mapping approaches involve generating and mapping mutants, thus limiting characterization to those genes for which mutants have been isolated. rDNA techniques provide a direct approach to genetic analysis, whereby a genomic library is created, from which overlapping clones are then assembled to construct genetic and physical maps for the entire genome. Ultimately, the entire genome is sequenced such that all genes are identified by both their location in the chromosome and their nucleotide sequence.

APPROACHES TO GENOME MAPPING

There are two basic approaches to mapping entire genomes.

■ BOTTOM-UP MAPPING

In the **bottom-up mapping** strategy, the chromosome is digested with restriction enzymes into small segments that are cloned into a vector. The individual clones vary in size between 10,000 bp and 1 Mb. The restriction fragments are characterized by restriction mapping, and with the aid of computers, ordered to identify overlapping clones. These clones are assembled into larger contiguous segments of DNA called **contigs** (see Figure 11–2). Because certain regions of a chromosome are not clonable, contig maps have gaps that require additional strategies to be used to fill them (e.g., using DNA probe techniques). Furthermore, contig mapping is suitable for analyzing DNA segments of less than 2 Mb. However, the use of **YACs (yeast artificial chromosomes)** allows larger DNA segments to be analyzed by this approach, and facilitate it by reducing the number

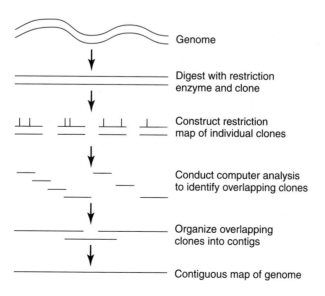

Genome

Digest with restriction enzyme and clone

Construct restriction map of individual clones

Conduct computer analysis to identify overlapping clones

Organize overlapping clones into contigs

Contiguous map of genome

FIGURE 11–2
The concept of bottom-up mapping.

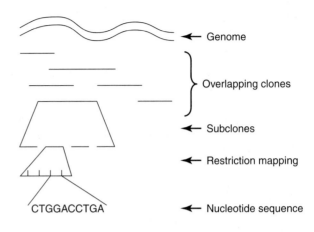

FIGURE 11–3
The concept of top-down mapping.

of clones to be ordered. The bottom-up approach was used in the *E. coli* genomic project, involving over 70 contigs and covering 4,700 kb of DNA.

■ TOP-DOWN MAPPING

In **top-down mapping,** very large (about 200 kb) randomly cloned DNA segments that cover the entire genome are isolated. Large DNA fragments are obtained by using restriction enzymes that cut less frequently. The pieces are ordered to create a **macrorestriction map.** Overlapping YAC clones can be mapped by *in situ* hybridization. BAC clones, as previously described, are more commonly used for cloning larger DNA fragments, while subclones are derived from one YAC for more detailed mapping (restriction mapping). Restriction fragments from the subclones are sequenced (see Figure 11–3). The macrorestriction map has more continuity than a contig map. However, the map resolution is lower. Because the fragments involved in this approach are large, separation by conventional gel electrophoresis is problematic. Pulse-field gel electrophoresis may be used in this regard for better results. The *Drosophila* genome project used the top-down approach to analyze the genome, involving 965 YACs.

■ CHROMOSOME WALKING

Chromosome walking is a technique that can be used to fill in gaps in genomic maps or to locate and clone specific genes. First, the approximate location of the gene of interest (e.g., disease gene) is ascertained through linkage analysis in which a marker is linked to within 1cM of the gene. The nearby sequence is then cloned and used to start the chromosome walking process. A probe is constructed from a genomic fragment identified from a genomic library as being the closest linked marker to the gene of interest. The end piece of the cloned DNA fragment is subcloned and then used to construct a probe, which will reprobe the genomic library to locate an overlapping clone. A restriction map analysis is conducted to determine the amount of overlap. A restriction fragment from one end of the overlapping clone near the gene of interest is used to reprobe the genomic library to search for subsequent overlapping clones. The process is repeated, clone by clone, until the other flanking marker is reached. The clones recovered in this exercise are sequenced to search for an open reading frame (a sequence of nucleotides that begins with a start codon followed by a set of amino acid-encoding codons and then capped by one or more stop codons) that putatively represents the gene of interest.

It may be necessary to sequence the putative gene in both a normal and mutant (e.g., diseased) individual for comparative analysis to confirm the open reading frame is indeed the suspected gene. Conclusive proof that the sequence represents the true gene for the disease is obtained by mutation analysis in which conversion of the mutant gene to the normal gene in a cell causes a reversion to normal phenotype. Chromosome walking is laborious to conduct. It has been utilized to isolate and clone important genes including those for cystic fibrosis and muscular dystrophy.

METHODS OF DNA SEQUENCING

Just like a word in a written language is spelled by arranging specific letters in a certain sequence, and usually only that specific sequence, a gene has a characteristic order of arrangement of the nucleotides which comprise it. The purpose of **DNA sequencing** is to determine the order (sequence) of bases in a DNA molecule. Classical mapping approaches involve generating and mapping mutants, thus limiting characterization to those genes for which mutants have been isolated. rDNA techniques provide a direct approach to genetic analysis, whereby a genomic library is created, from which overlapping clones are then assembled to construct genetic and physical maps for the entire genome. Ultimately, the entire genome is sequenced such that all genes are identified by both their location in the chromosome and their nucleotide sequence.

Allan Maxam and Walter Gilbert invented the **chemical degradation method** of DNA sequencing. Another method, the Sanger **enzymatic (dideoxy) method (chain terminating),** named after its inventor, is more commonly used. However, in these days of more advanced technology, there are high-throughput methods of DNA sequencing, making it more convenient for most researchers to contract sequencing jobs to outside companies. For the sake of instruction, one of these methods, the Sanger method, will be described briefly.

This enzymatic sequencing protocol uses triphosphate derivatives of deoxyribonucleotides to construct polynucleotides, a reaction catalyzed by DNA polymerase I. The dideoxy derivatives lack hydroxyl groups at both 2' and 3' carbons. Consequently, they can be added to a DNA strand undergoing synthesis, but since they lack a 3'-OH group, they cannot be extended. This leads to termination of synthesis and production of DNA fragments whose length depends upon the point at which a dideoxynucleotide was incorporated.

Four reaction mixtures are prepared, each containing all four of the dideoxy nucleotides (radioactive), DNA polymerase I, a short primer polynucleotide, and a single-stranded template fragment from the DNA to be sequenced (see Figure 11–4a). However, each mixture contains a different dideoxynucleotide triphosphate that acts as a chain terminator. Only a small amount of this dideoxy derivation (a nucleotide precursor) is added to the mixture. Consequently, the newly synthesized DNA strands are randomly terminated, resulting in the synthesis of DNA fragments of varying lengths (see Figure 11–4b).

Upon electrophoresis (PAGE) of the mixtures, autoradiography is conducted. The DNA sequence is read from bottom to top. However, it should be pointed out that the sequence read is that of the *complement of the template sequence*. This is because the fragments were synthesized off the template, with A directing the incorporation of T, and G directing the incorporation of C. As such, A at the 5' terminus on the gel indicates that T occurs at the 3'

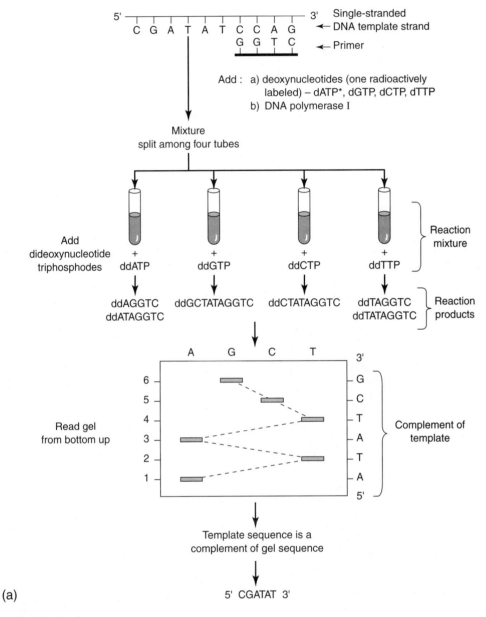

FIGURE 11–4

The Sanger method of DNA sequencing. (a) The procedure used for small projects. (b) The procedure for high-throughput applications. Four reaction mixtures are analyzed simultaneously, one for each nucleotide. In the high-throughput procedure, fluorescently labeled nucleotides are used to distinguish each nucleotide during data processing. It is critical to note that the produced sequence is a complement of the sequence of the actual DNA fragment sequenced.

terminus of the template sequence. For example, a determined sequence of 5′-ACCGT-TAC-3′ means the template of interest is 5′-GTAACGGT-3′.

Sequencing technology has undergone significant advancement since the discovery of the original technology. Third-generation technology is gel-less and incorporates such properties as enhanced fluorescent detection (the dideoxy is labeled with a tag that fluoresces a different color for each base) of individual labeled bases in flow cytometry, and direct reading of the sequence of the DNA strand. Each of the four dideoxynucleotides

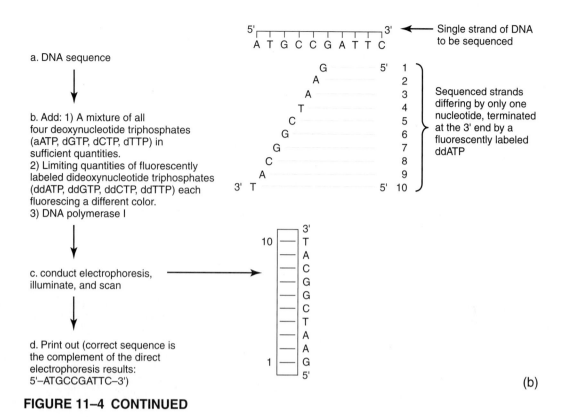

a. DNA sequence

b. Add: 1) A mixture of all
four deoxynucleotide triphosphates
(aATP, dGTP, dCTP, dTTP) in
sufficient quantities.
2) Limiting quantities of fluorescently
labeled dideoxynucleotide triphosphates
(ddATP, ddGTP, ddCTP, ddTTP) each
fluorescing a different color.
3) DNA polymerase I

c. conduct electrophoresis,
illuminate, and scan

d. Print out (correct sequence is
the complement of the direct
electrophoresis results:
5'–ATGCCGATTC–3')

(b)

FIGURE 11–4 CONTINUED

fluoresces a different color when illuminated by a laser beam and an automatic scanner provides a printout of the sequence (see Figure 11–5a). A DNA sequence is shown in Figure 11–5b.

Selected uses of sequencing include the following:

1. Study of the structure and organization of genes.
2. Study of gene regulation mechanisms.

PARTIAL VERSUS WHOLE GENOME SEQUENCING

Sequencing laboratories have a preference for their customized protocols, regarding the generation, isolation, and mapping of DNA fragments. This makes the correlating of mapping data from different programs problematic. A way of standardizing the process has been the strategy of partial genome sequencing in which partially sequenced unique regions of the genome (200 to 500 bp) called **sequenced tag sites (STSs)** are used as markers to facilitate mapping.

The methods of genome sequencing are discussed under the Human Genome Project where the two variations of shotgun sequencing (whole genome and hierarchical) are described.

AUTOMATED SEQUENCING

Capillary array electrophoresis (CAE) systems coupled with high-sensitivity detection provided by energy-transfer labeling reagents is now the accepted standard for high-throughput DNA sequencing facilities. Further advances are focused on the development

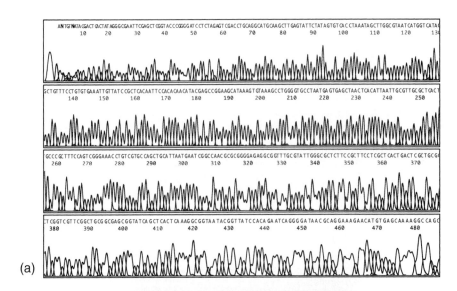

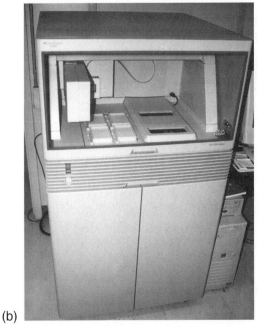

(a)

(b)

FIGURE 11–5
(a) A printout of a DNA sequence produced by an automated sequencer. (b) A DNA sequencer.

Source: Photo taken at the Oklahoma State University Nobel Research Center.

of capillary array systems capable of running more than 96 capillaries and the development of microfabricated CAE systems that provide higher throughput as well as the important ability to integrate microfluidic chemistries.

Sample purity and concentrations are two key factors that determine the success of automated sequencing reactions. Consequently, sequencing service providers recommend that customers purify their samples prior to submission for sequencing. Primer walk sequencing is the method of choice for sequencing DNA fragments between 1.3 and 7 kb. The DNA of interest may be either a plasmid insert or a PCR product.

HAPLOTYPE MAPPING

A **haplotype map (HapMap)** is a new type of genome map that some scientists believe will be pivotal in their quest to understand human disease. However, unlike its predecessor, the genome map, scientists are not yet unanimous in their enthusiasm about this new tool in the biotechnology toolkit. Humans are 99.9 percent genetically identical. Their differences are in observed phenotypes and other characteristics derived from the minute variations that occur at a rate of about one in every 1,000 bases, in which individuals inherit different nucleotides (ACTG). Some of these variations are called SNPs, as previously described. Scientists working on Crohn's disease discovered blocks of SNPs inherited as a block of about 60,000 bases. Each block is called a haplotype.

Instead of searching for individual SNPs, the HapMap focuses on the patterns of a few SNPs that define each haplotype. The objective is to determine which of these blocks is associated with specific diseases. For example, if a specific haplotype occurs more commonly in individuals with a certain disease, the mutation linked to that disease should be the same block of DNA. The assumption (which is debatable) behind this approach to understanding genes is that common mutations are responsible for most common diseases. Some argue that common diseases are caused by a combination of rarer mutations that the HapMaps are unlikely to reveal. Furthermore, many known diseases have not yet had their genetic underpinning elucidated, making the hypothesis of "common disease equals common mutation" unacceptable.

KEY CONCEPTS

1. Maps are generated in biological research for a variety of purposes.
2. Genomic maps depend on markers.
3. Genome sequencing provides the most complete picture of the genome.
4. Genome mapping entails determining the order of genes (or other genetic markers) and the spacing between them on each chromosome. Two categories of mapping strategies are employed in genome mapping: (a) genetic linkage mapping and (b) physical mapping.
5. Genetic maps consist of diagrams in which distances between markers are determined by meiotic recombinational frequencies between the markers (or genes).
6. Restriction fragment length polymorphisms (RFLPs) originate from natural variations in the genome that cause homologous chromosomes to produce different restriction products when digested with one restriction enzyme.
7. Physical maps are constructed from information obtained from the chemical characteristics of the DNA itself. This may involve DNA/chromosome fragmentation, histochemical staining and visualization, and DNA synthesis.
8. Physical maps may be grouped into two categories on the basis of degree of resolution (low resolution or high resolution).
9. There are two basic approaches to genomic mapping: bottom-up or top-down.
10. Chromosome walking is a technique that can be used to fill in gaps in genomic maps or to locate and clone specific genes.
11. Various organisms are used in biotechnology research as model organisms and in protocols.

OUTCOMES ASSESSMENT

1. What information does a DNA sequence provide scientists?
2. Describe the rationale for the Sanger dideoxy method of DNA sequencing.
3. Contrast top-down and bottom-up mapping strategies.
4. Discuss the importance of genetic mapping to science.
5. Compare a DNA sequence with the product of classical gene mapping.

INTERNET RESOURCES

Gene Mapping
1. Three-point test cross procedure: *http://www.ultranet.com/~jkimball/BiologyPages/T/ThreePointCross.html*
2. Genetic mapping problems: *http://esg-www.mit.edu:8001/esgbio/mg/linkage.html*
3. Human chromosomes: *http://www.accessexcellence.org/AB/GG/human.html*

Genome Sequencing
1. Strategy for vice genome sequencing at RGP: *http://rgp.dna.affrc.go.jp/genomicdata/seqstrategy/Seq-strategy.html*
2. Comparative scale of mapping: *http://www.accessexcellence.org/AB/GG/comparative.html*
3. Sequencing the genome: *http://vector.cshl.org/dnaftb/39/concept/index.html*
4. A stepwise presentation of the automated dideoxy method of genome sequencing: *http://www.ultranet.com/~jkimball/BiologyPages/D/DNAsequencing.html*

REFERENCES AND SUGGESTED READING

Acquaah, G. 1992. *Practical protein electrophoresis for genetic research.* Portland, OR: Dioscoredes Press.

Apley, L. 1997. *DNA sequencing: From experimental methods to bioinformatics.* New York: Springer-Verlag.

Beckmann, J. S., and T. C. Osborn (eds). 1992. *Plant genomes: Methods for genetic and physical mapping.* Dordrecht: Kluwer Academic Publishers.

Bostein, D., R. L. White, M. Skolnik, and R.W. Davis. 1980. Construction of a genetic linkage map in man using restriction fragment length polymorphisms. *American Journal of Human Genetics, 32*:314–331.

Lander, E. S., and D. Botstein. 1989. Mapping Mendelian factors underlying quantitative traits using RFLP linkage maps. *Genetics, 121*:185–199.

Lander, E. S., P. Green, I. Abrahamson, A. Barlow, M. J. Daly, S. I. Lincoln, and L. Newsbers. 1987. MAPMAKER: An interactive computer package for constructing primary genetic linkage maps of experimental populations. *Genomics, 1*:182–195.

Rosenberg, M., M. Przybylska, and D. Strauss. 1994. RFLP subtraction: A method for making libraries of polymorphic markers. *Proc. Nat. Acad. Sci. USA, 91*:6113–6117.

Williams, J. G. K., A. R. Kubelik, Rafalski, J. A., and S. V. Tingey. 1990. DNA polymorphisms amplified by arbitrary primers are useful as genetic markers. *Nucl. Acids Res., 18*:6531–6535.

Zuo, J., C. Robins, S. Baharloo, D. Cox, and R. Myers. 1993. Construction of cosmid contigs and high-resolution restriction mapping of the Huntington disease region of human chromosome 4. *Hum. Mol. Genet., 2*:889–899.

12 Storage and Retrieval of Genetic Information

A library, in the conventional sense, is a repository of information that has been catalogued for easy retrieval and use. Biotechnology is facilitated by the availability of a similar system for the storage and retrieval of genetic information. Just like the library, where information is stored in different formats (e.g., books, computer-based data, and microfiche), genetic information may be stored in different formats. The actual genes can be stored in physical form, or the information they contain can be decoded into DNA language comprising of four letters (sequenced) and stored in computer-based formats.

In this chapter, you will learn:

1. The nature of genetic libraries.
2. The types of genetic libraries.
3. How to retrieve data from genetic libraries.

Having a library of genetic information facilitates the work of scientists in a variety of ways. They do not always have to start from the beginning—extract DNA, isolate a gene, synthesize a DNA strand, or repeat some of the basic routines—when they conduct research. Copies of genetic material from previous work may be retrieved and used. Gene libraries (or gene banks) also facilitate the exchange and sharing of genetic materials and information.

STORING ACTUAL DNA

Pieces of DNA from an individual can be cloned. A set of cloned DNA segments belonging to an individual constitutes a **genetic library.** Furthermore, genetic libraries can be created according to themes: the entire genome, a single chromosome, or active genes in a cell.

■ GENOMIC LIBRARIES

A **genomic library** consists of cloned fragments of an entire organism. Genomic libraries of a number of prokaryotes and eukaryotes have been completed from various structural genomic projects. Because genomes can be large (e.g., 3.0×10^6 kb in humans), it is critical to the success of a genomic construction endeavor to select a cloning vector that will complete the job with the smallest number of clones. This is why viral vectors are advantageous in such projects.

145

To create a genomic library, the total DNA of the organism is extracted and subjected to a restriction enzyme digest. Each fragment is inserted into a vector, and each clone is identified, isolated, subcultured, and characterized. This is obviously a painstaking and tedious activity. There are sophisticated pieces of equipment and protocols available to facilitate the process, as will be discussed later (see structural genomics).

■ *cDNA LIBRARIES*

Complementary DNA (cDNA) is DNA synthesized from its mRNA by reverse transcription (i.e., against the central dogma of molecular biology). cDNA molecules can be synthesized to represent the genes that are active in a certain cell at a certain time. If you desire to construct a cDNA library for flowering genes, the best time to find gene products associated with flowering would be during flowering. First, the appropriate population of mRNA molecules is isolated and poly-A tailed at the 3′ end (see Figure 12–1). Poly dT is added as a primer to initiate synthesis of a DNA strand using **reverse transcriptase** as the enzyme. This activity creates an RNA-DNA double-stranded duplex molecule. To obtain only DNA, the RNA strand is removed by alkali treatment or by using the enzyme **ribonuclease H.** The resulting single-stranded DNA is used as a template to synthesize a complementary strand. The reaction is catalyzed by DNA polymerase I and produces a DNA duplex that is closed at the 3′ end, as a result of the single-strand DNA looping back upon itself to serve as a primer for synthesizing the complementary strand. The loop is broken by using the enzyme S₁ nuclease. The new double-stranded DNA can then be cloned as before.

Gene libraries may contain thousands to even millions of clones. The recognition sequence for *Eco*RI restriction endonuclease is GAATTC. Considering that there are four ni-

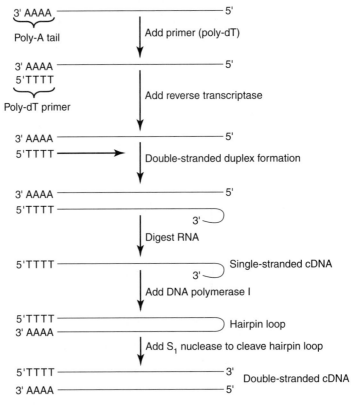

FIGURE 12–1
Steps in the production of cDNA.

trogeneous bases in DNA, the number of possible ways these bases (A, C, T, G) can be arranged to form GAATTC is $4^6 = 4,096$. This means that *Eco*RI will encounter its recognition sequence only once out of every 4,096 nucleotides. If *Eco*RI is used as a restriction enzyme for constructing a genomic library of *Arabidopsis* with the smallest genome (70 million nucleotide pairs) known in plants, the enzyme will cut the DNA $70,000,000/4,096 = 17,089$ times. This translates to a requirement of over 17,000 clones. The number of clones is astronomical for organisms with larger genomes.

cDNA libraries are always incomplete, because they represent only genes that have been reverse transcribed. In addition, cDNAs are smaller than the original genes because they lack introns.

RETRIEVING GENES FROM STORAGE (SCREENING GENE LIBRARIES)

How do scientists retrieve a specific gene from a gene library? It is a challenge to screen thousands (or millions) of clones to detect and select only a single clone, or several clones, containing a specific target DNA sequence. Several methods are available to detect a target nucleotide sequence from a library.

■ *DNA Colony Hybridization*

The DNA colony hybridization method depends on a **probe** (a polynucleotide that contains a base sequence complementary to all or part of the DNA sequence of interest). The length of a probe varies between 100 to 1,000 bp or more.

For visualization, a probe must have a tagging system. In one system, the probe is radioactively labeled and consequently visualized by autoradiography. A variation in radioactive probing is the method called **random primers.** However, nonisotopic methods may also be used in the detection of DNA sequences. A widely used method is biotin labeling in which biotin is attached to one of the four deoxyribonucleotides of the probe. To visualize this process, a coupling reaction that produces a chromogenic product (for direct visualization) or a chemiluminescent product (for autoradiographic visualization) is necessary.

Probes may be natural or synthetic in origin. Natural probes may be derived from close relatives of the organism. Such probes are called **heterologous probes** because natural differences exist between the target DNA and the probes, resulting in a lack of perfect matches during hybridization. A synthetic probe is designed based on the amino acid sequence obtained from the protein coded by the target gene.

To screen a genomic library by hybridization, a master plate of colonies is produced by plating out the selected clones. To do this, cells from the transformation reaction are plated out onto a solid medium that would permit only transformed cells to grow. A sample from each of the discrete colonies is transferred onto a nitrocellulose or nylon membrane, making sure to keep this pattern of the colonies on the master plate (see Figure 12–2). The cells are then lysed (broken up) to release the DNA, which is subsequently denatured and deproteinized. The product becomes irreversibly bound in the matrix (nitrocellulose or nylon). Next, an appropriately labeled probe is added to the matrix and then submitted to hybridization (in a hybridization chamber) (see Figure 12–3). After incubation, the matrix is washed with water to remove the nonhybridized probe molecules before visualizing by autoradiography. The colony with a positive response, as well as its corresponding colony, is identified on the master plate. This colony may contain the target DNA. Cells from the colony are subcultured.

Genetic libraries are frequently created from partial digestions. Consequently, more than one colony may show a positive signal in autoradiography. Further analysis is required to determine conclusively the clone that has the target gene. Techniques including

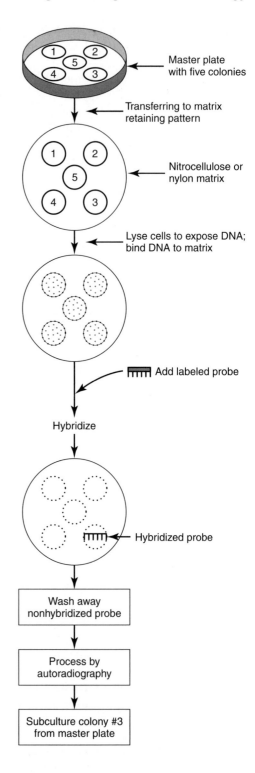

FIGURE 12–2
Steps in the technique of colony hybridization. The procedure is similar in principle to replica plating.

Master plate with five colonies

Transferring to matrix retaining pattern

Nitrocellulose or nylon matrix

Lyse cells to expose DNA; bind DNA to matrix

Add labeled probe

Hybridize

Hybridized probe

Wash away nonhybridized probe

Process by autoradiography

Subculture colony #3 from master plate

electrophoresis and restriction endonuclease mapping as well as DNA sequencing are among the tools employed in an attempt to identify the complete gene. An insert in one clone may be large enough to contain the complete gene, in which case DNA sequencing would bring about a quick resolution of the problem. Otherwise, additional cloning may be necessary or other libraries may have to be screened. It is for this reason that vectors capable of cloning larger fragments are desirable in some studies. They increase the probability that some clones in the library would carry the complete gene.

FIGURE 12–3
A hybridization chamber is
used to help a probe bind
to its target, which has
been immobilized by a
membrane.

◼ IMMUNOLOGICAL ASSAY

Screening by immunological assay is similar to screening by using a probe. Instead of a probe, an antibody that binds to the protein encoded by the target gene is used. The steps are similar to those described for probe hybridization except that the proteins are crucial to the immunological assay. Hence, after lysis of the cells, the matrix and proteins are treated with the antibody (primary antibody). After incubation, unbound antibody is washed away and another antibody (secondary antibody) specific for the primary antibody is applied. For visualization, an enzyme may be attached to the secondary antibody whose substrate is colorless. Upon introduction of the substrate, the enzyme hydrolyses it to produce a colored compound at the site of reaction, thereby identifying the clone that may harbor the target DNA.

COMPUTER-BASED STORAGE AND RETRIEVAL OF GENETIC INFORMATION

Computer-based storage, retrieval, and use of genetic information is currently one of the major activities in biotechnology. Thanks to the numerous genome sequencing projects that are in varying stages of completion, genetic information from various organisms (in sequence form) can be accessed from a variety of gene banks located in various parts of the world. This information is used in the modeling of proteins, drug discovery, and gene discovery, among other applications. This technology has given birth to a new scientific discipline called **bioinformatics,** the knowledge-based theoretical discipline that attempts to make predictions about biological functions using data from DNA sequence analysis. Bioinformatics is discussed later in the text.

KEY CONCEPTS

1. Genetic sequences (DNA, RNA, protein) can be stored.
2. DNA *per se* can be cloned and stored.
3. Gene libraries can be created according to themes.
4. Stored DNA can be retrieved by using probes, random primers, or immunological assays.
5. DNA sequences can be stored as databases on computers.

OUTCOMES ASSESSMENT

1. Give the importance of gene banks in biotechnology.
2. What is complementary DNA?
3. Contrast genomic and cDNA libraries.
4. Describe how a cDNA library is constructed.
5. Describe a method for screening a genetic library.
6. How is a genetic library similar to a book library?

INTERNET RESOURCES

1. Comprehensive listing of genomic and genetic resources at the Human Genome Research Institute: *http://www.nhgri.nih.gov/Data/*
2. Protein data: *http://www.expasy.ch/sprot/sprot-top.html*
3. Genome centers in the United States: *http://www.nhgri.nih.gov/Data/#us*
4. Chromosome maps of various organisms: *http://www.nhgri.nih.gov/Data/#other_orgs*

REFERENCES AND SUGGESTED READING

Apley, L. 1997. *DNA sequencing: From experimental methods to bioinformatics.* New York: Springer-Verlag.

PART III

Approaches to Biotechnology

Biotechnology is a rapidly evolving field, where new ways of doing things keep emerging as technologies advance. This part of the text is devoted to a discussion of the current conceptual approaches to conducting state-of-the-art research in biotechnology.

Thomas Roderick is credited with coining the term "genomics" in 1986 to describe the scientific discipline of mapping, sequencing, and analyzing genomes, as well as to provide a name for a new journal: *Genomics*. The genome is the totality of the genes in an organism. Eric Lander of MIT stated that the aim of genomics is to provide biologists with the equivalent of chemistry's Periodic Table. Like the Periodic Table (an inventory of all known elements), the proposed biological Periodic Table will represent an inventory of all genes involved in the assembling of an organism. Furthermore, the genes will need to be categorized according to certain properties to facilitate the use of such a table. Unlike the chemical Periodic Table that features only about 100 elements, a biological table will have to contend with the hundreds of thousands of genes present in certain species, making such an undertaking a daunting task. The biological table will be further complicated by the fact that it will not be two-dimensional, in order to account for similarities at the various levels of biological organization. These levels, Lander notes, include:

a. Primary DNA sequence in coding and regulatory regions
b. Polymorphic variations within a species or subspecies
c. Time and place of expression of RNAs during development, physiological response, and disease
d. Subcellular localization and intermolecular interaction of protein products

To accommodate the unfolding complexity of the problem, it is apparent that the conventional approach of exploring the genome piecemeal, one gene at a time, is a woefully inadequate strategy for accomplishing the task. Rather, there is the need to simultaneously examine all of the components of biological processes. In other words, a global approach is needed.

As previously stated, a genome is the full complement of the genes and chromosomes of an organism. Traditional genetics investigates single genes, one at a time, as "snapshots." Genomics is the approach of investigating the totality of genes in an individual as a dynamic system (not as a snapshot, but rather over time) and to determine from a global perspective how these genes interact and impact biological pathways and the general physiology of an organism, to obtain the "big picture."

There are currently two basic categories of genome analysis—structural and functional, each with its own set of tools and functions. These terms are widely accepted in the scientific community, but like the term biotechnology, there are different interpretations. Furthermore, subcategories are being coined at a feverish pace as the field of genomics advances.

Structural Genomics

PURPOSE AND EXPECTED OUTCOMES

Structural genomics, at its most basic level, is concerned with activities at the initial phase of genome analysis—mapping (the construction of high-resolution genetic, physical, and transcript maps of an organism). The ultimate physical map of an organism that can be achieved is the complete sequence of its total DNA (genomic sequence).

In this section, you will learn:

1. About the Human Genome Project.
2. About other genomics projects (model organisms) and their importance.
3. What lies ahead after genomics.

Genomic sequencing projects yield linear amino acid sequences. The recent explosion in genomic sequencing projects (the most notable being the Human Genome Project) has yielded an enormous storehouse of genetic data that needs to be converted into more useful information for understanding the function of proteins. Current usage of the term "structural genomics" appears to identify it with determining the three-dimensional structure of a protein. Unlike the genome sequence in which researchers can predict a definite end point, the duration of a protein structure project is not easy to predict, since new folds could be discovered.

GENOME SEQUENCING

The Human Genome Project (HGP), the most ambitious genome project to date, was jointly undertaken by multinational research teams and the private sector. Its primary purpose was to decipher the so-called "book of life," the totality of the human DNA that will be used to demonstrate what goes on in a genome sequencing project. This task was tackled at four levels: construction of genetic (linkage) maps, construction of a physical map, sequencing the genome, and comparative genomics.

■ *THE HUMAN GENOME PROJECT*

Brief Historical Background

Many people are surprised to learn that the HGP was initiated by the U.S. Department of Energy (DOE). However, it is much easier to understand DOE's role after learning how the project started. The two agencies that preceded the DOE (the Atomic Energy Commission and the Energy Research Administration) had been charged with studying the genetic and health effects of radiation and chemical by-products of energy production. These agencies determined that the best way to tackle the issue was to study the DNA directly. Between 1984 and 1986, the DOE and other institutions hosted a series of scientific meetings whose proceedings first proposed the idea of sequencing the complete human genome. A 1988 report produced by a committee appointed by the U.S. National Council for Research not only endorsed the concept but expanded the mandate to include the generation of physical and genetic maps as well as sequence maps of the human genome. Further aspects to investigate included the duplication of the human genome efforts in selected organisms (model organisms) such as bacteria, yeast, worms, and flies. The report also charged the scientific community with developing the requisite technologies to support the endeavors, and research into the ethical, legal, and social implications of the HGP.

The United States launched this challenge as a joint effort between the DOE and the National Institute of Health (NIH). In the UK, the key players were the UK Medical Research Council and the Wellcome Trust, while the Centre d'Etude du Polymorphisme Humain and the French Muscular Dystrophy Association were the principal players in France. Other international participants were the Science and Technology Agency and the Ministry of Education, Science, Sports, and Culture of Japan, as well as the European Community. The international HGP, operated largely by consensus and publicly funded, was launched in late 1990 (later participants included Germany and China).

The team first tackled the issue of developing genetic and physical maps of the human and mouse genomes, designed to identify disease genes and markers for the subsequent genome sequencing. Sequencing the yeast and worm genomes followed. Having tested their techniques on model organisms, a full-scale sequencing of the human genome commenced in 1997. Estimated to cost $3 billion, the HGP was slated to be completed in 2005. This date quickly changed when Craig Venter, a former NIH scientist turned entrepreneur, entered the arena as a private competitor and declared that he would complete the task in just three years, four years ahead of the public consortium's schedule. Fearing that, if successful, Venter's group would patent their product and sell it for profit (not to mention take all the credit), the HGP revised its strategy. Instead of a finished sequence, it proposed producing a draft sequence of 90 percent of the human genome by 2001, which coincided with the timing of Venter's group.

■ METHODOLOGIES

DNA Source

The public consortium collected samples from a large pool of people (blood from female volunteers, sperm from male volunteers). Only a few samples were eventually used, while steps were taken to make sure that source names were protected (i.e., no one knew whose genomes were actually sequenced). In the private effort established by Venter's Celera Genomics, samples were reportedly collected from five individuals who identified themselves as Hispanic, Asian, Caucasian, and African American. However, it came to light around April of 2002 that about one-third of the DNA actually sequenced was derived from Craig Venter. Coincidentally, Craig Venter resigned as CEO of the company in early 2002.

Sequence Technology

By consensus, the HGP team adopted the **hierarchical shotgun sequence** approach (also called **map-based, BAC-based,** or **clone-by-clone approach**). The methodology entailed generating and organizing a set (a BAC-library) of large-insert clones (about 100 to 200 kb long) to cover the entire genome, followed by conducting shotgun sequencing

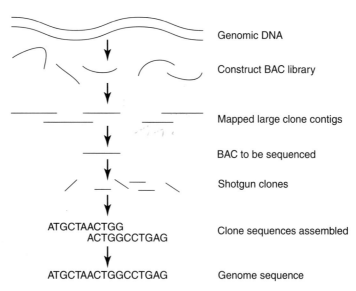

Genomic DNA

Construct BAC library

Mapped large clone contigs

BAC to be sequenced

Shotgun clones

ATGCTAACTGG
ACTGGCCTGAG   Clone sequences assembled

ATGCTAACTGGCCTGAG   Genome sequence

FIGURE 13–1
The strategy of shotgun genome sequencing entails the use of restriction enzymes to cut the genome into small fragments, cloning these fragments, sequencing them, and piecing them back together. The hierarchical shotgun method includes the production of a map.

of selected clones (see Figure 13–1). The sequence information was local, thereby eliminating the possibility of long-range misassembly and reducing the risk of short-range misassembly. In simple language, the clone-by-clone method of genome sequencing was likened to ripping a page out of the book of life (which constitutes a BAC), shredding it several times, and piecing it back together with the help of a computer by looking for overlapping ends of the strips. The next step is to form longer pieces called contigs. Subsequent clones were similarly treated. In actuality, each subclone was sequenced several times, the rationale being that the more times a base shows up at the same position in the subclones, the more certain the computer is that a correct identification has been made. To achieve an error rate of less than 1 in 1,000 for the completed rough draft, scientists had to sequence the genome four times over. For an error rate of 1 in 10,000, the genome had to be sequenced 8 to 11 times (i.e., 8- to 11-fold coverage).

The clone-by-clone approach was initially more expensive than the whole genome method because of the need to first create a map of clones as well as sequence the overlaps between clones. However, the work at the end of the project to resolve misassemblies in the finished sequence was more challenging with the whole genome methodology. The specific steps involved in the method of shotgun sequencing employed in the Rice Genome Research Program is summarized in Figure 13–2.

Assembling the Draft Genome

To put all the pieces together to form the rough draft of the human genome, the HGP group followed three steps: **filtering** (to remove contamination from nonhuman sequences and other artifacts), followed by a **layout** (to associate the sequenced clones with specific clones on the physical map). Finally, the sequences from overlapping sequenced clones were **merged** using the computer program GigAssembler. The rough draft as announced in 2001 consisted of BACs covering 85 percent of the gene-containing regions of the chromosome. Further, the BACs are in different stages of completeness, with about 24 percent of the sequence being classified as finished and highly accurate, 22 percent near finished, and 38 percent in draft form. The remaining 15 percent is being sequenced at the time of the announcement. Some 3 percent remains practically unclonable.

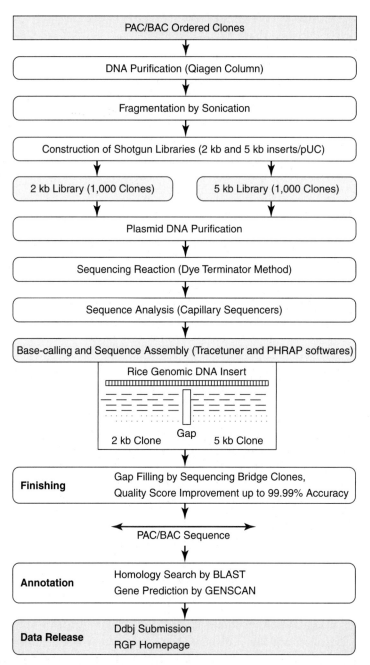

FIGURE 13–2

A summary of the shotgun sequencing strategy adopted by the Rice Genome Project in Japan for sequencing the rice genome.

Source: Image is courtesy of the International Rice Genome Project, Japan. Used with permission.

Technological Support for the Human Genome Project

The success of the HGP depended on the discovery (by both the project team and other inventors) and application of several key technologies for large-scale sequencing. The major ones included four-color fluorescence-based sequence detection, improved fluorescence dyes, dye-labeled terminators, custom-designed polymerases for sequencing, cycle sequencing, and capillary gel electrophoresis. Because of the role of computers in large-scale sequencing, the development of sophisticated software is critical to the success of such projects. The notable computer software that supported the human genome sequencing efforts were the GigAssembler (for merging the information from individual sequenced clones into a draft genome sequence), PHRED (for analyzing raw sequences to produce a "base call" with an associated "quality score" for each position in the sequence), and PHRAP (for assembling raw sequences into contigs and assigning to each position in the sequence a "quality score" on the basis of the PHRED scores). The development and use of robotics for automating various aspects of sequencing (e.g., sample preparation) was significant in facilitating the sequencing projects.

■ COMPLETENESS OF A SEQUENCED GENOME

In 2002, the *Drosophila* genome project announced the completion of its task. However, only 120 Mb of the 180 Mb genome were actually characterized. Was this a mistake, then? Not quite. Animal genomes are characterized by regions that cannot be cloned or assembled. These regions include telomeres, centromeres, and regions rich in sequence repeats. These regions contain heterochromatic DNA, which has few genes. Genomic projects hence focus first on the euchromatic regions. The definition of a "finished" sequence is variable from one project to another. For the Human Genome project, "finished" meant that fewer than one base in 10,000 was incorrectly assigned, more than 95 percent of the euchromatic regions were sequenced, and the gaps were smaller than 150 kb. At the time of the announcement of the draft genomes, the public project was considered to have more finished sequences of chromosomes 21 and 22, while the Celera project had more complete sequences of the other chromosomes.

The HGP is determined to produce a polished version of their draft by 2003. By June 2002, chromosomes 20, 21, 22, and Y were virtually finished, whereas chromosomes 6, 7, 13, and 14 were in the final stages of completion. Chromosomes 9 and 10 were about 85 percent completed, whereas chromosomes 1, 3, 17, and 18 were the farthest behind. The goal of the HGP is to sequence all parts of the genome, including the often-problematic regions where sequence repeats occur (the telomeres and centromeres). The duplicated regions may be 200,000 bases long, and comprise about 5 percent of the genome. Furthermore, any two duplications could be about 99 percent similar. This situation has posed significant challenges to assemblers. However, scientists have been able to write a more powerful computer program that is capable of fishing out such duplications. This will make assemblers more efficient in correctly ordering the DNA sequence.

■ THE PRIVATE SECTOR SEQUENCING STRATEGY

Craig Venter's group adopted the **whole-genome shotgun** approach that he pioneered while at NIH. This method entailed the sequencing of unmapped genomic clones (about 2 to 10 kb) followed by the use of linking information and computational analysis to avoid misassemblies (see Figure 13–3). This strategy may be likened to shredding not just a page at a time from the book of life, but the whole text (entire sets of encyclopedias) into millions of tiny overlapping pieces. Using supercomputers, these pieces are then reassembled together. It is believed (as the exact sequencing strategy has not been released) that genomic DNA from one sample was shredded into 2,000 base pieces, then 10,000, and finally 50,000, giving a threefold genome coverage. To increase the accuracy and fill in the gaps, parts of the genomes of three females and one male were sequenced. Because the public consortium had a policy of depositing its results into GenBank on a nightly basis, the Celera group took advantage of the HGP sequenced data, manipulating them to look like their own, and feeding both sets of data into their computer for comparison. This strategy created the appearance of a sixfold (or even greater) genome coverage, thereby increasing the accuracy and reducing the time of their project by one to two years.

FIGURE 13–3
The concept of the whole genome project (right column). This strategy excludes the construction of a physical map as part of the process (see map-based strategy on the left column).

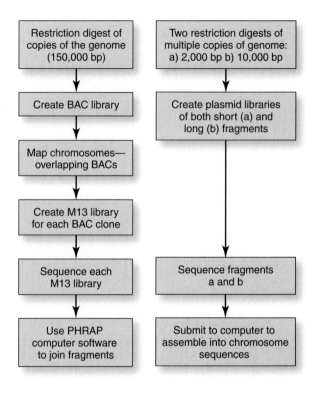

The contigs are linked together to form **scaffolds.** The contigs in a scaffold are ordered and oriented with respect to each other. The draft produced by the Celera group has some 200,000 gaps between and within scaffolds. However, their effort comes closer to covering all the euchromatic regions of the genome than the public draft. Furthermore, because Celera's assembly strategy is based only on overlaps and not supposedly the preestablished order of the pieces (as in the public consortium's project), more of the private sector draft of the genome is in the right place and in the right order.

■ SUMMARY OF INFORMATION FROM THE DRAFT OF THE HUMAN GENOME

The key information that can be gleaned from the rough draft of the human genome includes the following:

1. The human genome is the largest genome to be extensively sequenced to date. It contains 3,164.7 million nucleotide bases.
2. Of the total number of bases, 99.9 percent are exactly identical for all humans.
3. The human genome is now estimated to contain 30,000 to 40,000 protein-coding genes.
4. The average human gene is about 3,000 bases long (the longest-known sequence for a gene is 2.4 million bases for dystrophin).
5. Chromosome number 1 has the most genes (2,968), while chromosome Y has the fewest (231).
6. The human genome shows a wide variation in the distribution of features like genes, transposable elements, GC content, and CpG islands.
7. GC-poor regions strongly correlate with dark G-bands in karyotypes in cytogenetic analysis.
8. Only about 2 percent of the human genome is made up of protein coding.

9. Genes appear to be concentrated in random areas along the genome, interspersed with long noncoding regions.
10. At least 50 percent of the genome does not code for proteins (the so-called "junk DNA").
11. Over 1.4 million single nucleotide polymorphisms (SNPs) have been identified.
12. The human genome has a greater percentage of repeat sequences (50 percent) than the mustard weed (11 percent), the worm (7 percent), and the fly (35 percent).
13. Germline mutation rate is about two times as high in males as in females (i.e., most mutations occur in males).
14. Stretches of up to 30,000 C and G bases (CpG islands) repeating over and over often occur adjacent to gene-rich regions, forming a barrier between the genes and the noncoding junk DNA.

■ IMPACT OF THE HUMAN GENOME PROJECT

The impact of the HGP has and will have both direct and indirect benefits, including the following:

1. It will help scientists better understand the molecular mechanisms of disease, as well as the design of the rational diagnostics and therapeutics targeted at those mechanisms.
2. All human genes will be discovered; markers for these diseases will be developed for accurate diagnostics or inherited diseases.
3. By understanding the underlying biology of genome organization and gene regulation, scientists will be able to better understand how humans develop from single cells to adults, and also learn how and why mal-developments occur, and how changes occur with age.
4. The project will provide understanding of other genomes (animal models) that will help in understanding gene function in health and diseases.
5. New technologies, developed as a result of the project, can be applied to the study of other economically important crops and animals.
6. The project has increased public awareness of the development and applications of biotechnology.

■ AFTER SEQUENCING, WHAT NEXT?

After obtaining the genome sequence, scientists need to figure out its gene content, and then the functions of those genes. Analyzing genomes to discover the genes present and their function is called **annotation.** This is a very challenging task, considering the fact that the proportion of the protein-coding genes in the human genome is only 2 percent. This task is accomplished through the use of highly sophisticated software and powerful computers. There are two kinds of genes to search for: RNA genes and protein-coding genes.

Three basic gene prediction strategies are currently in use:

1. **Direct evidence** of transcription provided by expressed sequence tags (ESTs), or mRNAs. This approach is prone to artifacts from contaminating ESTs derived from unspliced mRNAs, genomic DNA contamination, and nongenic transcription-like material from a promoter of a transposable element.
2. **Indirect evidence** based on similarity to previously identified genes. This approach is not effective in identifying novel genes without sequence similarity to existing genes.
3. **Ab initio recognition** of groups of exons. This approach is based on hidden Mankov Models that combine statistical information about splice sites, coding bias, and exon and intron lengths. Examples of systems for this analysis include the Genscan, Genie, and FGENES.

COMPARATIVE GENOMICS

Biotechnology makes use of certain organisms as **models** for comparative research. This is because humans, farm animals, and many crop plants are genetically very complex and not readily, directly amenable to certain biotechnological procedures. Therefore, scientists develop models using plants and animals that are easy to manipulate. The information derived from such studies is used comparatively to develop and test procedures for studying the more complex organisms. The HGP focused on a few organisms (mouse, yeast, and worm).The common organisms used in model development in biotechnology research include those listed in Table 13–1.

■ GENERAL CHARACTERISTICS OF MODEL ORGANISMS

What makes scientists select certain organisms for developing models upon which other organisms are manipulated? Organisms selected and developed into model organisms for biotechnology research tend to have certain characteristics in common, including the following:

a. **Small size**

Small body size enables researchers to grow model plants and animals in large numbers in limited space (e.g., laboratory, greenhouse, growth chamber).

b. **Fast generation time**

A short gestation is desirable for rapid study of hereditary events, growth, and development, as well as for rapid multiplication.

TABLE 13–1

Examples of model organisms used in biotechnology research. Genome sequences for these organisms are either completed or in progress.

Organism	Scientific Name	Common Name
Plant		
	Arabidopsis thaliana	Arabidopsis
	Oryza sativa	Rice
	Zea mays	Corn
Animals		
Mammalian		
	Homo sapiens	Man
	Mus musculus	House mouse
	Rattus	Rat
Nonmammalian		
	Drosophila melanogaster	Fruit fly
	Escherichia coli	E. coli
	Caenorhabditis elegans	Roundworm
	Xenopus laevis	Frog
	Danio rerio	Zebrafish
	Saccharomyces cerevisiae	Yeast
	Anopheles spp.	Mosquito

c. **Small genome**

A small number of chromosomes facilitate genetic and cytogenetic studies. Low chromosome number is desirable for gene expression analysis, molecular cytogenetics, mapping, and other applications.

d. **Low levels of repetitive DNA**

■ *BRIEF STATUS OF MODEL ORGANISMS*

Genomic projects involving model organisms are in various stages of completion. The reader may consult the websites provided at the end of the chapter to check on the current status of a particular project.

Arabidopsis (*Arabidopsis thaliana*)

Arabidopsis thaliana belongs to the family Brassicaceae (mustard family). This family includes important species like cabbage, raddish, and canola. However, the plant has little agronomic value so far. *Arabidopsis* has 10 chromosomes and 125 Mb of genome, which contains 25,489 genes. This flowering plant (the first to be completely sequenced) has a short life cycle of just about six weeks from germination to seed maturity. It produces seed profusely. Because of its small size, arabidopsis can be cultured in trays or pots in limited space. The plant is often described as the plant equivalent of *Drosophila* in animal research.

Yeast (*Saccharomyces cerevisiae*)

Yeasts are heterotrophic and chlorophyll-lacking organisms. Budding yeasts are true fungi that belong to the phylum Ascomycetes and class Hemiascomycetes. Yeasts live symbiotically or parasitically on plants, the skin surfaces of animals, the intestinal tracts of warm-blooded animals, the vagina of humans, soil, and other places. They are capable of asexual reproduction (by budding or fission) or sexual reproduction (by asci). The most widely used species of yeast is the *Saccharomyces cerevisiae.*

S. cerevisiae has been used in the production of alcoholic beverages and in bakery products (for fermentation or leavening of dough prepared from cereals) since the times of ancient cultures. *S. cerevisiae* has 16 chromosomes and an estimated 6,000 genes. Its genomic DNA is about 12 Mb. It is a good model organism because the basic cellular mechanics of processes like replication, recombination, cell division, and metabolism are generally conserved between yeast and larger eukaryotes.

Yeast vectors are used in cloning large DNA fragments. A certain naturally occurring plasmid, **2-micron plasmid,** has been used to construct a number of yeast cloning vectors. A type of vector, called the yeast artificial chromosome (YAC), enables researchers to clone DNA sequences exceeding 1 Mb in size. This capacity makes them useful in large projects such as the Human Genome Project.

Mouse (*Mus musculus*)

The house mouse (*Mus musculus*) is one of the most commonly used mammals in medical and behavioral research because humans and mice share many of the same basic biological and behavioral processes. The mouse has 19 + X/Y chromosomes and a 3,059 Mb genome, with an estimated 80,000 genes in its genome. Because it is estimated that mice and humans are about 85 to 90 percent genetically identical, mice provide a good model for studying human behavior, determining predisposition to disease, predicting responses to environmental agents and drugs, and designing new drugs.

A mouse's genetic map based on DNA markers is also being developed. Researchers are using mouse models to study the effects of mutations that cause major human diseases like diabetes, muscular dystrophy, and certain cancers.

Fruit Fly (*Drosophila melanogaster*)

The fruit fly (*Drosophila melanogaster*) is one of the most important organisms, especially in genetics and developmental biology. The tiny insect has a short life cycle of about two weeks and is easy to maintain under laboratory conditions. The fruit fly has four pair of chromosomes with a genomic size of about 137 Mb and an estimated 14,000 genes. Chromosome number 4 is very tiny. Furthermore, the fly has thick polythene chromosomes, which are formed when chromosomes divide several hundreds of times but fail to separate. These unique chromosomes are used to aid in physically associating genes with chromosomal regions.

Many of the genes that define spatial patterns of cell types and body parts in *Drosophila*, as well as the regulatory pathways in which they operate, have counterparts in higher eukaryotes. Such findings have aided scientists in their study of human development.

Rat (*Rattus spp*)

Rats make up more than 25 percent of laboratory animals. Rat models are important in linking function to genes. There are over 250 rat disease models involving inbreds, mutants, transgenics, and genetic structures that are used in the study of human diseases and other related variables. Specific rat models are available for cardiovascular, pulmonary, immunology, endocrinology, diabetes, behavior, and many other models. In some instances, only rat models are suited to studying certain human diseases.

The size of the rat allows researchers to conduct certain studies that are not possible with smaller organisms. The size of rats allows invasive procedures such as intravenous cannulation for drug administration, blood collection, and tissue collection, as well as the use of multiple electrodes and transducers, and injection of neuroanatomical markers. Scientists are able to produce transgenic rats without much difficulty. However, the development of knockout rats by homologous recombination in embryonic stem cells is problematic.

Nematode (*Caenorhabditis elegans*)

The nematode (*Caenorhabditis elegans*) has six pair of chromosomes, with a genomic size of 97 Mb and an estimated 19,000 genes. This organism was the first multicellular organism to be completely sequenced. In addition, the developmental fate of each cell has been determined.

Escherichia Coli (*E. coli*)

The *E. coli* bacterium inhabits the intestinal tract of humans and plays a role in the digestion of food. However, *E. coli* is implicated in a number of severe disease outbreaks, often associated with animal food products. The bottom-up method was used in the *E. coli* genome project. Yuji Kohara and his colleagues created a genomic library of the K12 strain using lambda vectors. Seventy contigs covering 94 percent of the genome were identified. The resulting physical map covered the entire 4,700 kb of DNA in the *E. coli* genome. The clones in the contig library are sequenced.

Higher Plant Models

The two most important higher plants that are being developed as model plants are rice (*Oriza sativa*) and corn or maize (*Zea mays*). Genomes of all cereals (wheat, barley, rice, corn, and so on) are structurally similar. Corn is one of the higher plants for which the genetics have been well-developed. Rice is the most important cereal for human consumption. There are several subspecies of rice, the most widely eaten being the *indica*, followed by *japonica*. Two different groups, involving both the public and private sectors, are sequencing both subspecies, whose draft sequences were published in 2002. Japanese researchers led the international consortium called the International Rice Genome

Sequencing Project (IRGSP), where they focused on sequencing the *japonica* subspecies. The Monsanto Company of the United States also targeted this same subspecies, announcing a draft sequence in April of 2002 in conjunction with the University of Washington. These two research teams used the traditional approach of mapping the genome first and thereafter sequencing the DNA piece by piece (using the clone-by-clone or hierarchical shotgun method).

In May of 2000, the Beijing Genomics Institute (BGI) of China announced that it would utilize the whole-genome shotgun method to sequence a draft of the *indica* subspecies in only two years. In April of 2002, both the BGI and the Syngenta biotech company published draft sequences of the *indica* subspecies.

The 12 chromosomes of rice are estimated to contain between 32,000 and 55,000 genes and a genome of 430 Mb. Work is ongoing, as with the human genome, to complete the sequencing of rice genomes.

KEY CONCEPTS

1. The human genome is the largest genome to be extensively sequenced to date. It contains 3,164.7 million nucleotide bases.
2. The Human Genome Project is a consortium of international institutions.
3. The Celera Company pursued an independent human genome sequencing effort.
4. The public group used the hierarchical shotgun sequencing strategy; the private group used the whole-genome shotgun method in sequencing the human genome.
5. Some portions of the genome are difficult, if not impossible, to sequence; hence, eukaryotic genomes are seldom 100 percent completely sequenced.
6. Model organisms are sequenced and used in comparative studies.
7. Model organisms tend to have small genomes and quick generation times.
8. The HGP has numerous potential applications, including disease diagnosis and the development of efficient medicines.

OUTCOMES ASSESSMENT

1. A complete genome sequence does not necessarily mean a 100 percent genome coverage. Explain.
2. What is the importance of model organisms in biotechnology?
3. What makes certain species suitable as model organisms?
4. Contrast the public consortium and private company strategies for sequencing the human genome.
5. Give three specific key pieces of information that have been provided by the Human Genome Project.
6. Give three specific impacts (realized or potential) of the Human Genome Project.

INTERNET RESOURCES

1. General information about HGP: *http://www.ornl.gov/hgmis/resource/media.html*
2. Official site of the arabidopsis organization with general information about the plant: *www.arabidopsis.org*

3. Yeast and other fungi: *http://www.ultranet.com/~jkimball/BiologyPages/F/Fungi.html*
4. Roundworm: *http://www.ultranet.com/~jkimball/BiologyPages/C/Caen.elegans.html*
5. *E. coli* information: *http://www.ultranet.com/~jkimball/BiologyPages/E/Esch.coli.html*
6. The status of mapping of various organisms can be checked at the National Council for Biotech Information (NCBI): *http://www.ncbi.nlm.nih.gov/*
7. Genomic information: *http://www.celera.com/general*
8. The society for developmental biology maintains a Virtual Library with information on organisms (vertebrates, invertebrates, plants, unicellular/lower eukaryotes, and prokaryotes) indicating selected laboratories and scientists engaged in research involving these organisms: *http://sdb.bio.purdue.edu/other/VL_DB_Organisms.html*

REFERENCES AND SUGGESTED READING

Bostein, D., R. L. White, M. Skolnick, and R. W. Davis. 1980. Construction of a genetic linkage map in man using restriction fragment length polymorphisms. *Am. J. Hu. Genet., 32:*314–331.

Dennis, N., and E. Pennisi. 2002. Rice: Boiled down to bare essentials. *Science, 296:*32–33.

The International Human Genome Mapping Consortium. 2001. A physical map of the human genome. *Nature, 409:*934–941.

The International Human Genome Sequencing Consortium. 2002. Initial sequencing and analysis of the human genome. *Nature, 409:*860–921.

Pennisi, E. 1999. Academic sequencers challenge Celera in a sprint to the finish. *Science, 281:*1822–1823.

Pennisi, E. 2002. Genome centers push for polished draft. *Science, 296:*1600–1601.

Venter, J. C. et al. 2001. The sequence of the human genome. *Science, 291:*1304–1351.

Weber, J. L., and E. W. Myers. 1997. Human whole-genome shotgun sequencing. *Genome Res., 7:*401–409.

SECTION 2
Protein Structure Determination

PURPOSE AND EXPECTED OUTCOMES

Genes, as previously indicated, are expressed as proteins. The biological functions of proteins are determined by the nature and type of their structural folds. Scientists are interested, therefore, in studying protein structure.

In this section, you will learn:

1. How the three-dimensional structure of proteins is determined.
2. The benefits and application of protein structure modeling.

THE COMPLEXITY OF PROTEINS

In order to understand the biological function of a protein, its three-dimensional structure must be determined. The wealth of information relating to biological function that can be

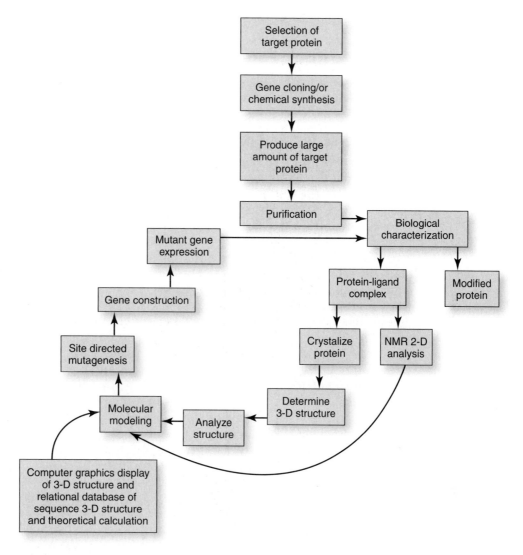

FIGURE 13–4

A summary of the steps in protein structure prediction.

obtained from three-dimensional structure is summarized in Figure 13–4. Accomplishing this task is not easy because of the complexity of proteins as summarized below:

1. Many proteins change their shapes in the process of performing their functions. Proteins and nucleic acids have the capacity for conformational change as they respond to binding molecules.
2. Proteins often elicit help from cofactors in performing certain roles that may range in complexity from a single ion (e.g., Zn^{2+}) to a complex vitamin (e.g., B_{12}).
3. Proteins are subject to post-translational modifications, reversible or permanent, which impact their properties. Such modifications may be as simple as the formation of disulfide bridges when amino acid side chains fuse together, forming complex polysaccharide antennae attached to the side chains.
4. Proteins have a tendency to form aggregates that may be as simple as dimers or complexes involving hundreds of subunits.

These and other factors pose considerable challenges to scientists pursuing the task of fully unraveling the structure of biomolecules like proteins.

OPERATIONAL STRATEGY PROTEIN STRUCTURE DETERMINATION

Determining protein structure is accomplished by a two-pronged approach: **experimental** and **computational** (prediction). These activities include PCR amplification of the coding sequence from a genomic cDNA, cloning into appropriate expression vectors, protein expression at sufficiently high levels, protein purification, characterization, crystallization, crystal drop inspection, crystal mounting, model building, and NMR spectral inspection.

As previously stated, the aim of structural genomics is to structurally characterize most protein sequences by an efficient combination of experimentation and prediction. Target protein selection is fundamental to this process. Various schemes exist for target selection. One scheme focuses on selecting all proteins in a model genome. To select a target protein, a search criterion needs to be developed (e.g., disease-associated genes, or the proteins that are common to most organisms). The Protein Data Bank is then interrogated, utilizing sequence comparison and fold recognition. Another scheme focuses on only novel folds. Model-based schemes select targets such that most of the remaining sequences can be modeled accurately by comparative modeling. Just like the strategies of large-scale genome sequencing endeavors, where the ultimate goal is to map all genes in the genome, the ultimate goal of structural genomics is to determine the structure of all proteins in an organism. This suggests a departure from hypothesis-based research, where a specific biological justification precedes structural analysis of a protein, to one in which structures are determined first before other questions are asked.

After identifying the target protein, large amounts of it must be produced and purified for analysis. However, with advances in technology that have increased the speed of macromolecular structure determination, smaller amounts of protein and fewer crystals are needed these days for analysis. Such new technologies include the use of selenomethionine derivatives, cryo-freezing, robotic crystallization, and synchrotron radiation source in X-ray crystallography. New technologies in nuclear magnetic resonance (NMR) include new and improved magnets and probe technology and experimental methods (e.g., TROSY) that have enabled a wider range of proteins to be amenable to structure determination. The selected protein must be expressed at high levels in an expression system. Many proteins, especially the large proteins and those with complex cofactors like iron sulfur tungsten clusters or unusual post-translational modifications like a main chain oxygen molecule replaced by sulfur, are often difficult to express.

BASIC METHOD OF PROTEIN STRUCTURE DETERMINATION

There are currently two basic methods of protein structure determination: by X-ray crystallography or NMR. The method of crystallography requires a crystal to be produced (which is not feasible in all cases). The protein can be big, and the methods provide high resolution (1.2 Å). However, there is the possibility that crystal packing could modify the protein structure. In the case of NMR, the protein needs to be soluble and not too big. The method is therefore less accurate, but it can capture motion. The quality of protein should be more than 95 percent pure in order for the best results to be obtained from either technique.

It is important that, once resolved, the solved structure be placed in the appropriate genomic context and annotated to facilitate the prediction of functional details. Just like genomic sequences, structural genomics needs a systematic and coordinated approach if this monumental task is to be resolved cost effectively and quickly. Different groups of workers could work on the same structure, or focus on different organisms or classes of proteins. Coordination of target selection would enable scientists to achieve reasonable coverage of protein fold quickly.

■ *X-ray Crystallography*

Purified protein can be induced to crystallize by a variety of methods, the common ones being batch methods and vapor diffusion. The protein molecules in solution are induced to form a supersaturated solution through the addition of precipitants (e.g., polyethylene glycol). Factors in the crystallization environment that are varied according to the nature of the protein include pH, temperature, precipitant, and protein concentration. The best set of conditions is usually arrived at after experimentation.

After crystallization has proceeded to a desirable extent, the crystals are mounted and snap frozen (e.g., by immersion in cryogenic liquids) to prevent ice lattice formation. It should be pointed out that freezing may destroy the macromolecular structure and render the crystal unfit for analysis by crystallography.

Crystallography entails the exposure of protein crystals to X-rays to determine their secondary and tertiary structures. The X-rays are scattered in a pattern that varies according to the electron densities in different portions of the protein. The resulting images are translated into electron density maps. These maps are superimposed on one another (manually or by computers) and manipulated to construct a model protein.

Crystallography is a lengthy process that requires special skills to conduct. The results provide very accurate structural information that is valuable to other applications (e.g., studying protein interactions, drug design).

■ *Nuclear Magnetic Resonance*

The use of the term "nuclear" in the name of NMR is misleading. It is not directly related to radioactive nuclear decay, but simply relates to the properties of the atomic nucleus, similar to the common phenomenon of magnetism. Some stable isotopes have nuclear magnetism, as do radioactive isotopes. Almost all NMR spectroscopy involves the use of stable isotopes.

The modern NMR spectrometer is essentially a computer-controlled radio station with the antenna installed in the core of a magnet. When a sample is introduced, the nuclear dipoles in the sample align in the magnetic field where it can absorb energy and change its orientation back and forth. The stronger the dipole in the sample, the greater the alignment. Furthermore, the stronger the dipole, the greater the energy associated with the alignment. The computer directs a transmitter to pulse radio waves to the sample inside the antenna. The sample absorbs some of the pulse and, after a period, readmits radio signals that are subsequently amplified by a receiver and stored on the computer. Nuclei in different chemical environments on molecules radiate different energies that can be differentially analyzed by researchers using appropriate software.

The application of NMR to the study of biological samples rests on the hydrogen nucleus. The hydrogen atom is the most abundant in organic molecules and also has one of the strongest nuclear dipoles in nature. In biological NMR spectroscopy, researchers use pulse sequences of the hydrogen nucleus to map the chemical bond connectivity, spatial orientation, and distance geometry of large biomolecules. This information is used in molecular modeling.

NMR spectrometry has certain advantages over X-ray crystallography, the first being that the often-problematic crystallization of the protein is not a requirement in NMR studies. The technique can also be used to elucidate details about specific sites of molecules without the need to solve their entire structure. NMR spectroscopy is sensitive to the motions on the time scale of most chemical events, allowing the researcher to capture motion in very minute details. A modern variation of the technique is called the Transfer Nuclear Overhauser Spectroscopy (TrNOESY), which enables scientists to determine the shape of small molecules bound to large ones in order to understand and define the binding pocket of the macromolecule.

Current NMR spectrometers are able to determine the structure of small proteins (about 10,000 Dalton molecular weight) without any problems. Larger proteins are more technically challenging.

STEPS IN PROTEIN STRUCTURE PREDICTION

Experimental structural determination methods continue to yield high-resolution structure information about a subset of the proteins. It is expected that most of the proteins will be modeled rather than determined by empirical methods. There are two categories of prediction models:

1. Comparative structure prediction.
2. *De novo* structure prediction.

   Computational structure prediction by comparative methods follows four basic steps:

1. Finding a template (a known structure related to the sequence being modeled).
2. Sequence-template alignment.
3. Model-building.
4. Model assessment.

A flowchart for protein structure prediction is suggested in Figure 13–4.

■ GENERAL CONSIDERATIONS OF PROTEIN MODELING

Bioinformatics (discussed in the next chapter) provides tools for protein modeling. Protein modeling is a challenging task, involving numerous steps from the time a gene locus is identified to when a three-dimensional model of the corresponding protein it encodes is achieved. The basic hurdles to overcome in protein modeling include:

1. The **transcription start/stop location** must be identified. It is important that this location on the DNA sequence be accurately determined.
2. The **location of the translation start/stop sequence** must be identified. The (nearly) universal start codon on the mRNA is AUG. However, there are six possible reading frames for a given DNA sequence (three on each strand). Because genes are usually transcribed away from their promoters, a definitive location of this element can reduce the number of possible frames to three. An incorrect reading frame would usually result in a predicted peptide sequence of reduced length.
3. In eukaryotes, there is a need to **detect the intron/exon splice sites.** The presence of introns makes eukaryotic reading frames discontinuous. These introns must be spliced out of the sequences (unless a cDNA sequence is being analyzed) and the exons pieced together to produce the actual coding sequence for the protein. This splicing is made a little easier by the fact that most introns begin with the nucleotides GT and end with the nucleotides AG.
4. The final step in the modeling process is the **prediction of the three-dimensional structure** of the protein. The completion of step 3 would produce the primary amino acid sequence. This usually entails the use of information from a variety of sources including pattern analysis (alignment to known homologues with more secure conformation), X-ray diffraction data (which is best when some data is available on the protein of interest), and physical forces/energy states (biophysical data and analyses of an amino acid sequence that helps to predict how it folds). The combined use of these sources of information would indicate the probable locations of the atoms of the protein in space and bond angles. The final step in the modeling involves the use of computer graphics programs to render the data in a three-dimensional format. The

resultant model is usually a rough model since several other probable conformations may exist.

■ *COMPARATIVE STRUCTURE PREDICTION*

Of the 13,000 entries in the Protein Data Bank as of 2000, only about 5,000 were estimated to represent distinct experimental templates for comparative protein structure modeling. Templates may be found by employing sequence comparison methods like PSI-BLAST or by sequence-structure threading methods that are sometimes more effective at revealing more distant relationships than sequence methods. Sequence threading entails the threading of a sequence through each of the structures in a library of known folds. Alignment is not based on sequence similarity but rather energy of the corresponding coarse models. The product of comparative modeling is an all-atom model sequence based on its alignment to one or more related protein structures. In its original form, a comparative model is constructed sequentially or simultaneously from a few core template regions, loops, and side chains from aligned or unrelated structures. There are also other methods of modeling proteins.

The accuracy of a comparative model is related to the percentage of sequence identity on which it is based, correlating with the relationship between the structural sequence similarities of two proteins. The scope of modeling is limited to the content of the protein structure database. Furthermore, only a portion of a given protein can usually be modeled. When models achieve more than 50 percent sequence identity to their templates, they are classified as high-accuracy comparative models. The errors at this level of identity occur primarily in the area of mistakes in side-chain packing and small shifts or distortions in the core main-chain regions. Medium-accuracy models are based on 30 to 50 percent sequence identity while low-accuracy models are based on less than 30 percent sequence identity. It is estimated that, based on 30-percent sequence identity cut off for successful modeling, if structural genomics projects focused on 10,000 to 20,000 targets, there would be enough protein structures to support homology modeling of every globular segment of every protein in nature. It is estimated that 1,000 to 5,000 distinct stable polypeptide chain folds occur in nature.

■ *DE NOVO STRUCTURE PREDICTION*

Unlike comparative modeling that is limited to protein families with at least one known structure, *de novo* prediction is without such a restriction. Starting on the premise that the native state of a protein is at the global-free energy minimum, the researcher conducts a search of conformational space for tertiary structures that are particularly low in free energy for the given amino acid sequence. Recent advancement in this area introduces a toolkit, Rosetta, developed by Norwegian researchers for analyzing tabular data within the framework of rough set theory. Rosetta language provides modeling support for different design domains, employing semantics and syntax appropriate for each.

APPLICATIONS AND BENEFITS OF PROTEIN STRUCTURE MODELS

Information from structural genomics may be of academic interest and facilitate the further understanding of the structure and function of proteins. In terms of fold and function, four outcomes are possible from the determination of a target protein structure. The fold may be new or old and the function known or unknown. This yields four possible combinations of new *x* known, new *x* unknown, old *x* known, and old *x* unknown. At the very least, each new structure will allow scientists to develop a model for a protein family for which structural information did not previously exist. Each of these homology models may be further investigated in a variety of ways (site-directed mutagenesis, ligand binding sites, protein-protein interaction). In a case where the outcome is new-known, the new structure could be compared with structurally distinct but functionally similar proteins to identify the regions of the protein responsible for function.

High-and-medium-accuracy comparative models are useful in refining the functional predictions that have been based on sequence match alone. This is so because ligand binding is more directly determined by the structure of the binding site than by its sequence. Frequently, the researcher can correctly predict features of the target protein that do not occur in the template structure. Structural genomics of single proteins or their domains, combined with protein structure prediction, may help scientists efficiently characterize the structure of large macromolecular assemblies.

In terms of industrial applications to benefit society, structural genomics could impact the genetic engineering of industrial enzymes through the provision of a large number of structures of thermostable proteins. Potential benefits to medicine are even more significant. Some of the newly characterized proteins may be protein pharmaceuticals. Structural genomics endeavors involving pathogenic organisms provide opportunities for structure-based cloning designs, while providing expression systems for high-throughput screening. Success with understanding protein function from structural genomics data may be illustrated by the case involving the *tub* gene. The gene, responsible for obesity in mice (the Tubby mouse), has been implicated as a transcription regulator. Scientists used the three-dimensional model of the core domain of murine Tubby to reveal the presence of a positively charged groove, suggesting a role as a DNA-binding protein. It was subsequently discovered that mutations in the TVLPI gene (human homology of Tubby implicated in human retinitis pigmentosa) cluster to this groove. The protein is located in the nucleus. The Tubby protein regulates gene expression in cell-based assays.

In the area of drug discovery, pharmaceutical industry has certain favorable traditional classes of proteins that are deemed best drug targets. These include GPCRs, ion channels, nuclear hormone receptors, proteases, kinases, integrins, and DNA processing enzymes (e.g., helicases and gyrases). Many of these are soluble proteins and hence can be subjected to a structural genomics procedure. However, the GPCRs and ion channels that currently make up more than 50 percent of human drug targets are integral membrane proteins. These proteins have proven problematic for NMR and crystallography even though some headway appears to be made.

KEY CONCEPTS

1. The three-dimensional structure is critical to understanding the protein function.
2. Protein structure determination is complicated because of their complex nature.
3. Protein structure can be determined experimentally or by computation methods (prediction).
4. There are several steps in protein structure determination: PCR amplification of the coding sequence from a genomic cDNA, cloning into an appropriate expression vector, protein expression at sufficiently high levels, protein purification, characterization, crystallization, crystal drop inspection, crystal mounting, model building, and NMR spectral inspection.
5. Entries in the Protein Data Bank are used for comparative structure prediction.
6. Applications of protein structure determination include engineering of industrial enzymes and drug discovery.

OUTCOMES ASSESSMENT

1. Explain why protein structure determination is such a challenging task.
2. Discuss the role of Protein Data Banks in protein structure determination.
3. Give specific applications of protein structure determination.
4. Describe the method of comparative structure prediction in protein structure determination.

INTERNET RESOURCES

1. Intro to protein structure determination: *http://www.bmm.icnet.uk/people/rob/CCP11BBS/*
2. Principles of protein structure—excellent course: *http://www.cryst.bbk.ac.uk/PPS2/course/index.html*
3. Oak Ridge National laboratory protein research program: *http://compbio.ornl.gov/structure/*
4. Use of NMR: *http://www.nmrfam.wisc.edu/~volkman/LinuxNMR/*
5. Three-dimensional protein structural analysis: *http://www.hgmp.mrc.ac.uk/GenomeWeb/prot-3-struct.html*
6. Slide presentation on protein structure determination: *http://www.soi.city.ac.uk/~drg/courses/P227/slides/slides5/sld001.htm*
7. X-ray diffraction apparatus: *http://crystal.uah.edu/~carter/protein/xray.htm*

REFERENCES AND SUGGESTED READING

Baker, D., and A. Sali. 2001. Protein structure prediction and structural genomics. *Science, 29*:93–96.

Blundell, T. L., B. L. Sibanda, M. J. E. Sternberg, and J. M. Thornton. 1987. Knowledge-based prediction of protein structures and design of novel molecules. *Nature, 323*:347–352.

Greer, J. 1974. Three-dimensional pattern recognition: An approach to automated interpretation of electron density maps of proteins. *J. Molecular Biology, 82*:279–301.

Jones, D. T., and J. M. Thornton. 1996. Potential energy functions for threading. *Curr. Opin. Struct. Biol., 6*:210–216.

Sanchez, R., and A. Sali. 1997. Advances in comparative protein-structure modeling. *Curr. Opin. Struct. Biol., 7*:206–214.

Suelter, C. H. 1991. *Protein structure determination.* New York: John Wiley and Sons.

14 Functional Genomics

SECTION 1
Bioinformatics

PURPOSE AND EXPECTED OUTCOMES

In the last two decades, numerous research undertakings in both the public and private sectors have yielded complete structural genomic analysis of several important species, as previously discussed. A number of projects devoted to sequencing major crops and animals are currently underway or in various stages of completion. Databases are literally busting at the seams with gene sequences. The next important step and bigger challenge to scientists is to decipher the functions of these genes. The challenge is greater for plant scientists who tend to handle genomes that are bigger and more complex than those encountered by animal researchers.

In this section, you will learn:

1. How computer data storage and retrieval systems facilitate biotechnology.
2. The types and sources of databases used in biotechnology.
3. How to retrieve data from genetic databases.

WHAT IS BIOINFORMATICS?

The effort at understanding the functions of gene sequences is generally called **functional genomics**, as opposed to **structural genomics**, which focuses on genome sequencing. Because most genes are expressed as proteins, one of the common ways of understanding gene function is by tracking protein expression by cells, called proteomics. The genome is essentially a set of instructions for making various kinds of proteins. It is important to identify what protein a gene produces, because genes do not cause disease; the proteins they encode do.

Genes may provide instructions for making specific proteins. However, in the process of carrying out these instructions, additional proteins can be produced, as previously noted. These uncertainties make linking genes to function a complex undertaking. Nonetheless, understanding the genome structure alone is insufficient; it is critical to identify the proteins the genes encode. A case in point is Alzheimer's disease. Several genes that increase the risk of having the disease have been identified. However, a conclusive diagnosis of the disease comes from the presence of beta amyloid protein fragments. Unfortunately, there is no beta amyloid gene, and consequently Alzheimer's disease cannot be identified through a DNA chip (as discussed later). A protein chip might be the answer. Many techniques have been developed, and continue to be developed, for deciphering gene function.

Bioinformatics may be defined as a knowledge-based theoretical discipline that attempts to make predictions about biological function using data from DNA sequence analysis. It uses supercomputers and sophisticated software to search and analyze databases accumulated from genome sequencing projects and other similar efforts. Practitioners should be well-versed in both computer science and biology. At the present time, scientists are unable to deduce protein function from an amino acid sequence. What bioinformatics allows scientists to do is make predictions based on previous experiences with biological reality. The biological information bank is searched to find sequences with a known function that resemble the unknown sequence and thereby predict the function of the unknown sequence. The general steps in a bioinformatics project are outlined in Figure 14–1.

THE PRINCIPLES OF COMPUTATIONAL BIOLOGY

Bioinformatics, in essence, is an application of information science to biology. The most fundamental, and perhaps simplest, task in bioinformatics is the creation and maintenance of biological information databases. This task of collecting, organizing, and indexing sequence information into a database is nonetheless very challenging. Most of the databases comprise nucleic acid sequences and the protein sequences derived from them. A critical aspect of the design and storage of such biological information is the development of an interface that enables researchers to readily access existing information as well as contribute new entries to the repository.

The information *per se* has limited uses, unless it is analyzed to reveal the functionality of the sequences. A sequence of DNA does not necessarily constitute a gene. It may represent only a fragment of a gene or possibly comprise several genes. **Computation biology** is the science of analyzing sequence information, and entails the following activities:

a. Finding the genes in DNA sequences. DNA sequence repositories exist for a variety of organisms.
b. Developing methods to predict the structure and/or function of newly discovered proteins and structural RNA sequences.
c. Grouping or clustering protein sequences into families of related sequences, and the development of protein models.
d. Aligning similar proteins and generating phylogenetic trees to examine evolutionary relationships.

RESOURCES FOR BIOINFORMATICS PROJECTS

There is a vast amount of biological information residing on numerous supercomputers distributed all over the world and interconnected through the World Wide Web to facilitate information exchange. The Internet and World Wide Web communication resources are indispensable in the use of bioinformatics, as they facilitate networking and resource sharing. Access to the databases may be free or paid, depending on the kinds of information and the custodians of the databases. Similarly, software packages for accessing and analyzing the information may be in the public domain or may be available for use at a fee.

The first-time users may take advantage of guidelines and the variety of resources that have been compiled by other previous users. These include links to major bioinformatics websites and web addresses of entities engaged in genome research. Newsgroups provide opportunities for discussions and the exchange of ideas among active users of the technology. Because biotechnology is a very vibrant and rapidly evolving discipline, vast amounts of information are generated on a regular basis. Consequently, it is important for researchers to update themselves on the current state of the science through, for example, frequent visits to major bioinformatics websites.

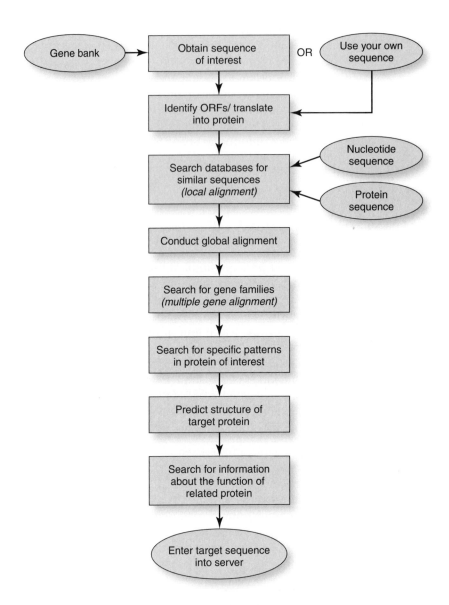

FIGURE 14–1

A summary of the steps in bioinformatics. The researcher may utilize an original sequence or obtain one from the huge amount of genetic data being managed by numerous repositories around the world. The specific steps may vary depending on the purpose of the project, the starting material, and the software being used to conduct the search. Gene-seeking computer programs have four general characteristics: (1) algorithms for pattern recognition use statistical probability analysis to determine the similarity between two sequences; (2) data tables contain information on consensus sequences for various genetic elements; (3) taxonomic differences are included because consensus sequences vary between different taxonomic classes to facilitate analysis and minimize errors; and (4) specific instructions describe how the algorithms should be applied in an analysis and how the results should be interpreted.

TYPES OF BIOINFORMATICS DATABASES

Two general categories of information are used in bioinformatics research.

1. **Primary databases**   These databases consist of original biological data such as raw DNA sequences and protein structure information from crystallography.
2. **Secondary databases**   These databases contain original data that have been processed for value added to suit certain specific applications.

A good database, primary or secondary, should have two critical parts: (1) the **original sequence**, and (2) an **annotation description** of the biological context of the data. It is critical that each entry be accompanied by a detailed and complete annotation, without which a bioinformatics search becomes an exercise in futility since it would be difficult to assign valid meaning to any relationships discovered. Some databases include taxonomic information such as the structural and biochemical characteristics of organisms. Most of these biological databases consist of long strings of nucleotides and/or amino acids. These sequences are presented in universally accepted biological shorthand (e.g., A,C,T,G) for DNA (see Table 14–1). This coding decreases the space for storing the continually growing data and also facilitates data processing and analysis.

DNA sequences constitute the majority of primary data in the bioinformatics system. There are three major worldwide entities that are collaboratively responsible for maintaining these DNA databases: the European Molecular Biology Lab (EMBL) of Cambridge, UK; GenBank of the National Center for Biotechnology Information (NCBI), which is affiliated with the National Institute of Health in the United States; and the DNA Databank of Japan. No universal format is currently available for sequence annotation. As would be expected, the databases are growing very rapidly. The top five species whose DNA is represented in these databases are *Homo sapiens* (humans), *C. elegans* (earthworm), *S. cerevisiae* (yeast), *Mus musculus* (mouse), and *Arabidopsis thaliana* (arabidopsis). As another note of caution, it is important that scientists who submit data to these banks quote the particulars of the source (e.g., the release date) from which a match was found. To facilitate the search by users, the managers of these sequence repositories have tried to organize the databases into smaller, meaningful categories (e.g., rodents, organelles, primates, bacteriophages, and so on).

TABLE 14–1

Standard codes for various amino acids. An example of a sequence of DNA may be 5'-CGGGTAACTT-3', whereas an example of a protein sequence could be NH$_2$-GGALMQEEQ-OH.

Letter		Full Name	Abbreviation	Letter		Full Name	Abbreviation
G	=	Glycine	Gly	P	=	Proline	Pro
A	=	Alanine	Ala	V	=	Valine	Val
L	=	Leucine	Leu	I	=	Isoleucine	Ile
M	=	Methionine	Met	C	=	Cysteine	Cys
F	=	Phenylalanine	Phe	Y	=	Tyrosine	Tyr
W	=	Tryptophan	Trp	H	=	Histidine	His
K	=	Lysine	Lys	R	=	Arginine	Arg
Q	=	Glutamine	Gln	N	=	Asparagine	Asn
E	=	Glutamic acid	Glu	D	=	Aspartic acid	Asp
S	=	Serine	Ser	T	=	Threonine	Thr

Next to DNA sequences, the second-most important source of primary data is material from genome projects. Whereas some of these genome projects have been completed, others are in various stages of completion. Still, there are nonetheless plans to start new genomics projects.

ACCESSING AND USING WEB-BASED RESOURCES

The protocol for submitting a query is fairly straightforward. The unknown sequence is first loaded into a word processor. The resource to be accessed is then located on the Internet via the normal protocols specified by the Internet access provider. Next, the sequence is highlighted in the word processor and copied into a buffer. It is then pasted in the appropriate textbox on the web page associated with the bioinformatics data repository being accessed. Because bioinformatics analysis involves searching and manipulating huge databases, it is often impractical to run specific searches in real time. This is especially true during the peak search time. An alternative is to submit a search via e-mail. This way, search jobs can be placed in a queue and executed as soon as "the system is less busy."

PROTEIN DATABASES

Various organizations maintain databases for both protein sequences and structure. The Department of Medical Biochemistry (University of Geneva) and European Bioinformatics Institute collaboratively maintain properly annotated translations of sequences in the EMBL databases. This is called the SwissProt. TREMBL (translated EMBL) is another protein database consisting solely of protein-coding regions of the EMBL database. The NCBI (National Council for Biotechnology Information) also maintains a database of the translations of the GenBank. Another kind of protein database consisting of experimentally derived three-dimensional structures of proteins is kept at the protein databank, where these structures are determined by X-ray diffraction and nuclear magnetic resonance.

SEQUENCE RETRIEVAL SYSTEMS

Newly produced sequences are deposited in one of the repositories previously described. Scientists are obligated to search one of these reputed sequence banks for their work. In fact, some professional journals will not accept results for publication based on data that have not been previously submitted to one of these information centers. Upon submission, an entry receives an accession number that can be referenced in a publication. Retrieval of sequences is usually done on the basis of accession numbers. The most commonly used information retrieval systems in bioinformatics are the U.S.-developed Entrez and the English-developed Sequence Retrieval System (SRS).

■ THE PROPERTIES OF A GENE-SEEKING COMPUTER SOFTWARE

Searching for genes is a more challenging task than assembling and organizing the biological information. As previously indicated, a DNA sequence does not automatically constitute a gene. Gene search is facilitated by the fact that organic evolution has revealed that all genes share certain characteristics. With this knowledge, scientists have been able to construct what are called **consensus sequences** for many genetic elements that are representative of a given class of organisms (e.g., prokaryotes, eukaryotes). The commonly occurring genetic elements shared among organisms include promoters, enhancers, polyadenylation signal sequences, and protein binding sites. Because these genetic elements share common sequences, computer scientists are able to apply mathematical algorithms to the analysis of sequence data.

A typical gene-seeking computer program will have certain basic elements:

1. **Algorithms for pattern recognition**   This property entails the use of probability concepts to determine if two sequences are statistically similar.
2. **Data tables**   These are tables containing information on consensus sequences for various genetic elements. By including these differences, a researcher is able to speed up the analysis and minimize error.
3. **Taxonomic classes**   This property is critical because consensus sequences vary among taxonomic classes of organisms.
4. **Analysis rules**   Like all computer software, the programming instructions determine how the algorithms are applied. In this case, the program will define the acceptable degree of similarity to declare a match. It will also determine whether entire sequences or merely fragments of sequences will be included in the analysis. It is desirable for the researcher to be able to manipulate these variables.

■ SEQUENTIAL ALIGNMENT AND SEARCHING OF DATABASES

A purpose of searching a bioinformatics database is to determine if the researcher's unknown sequence, DNA or protein, matches any sequence in the database in terms of structure or function. This requires the proper choice and skillful use of software to align the unknown sequence with the known. This search involves two key activities:

1. **Sequence alignment scoring matrices**   Computer software is used to align the unknown sequence with those sequences in the bank. Scores are assigned on the basis of the sequence homology detected. It is most useful to align sequences such that the largest scores are assigned to the most biologically significant matches. There are certain international codes (IUB/IUPAC) for nucleotides and common amino acids (see Table 14–1). In the following example, two different alignments are displayed for an unknown sequence TGYAPPPWS:

a. TTYGAPPWCS   b. TTYGAPPWCS
  TGYAPPPWS   TGYAPPPWS
   *   * **   *   *   *   * * *

In match (a) the alignment favors relatively common amino acid residues (A, P, S, and T), whereas the alignment in (b) conserves the less-common amino acid residues like W (tryptophan) and Y (tyrosine). A C-to-C (cysteine to cysteine) match in an alignment is more important than an S-to-S (serine to serine) match because cysteine is relatively more rare than serine. By the same token, a D-to-E match should receive a higher score because the two amino acid residues are chemically similar and could be functionally related in the two proteins being compared. However, a V-to-K match should score negatively because these amino acid residues are so dissimilar that they could hardly be functionally related in the two proteins.

The scoring matrices for DNA alignments are relatively simple and straightforward:

	A	C	G	T
A	0.9	− 0.1	− 0.1	− 0.1
C	− 0.1	0.9	− 0.1	− 0.1
G	− 0.1	− 0.1	0.9	− 0.1
T	− 0.1	− 0.1	− 0.1	0.9

A base-pair match receives an alignment score of 0.9 while a mismatch receives − 0.1. The sequence alignment for the example below is 4.3 (obtained as $(5 \times 0.9) + (2 \times − 0.1)$).

```
GCGCCTC
GCGGGTC
* * *   * *
```

There is no universal scoring matrix for proteins as yet. A general purpose one is BLOSUM62; there are also others (e.g., PAM40 and PAM120).

The Needleman-Wunch algorithm is considered one of the very best for calculating pairwise sequence alignment for both proteins and nucleic acids. However, the results do not necessarily reflect biological relationships between the compared sequences. Multiple sequences can be aligned using computer software such as CINEMA.

■ *Comparing Sequences against a Database*

One of the most common searches of bioinformatics databases is to compare an unknown sequence against those in the database to discover similarities. Typical homology search algorithms are used in this activity. The most widely used software for this search is the Basic Local Alignment Search Tool (BLAST) and FASTA. BLAST uses a strategy based on short sequence fragments between the unknown sequence and those found in the database. It is designed to match only continuous sequences (with no gaps from deletion or insertion mutations taken into account). To overcome some of the weaknesses of BLAST, there is other software such as BEAUTY that may be used to supplement BLAST. FASTA focuses on alignments that are likely to be significant; hence, it sometimes misses weak but significant scores. BLITZ is a more sophisticated and powerful search that requires more computing power. It is more expensive to use but searches more quickly and extensively than FASTA or BLAST. There are various versions of these software with different functions. New and more powerful software continue to be developed for bioinformatics.

■ *What the Search Results Mean*

Finding a match is one thing; determining if it is biologically significant is another. Generally, for protein global alignment, any sequence that shows 25 percent or more identity over a stretch of at least 80 amino acids should have the same basic fold. Furthermore, at least 75 percent sequence identity should be observed before a match can be suspected of being significant.

How valuable a tool is bioinformatics to biotechnology? The limitations of bioinformatics is that the chances of finding a significant match depend on the information in the database and what is known about those genes. In the case of animals, model organisms with both small and large genomes have been completely sequenced. In addition, these organisms have been studied in more detail in other respects. Plant model organisms are playing catch-up.

KEY CONCEPTS

1. Understanding the genome structure alone is insufficient; it is critical to identify the proteins it encodes.
2. Bioinformatics is a knowledge-based theoretical discipline that attempts to make predictions about biological functions using data from DNA sequence analysis. It uses supercomputers and sophisticated software to search and analyze databases accumulated from genome sequencing projects and other similar efforts.
3. Two general categories of information are used in bioinformatics research: primary databases (which consist of original biological data such as raw DNA sequences and protein structure information from crystallography) and secondary databases (which contain original data that have been processed for value added to suit certain specific applications).
4. A good database, primary or secondary, should have two critical parts: (1) the original sequence, and (2) an annotation description of the biological context of the data.
5. One of the most common searches of bioinformatics databases is to compare an unknown sequence against those in the database to discover similarities.

OUTCOMES ASSESSMENT

1. What is bioinformatics, and what is its role in biotechnology?
2. Discuss the kinds of data used in bioinformatics.
3. Discuss the composition of genetic databanks used in bioinformatics.
4. What does an annotation of a database entail?
5. Give three examples of the primary database in bioinformatics.
6. Give two examples of protein databases.
7. Discuss how sequential alignment is used in searching bioinformatic databases.

INTERNET RESOURCES

1. Tool for analyzing the functions of proteins: *http://vector.cshl.org/dnaftb/41/concept/index.html*
2. Searching a database with BLAST: *http://vector.cshl.org/dnaftb/40/concept/index.html*
3. Comprehensive listing/links to worldwide sites for all aspects of bioinformatics data: *http://zlab.bu.edu/~mfrith/tools.shtml*
4. Links to resources. U.S. Department of Commerce/NOAA: *http://research.nwfsc.noaa.gov/bioinformatics.html*
5. Basics of bioinformatics: *http://biotech.icmb.utexas.edu*
6. Comprehensive listing of genetic databases: *http://www.nhgri.nih.gov/Data/*
7. GenBank: *http://www.ncbi.nlm.nih.gov/Genbank/GenbankOverview.html*
8. ELMB site: *http://www.embl-heidelberg.de/*

REFERENCES AND SUGGESTED READING

Apley, L. 1997. *DNA sequencing: From experimental methods to bioinformatics.* New York: Springer-Verlag.

Lipman, D. J., and W. B. Pearson. 1985. Rapid and sensitive protein similarity searches. *Science, 227*:1435–1441.

Section 2
DNA Microarrays

PURPOSE AND EXPECTED OUTCOMES

During a physiological change, some genes become more active while others become less active (i.e., differences occur in the levels of gene expression). More than one gene is often involved in most biological processes. To understand how an organism functions, it would be best to examine many of its numerous life processes simultaneously to see how they respond to changes over time. Conventional biological studies are designed to study genome function in a piecemeal fashion, one gene at a time.

In this section, you will learn:

1. The use of array technology for global studies of the genome.
2. The rationale of the microarray technology.
3. The design of DNA microarrays.
4. The application of DNA microarrays in biotechnology.

WHAT ARE DNA MICROARRAYS?

DNA microarrays (also called **DNA chips, genome chips, gene arrays**, or **biochips**) technology tremendously enhances throughput in genetic and other biological experimentation. The rationale for this technology is that a large number of genes and their products (RNA, proteins) work together in a complex fashion to make an organism function as an integral whole. DNA microarray technology allows researchers to adopt the "whole picture" approach in biological experimentation. That is, an ordered array allows the sum of all interactions across the full set of gene sequences to be measured simultaneously and calculated instantly. Array experimentation is not a novel approach to research. Microplate assays have provided an opportunity to observe patterns in certain biological studies for a long time.

The microarray technology provides a powerful and universal tool for researchers to use in exploring the genome in a systematic and comprehensive fashion (not driven by hypothesis) to survey DNA and RNA variations. The underlying principle of the technology is specificity and affinity of complementary base pairs. Microarrays can be used to explore and exploit the properties of genes in a variety of ways. The challenge in doing so rests in finding an experimental method that will allow the property of interest to be turned into the basis for differential fractionation of DNA or RNA sequences. Gene attributes of interest include those that are amenable to microarray research, as well as their transcription, translation, subcellular location of the products, genotype, and mutant phenotypes. The experimental methods for exploring attributes vary in complexity. Studying the differential expression at the mRNA level is relatively straightforward. Measuring the differential hybridization of a DNA microarray to fluorescently labeled cDNAs prepared from the two mRNA samples can be used to compare the relative abundance of mRNA from each gene.

DESIGN FORMATS

Technically, the first array to be developed was the Southern blot, in which labeled nucleic acid molecules were used to interrogate nucleic acid molecules immobilized on a solid support. As with most fledgling technologies, there are variant forms of DNA microarray technology. However, there are two basic design formats.

1. **Stanford University format**

     This format, pioneered by scientists at Stanford University, is traditionally called the **DNA microarray technology**. DNA probes of between 500 and 5,000 bp long are immobilized on a nonporous surface (e.g., glass) through robotic spotting. A single-stranded target (unknown) sequence (or a mixture of several target sequences) is poured over the plate. After rinsing, spots where the probes found their complements are visualized.

2. **Affymetrix format**

     Affymetrix, Inc. developed this format, historically called the **DNA chip technology**. An array of oligonucleotides (20- to 80-mer oligos) or peptide nucleic acid probes are either synthesized *in situ* (on the chip) or by conventional methods and deposited on the chip. The unknown DNA sample is labeled, then applied to the chip for hybridization to complementary sequences. The identity and abundance of the complementary sequences are then determined. In fact, Affymetrix Inc. markets its DNA chips under the trademark name GeneChip®, and hence opposes the alternative reference of DNA microarrays as gene chips.

MICROARRAY FABRICATION

DNA microarrays are essentially ordered sets of DNA molecules of known sequences. Arrays are usually rectangular (e.g., 60 × 40, 300 × 500). A spot of the DNA may be less than 200 microns in diameter and placed at a precisely desired location. Each spot contains a specific sequence. The test material usually consists of RNA that has been amplified by using PCR methodology. The key components of microarrays are:

a. Media or material
b. Spotter
c. Labeling and detection
d. Analytical software

Most of the microarrays are offered in standard microscope slide configurations (e.g., polylysine-coated glass or CMT-GAPS amino-salanized slides). Labeling and detection are usually fluorescence-based systems. Each probe is labeled with a different-colored fluor that is different enough to be distinguishable by "reading" devices equipped with optical filters. Readers (scanners) commonly use high-intensity white light or laser-induced fluorescence that may or may not be confocally focused.

Fabrication of DNA microarrays is amenable to automation, involving the use of high-speed robotics. Three main fabrication technologies are commonly used. A key difference between microarrays and the traditional hybridization assays is that microarrays generally use nonporous substrates. The **photolithographic** fabrication technology, distributed by Affymetrix Inc., uses semiconductor technology in chip fabrication, and is applicable to only oligos and has length restrictions. Affymetrix Inc. uses this technology in fabricating its GeneChip®. The **piezoelectric** technology is an adaptation of the inkjet printing technology designed to deliver extremely low volumes of reagents to specific spots. Electricity is used to deliver the DNA bases, cDNAs, and other molecules. There is no length restriction. Many companies offer this product.

The third fabrication technology is **microspotting**. Unlike piezoelectric technology, in which there is no direct surface contact, microspotting relies on direct surface contact. The technology is amenable to oligos and other biomolecules as well. All of these three fabrication technologies can be used to create single chips capable of holding entire genomes. This means a chip can have tens of thousands of spots.

DESIGN AND IMPLEMENTATION OF A DNA MICROARRAY EXPERIMENT

There are six basic components of a DNA microarray experiment—probe, chip, target, assay, readout, and computer software (see Figure 14–2). The experimenter first selects the genetic material of known identity (e.g., cDNAs or small oligos) to be used as probes. The next decision is the fabrication format (i.e., how the probes will be arranged on the chip). The sample (cDNA, RNA) to be used to interrogate the spotted and immobilized probes is prepared and fluorescently labeled (see Figure 14–3). The assay is then conducted and the results read by, for example, electronic devices. The data may be submitted to a variety of data management systems to obtain useful and desired information according to the objectives of the researcher.

FIGURE 14–2
A summary of the general steps in a microarray research project. A probe may be cDNA or an oligo with a known identity. Fabrication processes differ and may include photolithographic or piezoelectric methods. The ways of assaying in a microarray experiment include hybridization, electrophoresis, and flow cyclometry. The readout may be fluorescent or electronic. Informatics includes data mining, visualizing, and image processing.

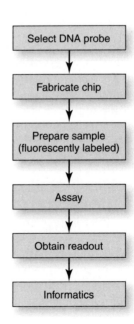

(a)

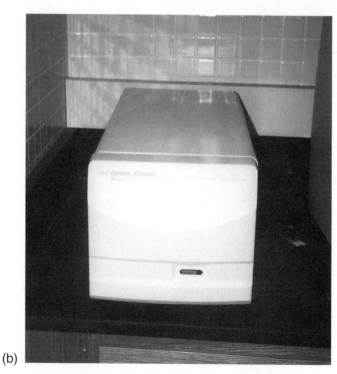

(b)

FIGURE 14–3

A DNA microarrayer. (a) A key component of a microarray unit is the microarray robot. This device moves in the x-, y-, and z-axes to print microarray slides using an attached printhead. Printheads differ in capacity. (b) Another key component of a microarray setup is the scanner used to record the patterns resulting from the array assay.

Source: Photo taken at the Oklahoma State University Noble Research Center.

As previously indicated, DNA microarrays are not as yet commonplace in research. The cost of an arrayer and scanner is estimated at about $60,000. This price will certainly decrease steadily as more researchers use the technology and companies that manufacture the equipment start having to compete for customers. As an alternative, researchers may be able to purchase prefabricated microarrays of genomes of interest.

APPLICATIONS OF DNA MICROARRAYS

DNA microarrays are used in two primary applications: (1) gene expression (level or abundance of gene expression) and (2) sequence identification (normal genes and detection of mutations).

■ GENE EXPRESSION

The expression pattern of a gene provides indirect information about its function. One of the attributes of genes that is of great interest to microarray researchers is their expression. Several reasons may be cited for this focus on understanding the regulation of gene expression at the level of transcript abundance.

1. The technology of microarrays is readily adaptable to measuring the transcript for every gene in the genome at once.
2. The function of a gene product and its expression pattern are strongly linked. It is known that a gene is expressed in specific cells and under the specific conditions in which its product impacts fitness.
3. Promoters of genes essentially operate like transducers. They change the level of transcription for specific genes in response to changes in the internal state of the cells and the cellular environment. By knowing the information a promoter transduces, scientists can use microarray technology to read the information from the profile of gene transcripts.
4. Because the set of genes expressed in a cell determines the attributes of the cell (e.g., what it is made of, its biochemical and regulatory systems, its structure, and its functions), genome-wide technologies like DNA microarrays are suited to exploring the total dynamic molecular picture of a living cell.

■ STEPS IN GENE EXPRESSION ANALYSIS WITH DNA MICROARRAYS

The gene expression methodology may be illustrated by the following example in which gene expression between two samples, A and B, is being compared (see Figure 14–4).

1. Prepare fluorescently labeled cDNA of the total pool of mRNA from each cell population by reverse transcription in the presence of fluorescently labeled cDNA precursors. Use different fluors to allow distinction between their effects.
2. Mix two fluorescently labeled cDNAs.
3. Hybridize with a DNA microarray in which a distinct spot of DNA represents each gene. The cDNA sequences representing each individual transcript will hybridize with only the corresponding gene sequence in the array, regardless of the fluorescent labels.

The result of this methodology is a pattern in which the relative abundance of the transcripts from each gene corresponds to the ratio of the two fluors used. It should be mentioned that gene expression data have certain limitations. For example, mRNA levels do not always reflect protein levels, and also, the expression of a protein may not always have a physiological consequence.

■ DRUG DISCOVERY AND DEVELOPMENT

One of the bright future applications of microarray technology is in **pharmacogenomics**, a marriage between genomics (gene and gene function) and pharmacology (devoted to understanding the operation, use, and effects of drugs). The rationale of pharmacogenomics is that genomics provides a complete picture of all genes in an individual. These

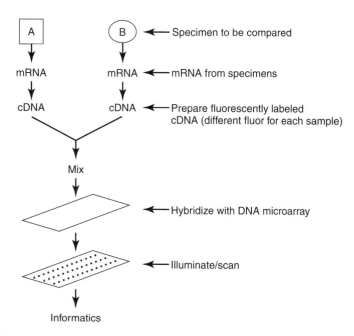

FIGURE 14–4
An example of DNA microarray research. The mRNA from the specimens to be analyzed is first extracted. Fluorescently labeled cDNA is prepared from the specimens and mixed before hybridizing a DNA microarray. The relative abundance of the fluorescence as scanned is used in analysis.

genes direct the production of all the gene products (proteins) in an individual. Most drugs act at the protein level (to disrupt or alter protein function). A genomic sequence can be used to identify all the potential drug targets in an organism. Microarray information will accelerate and reduce the cost of drug development. By extension, knowing protein expression differences among individuals would allow doctors to tailor prescriptions to individual needs (instead of generic dosages).

1. **Conventional method of drug discovery**

    The conventional process of drug discovery and development starts with the identification of a biochemical pathway that is associated with a pathophysiological process. The next step is to identify an enzyme in the pathway (e.g., the rate limiting step in the pathway). This enzyme is subsequently isolated, characterized, and purified (usually from animal tissue). The next step is to screen the purified enzyme against collections of different small molecules to determine compatible compounds. When the search has been narrowed down to a few "druggable" candidates, medicinal chemists take over to tweak the properties of these compounds, removing undesirable properties (e.g., poor specificity for a target enzyme) while increasing its bioavailability.

2. **Molecular approaches to drug discovery and development**

    Drug discovery has been facilitated through the use of genetic engineering, a technique that enables targets that are limited, not readily accessible, or risky to handle to be cloned. The technique of site-directed mutagenesis may be used to test hypotheses about drug target interaction and thereby facilitate the drug discovery process. Cross-hybridization with cloned sequences provides a quick way of identifying related targets. The technology of DNA microarrays has the potential for use in identification and validation (linking targets with therapeutic utility) of therapeutic targets. It is known that a deviation from normal physiology in an organism is often associated with a variety of histological and biochemical changes such as changes in gene expression. Disease condition may occur as a result of either upregulation or downregulation of a gene activity. Microarrays provide an opportunity to compare the expression of thousands of genes obtained from normal and diseased individuals, thereby allowing multiple potential targets to be discovered. In trying to find a therapeutic solution to disease, researchers may target products of disease-causing genes. Alternatively, they may interfere with the genes themselves and hopefully achieve the alleviation of disease symptoms.

Microarray technology can also be used to study disease from the perspective of the pathogen by identifying genes that are turned on *in vitro* but not at the site of infection *in vivo* (and vice versa), and also genes that are only turned on during infection *in vivo*. This will help identify virulent genes.

Another application of microarrays in drug discovery and development is the study of the mechanisms of drug action. The effect and dose of a drug can be followed in a clinical setting. Sentinel genes can be analysized to help determine the mechanism or action of a drug or toxin. Numerous events are triggered by the initial action of a drug. The microarray technology may be used to screen thousands of genes simultaneously to identify multiple potential drug effectors.

■ SEQUENCE IDENTIFICATION

Researchers may use DNA microarrays to detect mutations or polymorphisms in a gene sequence. This strategy enables researchers to rapidly diagnose diseases for which a gene mutation has been identified. The target DNA differs from one spot to another in the microarray by one or a few specific nucleotides. Such a target DNA is SNP (single nucleotide polymorphism). To use this technique, researchers first have to establish a SNP pattern associated with a specific disease. Genomic DNA from the subject is obtained and hybridized to a microarray spotter with various SNPs to determine if the individual is susceptible or at risk of developing the disease. This determination is made based on the frequency of hybridization of the sample DNA with specific SNPs associated with the individual being tested.

KEY CONCEPTS

1. Microarrays are tools designed for a global approach to biological experimentation.
2. DNA microarrays are also called DNA chips, gene chips, biochips, and gene arrays.
3. The underlying principle in the technology is the specificity and affinity of complementary base pairs.
4. There are two basic design concepts: the Stanford and Affymetrix design formats.
5. The common methods of microarray fabrication are photolithographic, piezoelectric, and microspotting.
6. DNA microarray technology is used in two primary applications: (1) sequence identification (normal genes and detection of mutations) and (2) gene expression (level or abundance of gene expression).
7. The key components of microarrays are media or material, spotter, labeling and detection, and analytical software.

OUTCOMES ASSESSMENT

1. Discuss the rationale of the concept of array technology.
2. Discuss the basic types of microarray design.
3. Discuss the potential of DNA microarrays in drug discovery and development.

4. Give reasons why gene expression is one of the major applications of DNA microarrays.
5. Describe how DNA microarrays may be used in a gene expression study.

INTERNET RESOURCES

1. Microarray technology: *http://www.accessexcellence.org/AB/GG/microArray.html*
2. A description of the microarray technology: *http://www.e-proteomics.net/tech.html#microarray*
3. Links to software for microarrays: *http://www.deathstarinc.com/science/biology/chips.html*
4. General information: *http://www.unil.ch/ibpv/microarrays.htm*
5. About gene chips: *http://www.gene-chips.com/*

REFERENCES AND SUGGESTED READING

Brown, P. O., and D. Bostein. 1999. Exploring the new world of the genome with DNA microarrays. *Nature Genet. 21:*33–37.

Debouck, C., and P. Goodfellow. 1999. DNA microarrays in drug discovery and development. *Nature Genet. 21:*48–50.

Lander, E. S. 1996. The new genomics: Global views of biology. *Science, 274:*536–539.

Lander, E. S. 1999. Array of hope. *Nature Genet.* (supplementary), *21:*3–4.

Lemieux, B., A. Aharoni, and M. Schena. 1998. Overview of DNA chip technology. *Molecular Breeding, 4:*277–289.

Lockhart, D. J., and E. A. Winzeler. 2000. Genomics, gene expression and DNA arrays. *Nature, 405*(6788):827–836.

Schena, M., and R. W. Davis. 1999. Genes, genomes and chips. In *DNA Microarrays: A practical approach* (ed. M. Schena). Oxford, UK: Oxford University Press.

Southern, E. M. 1975. Detection of specific sequences among DNA fragments separated by gel electrophoresis. *J. Mol. Biol., 98:*503–517.

Section 3
Proteomics

PURPOSE AND EXPECTED OUTCOMES

Microarrays may help scientists, through comparative or parallel gene expression assays, to place genes in the same metabolic pathway. However, just because a gene is expressed does not indicate conclusively that its mRNA is actually making the ultimate protein product. The conclusive evidence of gene function comes from identifying the protein products (not mRNA expression) actually made by the gene.

In this section, you will learn:

1. The goal of proteomics.
2. The key technologies employed in proteomics.

WHAT IS PROTEOMICS

The term "genome" refers to the full complement of genes in an organism's cell. Similarly, "proteome" (a contraction of protein and genome) is the term used to describe all the proteins produced by the genome. **Proteomics is** the science of analyzing and cataloging all the proteins encoded by a genome. Two generalized categories of analysis are identified in proteomics: (1) **expression proteomics** (which is involved in studying global changes in protein expression) and (2) **cell-map proteomics** (which entails a systematic study of protein-protein interactions through the isolation of protein complexes). Proteomics tracks protein expression by cells, beginning with the functional protein and backtracking to the gene that encoded it.

Unlike genomics, which entails a process in which there is a well-defined ultimate end point (i.e., the complete sequence of an organism's DNA), such a finality is impossible to define for proteomics. This is because, unlike DNA that is fixed, the proteome depends on when the sample is taken from the organism and the kinds of tools used for the investigation. Proteins are part of the dynamics of a living system, appearing and disappearing according to the needs of the cell. Proteomics is hence the study of global protein expression patterns that define cells in specific biological states. Furthermore, because proteins vary widely in their chemical behavior, it is difficult to prescribe one best technique to study all proteins. Unlike genomics, in which robotic gene sequencers were the workhorses of the large-scale projects, proteomics depend on a variety of key technologies, including the yeast two-hybrid system, mass spectrometry, two-dimensional gel electrophoresis, and DNA microarray hybridization.

Proteomics may also be described as protein expression profiling. The recent surge in large-scale protein study is attributable to the large amount of genomics data produced by genomics projects, especially the Human Genome Project, coupled with advances in computing power and robotics that enable scientists to handle and manipulate huge databases. Most of all, scientists have not yet found out the cellular functions of the known and predicted proteins. Raw DNA and protein sequences operated by genomics projects provide little information. It is hoped that proteomics will be able to provide information about protein function to help scientists understand cellular metabolism more fully.

The basic process in determining gene function is as follows:

a. A protein of interest is first extracted from a tissue sample.
b. A fragment of the protein's amino acid sequence may be translated into its DNA base sequence.
c. With this information, a computer database is searched to identify the gene that encodes the protein as well as the complete protein.

It is not always necessary to convert the amino acid sequence into DNA. The amino acid sequence may be used to query ORF databases that have translated DNA sequences (e.g., BLASTP), or amino acid sequences could be used in a TBLASTN search of a DNA database.

THE CHALLENGES OF STUDYING PROTEINS

Proteins are very complex biomolecules. The information produced from genomic analyses do not predict the post-translational modifications that most proteins undergo after synthesis. Proteins are more difficult to work with than nucleic acids for several reasons. Because proteins cannot be amplified like DNA (using PCR), less-abundant sequences are difficult to detect. Often the secondary and tertiary structures of protein must be maintained during their analysis. A variety of environmental factors and certain treatments (e.g., heat, enzymes, light, and excessive physical mixing) may denature proteins. Proteins with poor solubility are difficult to analyze.

However, even though nucleic acids are relatively easier to work with, there are significant limitations to the types of information their analysis may yield. For example, DNA sequence analysis is unable to predict whether a protein it encodes would be in an active form. The phenomenon of alternative gene splicing and post-translational modification may lead to multiple proteins from a single gene. Other shortcomings of DNA/RNA analyses include their inability to predict the amount of gene product that is made, if and when a gene will be translated, the type and amount of post-translational modification, or events involving multiple genes (e.g., aging, stress responses, and drug responses).

The difference in ease of analysis and depth of information produced notwithstanding, both genomics and proteomics are necessary endeavors that complement each other. Proteomics provides additional understanding to gene function.

FRONTLINE TECHNOLOGIES OF PROTEOMICS

Technologies used to separate mixtures of proteins to identify those of interest include the two-dimensional gel electrophoresis and HPLC (see Figure 14–5). Other methods of finding out about protein function include studying how they interact with other proteins, and how the genes that encode proteins behave upon disruption. In developing technologies, researchers are always searching for those that are amenable to high throughput and system-wide applications. Considering the astronomical amount of data spewed out by genome sequencing projects, proteomics needs technologies that can allow large numbers of genes to be studied simultaneously and rapidly.

FIGURE 14–5
A high-performance liquid chromatograph (HPLC).
Source: Photo taken at the Oklahoma State University Nobel Research Center.

■ *TWO-DIMENSIONAL GEL ELECTROPHORESIS*

The two-dimensional gel electrophoresis technique, discussed previously, is used to separate proteins and observe their interactions. The protein of interest is cut from the gel, purified, and then fragmented. The fragments are submitted to analysis by a mass spectrometer and associated computer, which identifies proteins based on atomic mass. The two-dimensional gel electrophoresis is the workhorse for this task for key players like Oxford Glyco Sciences and Large Scale Biology Corporation of California. The two-dimensional gel electrophoresis may be used to compare tissues to detect differences in protein expression (e.g., between a diseased tissue and a healthy one).

Two-dimensional electrophoresis is ineffective in separating proteins that are either very small or very large. Separation of proteins that are localized in membranes are also problematic with this approach. Unfortunately, these are the proteins that have great potential to become drug targets. Furthermore, mass spectrometers that are part of this methodology tend not to identify proteins that are expressed in minute quantities. Nonetheless, some reseachers use the HPLC protein separation technology because they believe it is very effective for small proteins.

■ *YEAST TWO-HYBRID ASSAY*

This relatively straightforward approach for understanding protein function consists of different two-hybrid systems, although the basic application is to learn about the function of a given protein by isolating proteins that interact with it (i.e., to detect protein-protein interactions). Scientists use known proteins (called **baits**) to bind to unknown proteins (called **preys**). First, a gene for a bait protein is inserted into yeast alongside DNA for half of an "activator" protein. The other half of the activator DNA is inserted alongside DNA for a random, unknown protein (prey). The yeast cells are cultured and the proteins allowed to interact. The rationale is that if the bait and prey successfully bind, the two halves of the activator will interact, causing yeast genes to be turned on. The cell then turns blue as a sign that a match has occurred.

The technology of yeast two-hybrid exploits the fact that transcriptional activators are modular in nature. Two physically distinct functional domains are necessary to obtain transcription. One, a DNA binding domain, binds to the DNA of the promoter, and the other, an **activation domain**, binds to the basal transcription apparatus and activates the transcription. The known gene that codes for a known protein, X, is cloned into the bait vector, placing it into a plasmid next to the gene encoding a DNA binding domain from a transcription factor. If, for example, the vector pHybLex/Zeo is used for cloning, the X would be expressed as a fusion protein containing bacterially derived LexA DNA binding domain. A second gene (or a library of cDNAs to be screened), Y, is cloned in a frame adjacent to an activation domain of a different transcription factor. For example, Y could be inserted next to the DNA encoding the B42 activation domain in a prey vector (e.g., pYESTrp2). If an unknown protein combines with the known protein (X), it will bring the activation domain over to the DNA binding domain, and consequently activate the transcription.

The next step in the two-hybrid technology is for the researcher to place plasmids containing the bait and the prey into a yeast strain along with a selectable marker gene that has a promoter containing the sequence bound by the bait protein DNA binding domain. Both bait and prey must unite in order for a working transcription factor to be formed. This will happen only if the known protein (X) combines with an unknown protein (Y) that is carrying the activation domain. In this example, the interaction of X and Y proteins will bring the B42 active domain into the proximity of the LexA DNA binding domain, causing the reporter gene to be activated (see Figure 14–6). Depending on the nature of the reporter gene, the transcriptional product may be assayed enzymatically, or visualized by some other methods.

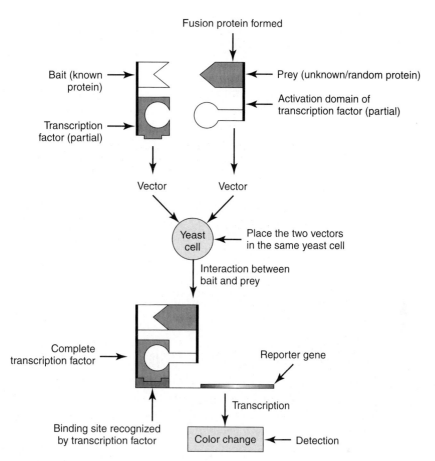

FIGURE 14–6
A summary of the steps involved in the yeast two-hybrid system for detection of protein-protein interaction. A bait (known protein) is attached to a part of a transcription factor and the product is inserted into a vector. Another fusion product is formed between an unknown protein and the activation domain of the transcription factor and inserted into a different vector. The two vectors are placed into a yeast cell. If an interaction does occur between bait and prey, the event is indicated by the transcription of the reporter gene that results in a color change.

The yeast two-hybrid method is capable of detecting proteins that interact when placed inside yeast cells and outside their natural environment. There is a concern that, whereas these proteins may have value in basic sciences, only a few are likely to be related to disease to become drug targets. However, this occurrence is not unique to this technology. The other concern is that because the interaction and abundance of specific proteins change over time, mapping the links between proteins is not enough to explain the biology of a cell. The yeast two-hybrid technology is slow and not practical for surveying an entire proteome.

■ X-RAY CRYSTALLOGRAPHY

The X-ray crystallography approach of proteomics, described in Chapter 13, uses high-speed X-ray crystallography to map the atomic landscape of the proteins.

FIGURE 14–7
A mass spectrophotometer.
Source: Photo taken at the Oklahoma State University Nobel Research Center.

■ *MASS SPECTROMETRY OF BIOMOLECULES*

Protein sequencing in the post-genomic era has become a fierce and competitive race to claim as much intellectual property as possible. One of the leaders in this charge, GeneProt of Switzerland, invested in 51 mass spectrometers from the onset, as well as several dozen robots, all supported by a supercomputer (see Figure 14–7). The availability of genome-wide nucleotide sequences from genome projects has simplified the approach to identifying all possible open reading frames as well as deducing the sequence of proteins they encode. The remaining challenge is how to determine which of the protein sequences correspond to a given protein spot on a two-dimensional gel produced from a sample. Mass spectrometry can determine the mass of a peptide with high accuracy. It can also be used to calculate a protein spot's probable amino acid composition. When the spot is recovered, cleaved into short peptide fragments, and analyzed, its mass as well as its amino acid sequence can be determined with high accuracy.

How Mass Spectrometry Works

In mass spectrometry, an ionized molecule is accelerated in an electric field toward a detector, with large molecules traveling slower than smaller molecules. Similarly, molecules with a larger charge (e.g., 2^+) will travel faster than molecules with a smaller charge (e.g., $+$). The time of flight (the time it takes for a molecule to travel from the point of ionization to the detector) is a function of the mass-to-charge ratio (m/z) of a particle.

Two ion-producing methods are commonly used in the mass spectrometry of biomolecules: **electrospray ionization** (ESI), and the **matrix-assisted laser desorption ionization** (MALDI). The MALDI is the preferred method for the high-throughput sequence determination needed for genome-wide analysis. The use of mass spectrometry in protein sequencing is summarized in Figure 14–8. Computational analysis utilizes a database from a variety of sources for accurate protein identification. Full molecular weight of the protein can be determined if intact protein mass is delivered to the detector. However, to obtain amino acid sequence information, the intact protein must be fragmented into peptides. This may be done chemically or enzymatically. Peptide mass fingerprinting and mass analysis of fragments from individual peptides is conducted sequentially to yield information for the conclusive identification of a protein.

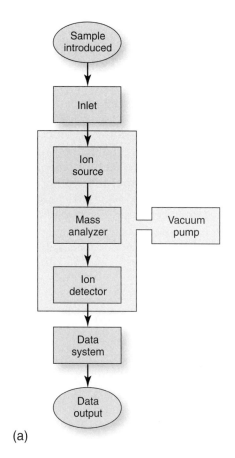

(a)

FIGURE 14–8
(a) A summary of the steps in using a mass spectrophotometer.
(b) A typical printout of a mass spectrophotometer showing the characteristic spikes that indicate the intensities of specific compounds.

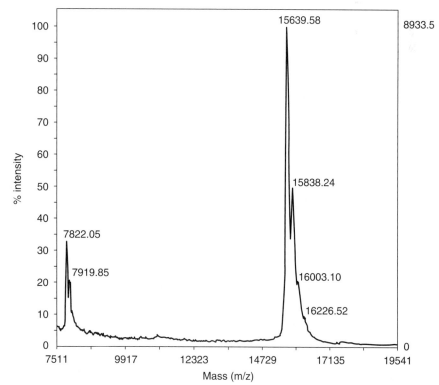

(b)

Computation is a key part of protein sequencing. Analysis utilizes tissue or species-specific databases for each sequence determination. Pieces of information, including the isoelectric points and the mass measurements of the intact protein, its peptide fragments, and their degradation products, are used as screening criteria for the identification of the corresponding open reading frame in the genome of the species. In addition, mass spectral patterns from mass spectrometry may also be compared against a peptide fingerprint database to identify unknown samples. Sources of such databases include those at SwissProt, Procol, Mascot, Protein Prospector, and PIR.

Using mass spectrometry coupled with robotics and high power computation, scientists are able to analyze tens of thousands (over 80,000) of protein spots per day. Post-translational modification problems (e.g., glycosylation or phosphorylation) may be flagged in several ways. For example, all glycopeptides contain a hexose N-acetylhexosamine, and/or neuraminic acid isomers, which produce characteristic fragmentations that facilitate their tagging. Scientists may also match an unknown peptide fragment's fingerprint to that of a previously characterized fingerprint in a database to help clarify peptide patterns.

OTHER FUNCTIONAL GENOMICS TOOLS

New techniques for determining gene function are being developed. These include metanomics, gene knockouts and knockins, reverse genetics, and synteny. Some of these are more advanced in development and application than others.

◼ METANOMICS

Scientists in different disciplines frequently borrow from each other, adapting technologies developed in one discipline for application in another. Toxicologists use gas chromatography to separate components of a liquid, followed by mass spectrometry for the determination and quantification of the components. These techniques are being adapted to help understand the functions of genes by analyzing the metabolic profiles of organisms and tissues. Proteins expressed by an individual change under varying environmental conditions, growth, disease, and death of cells and tissues. Proteins with general functions like kinases are difficult to study because as protein patterns change, it is not always easy to determine precisely what role a specific protein plays in a metabolic pathway. The strategy with metanomics is to create metabolic profiles (or maps) for an organism or tissue under different conditions (with new genes or mutated genes). These profiles can be compared to allow scientists to learn how the organism changes or adapts.

Metanomics, the brainchild of biochemist Richard Trethewey at the Max Planck Institute for Molecular Plant Physiology in Golm, Germany, is one of the good fortunes of an experiment that apparently failed. Trethewey and his colleagues set out to reduce the sugar levels and increase the starch content of potatoes, in order to satisfy the demands of the potato chip market. Transgenic potatoes carry the yeast gene for the enzyme invertase and a bacterial gene for the glucokinase enzyme. The strategy was to cause the invertase to convert sucrose (a disaccharide) into its nomomers (fructose and glucose), and then, hopefully, convert these products to starch by glucokinase. This would accomplish the research goal in one process. Unexpectedly, the transgenic potato decreased in both sugar and starch content. Following an extensive analysis, the team determined that the additional invertase and sugar molecules produced by the transgene products were diverted to producing novel metabolites that had no impact on starch formation. The team got the idea that gene function could be better understood by tracking metabolite expression by cells. Subsequently, they founded a company called Metanomics in Berlin that is currently embarking on cataloging metabolic profiles of normal and mutant versions of *Arabidopsis.*

■ GENE KNOCKOUT AND KNOCKIN TECHNOLOGIES

Transgenic technology entails the expression of foreign genes in an organism. Gene **knockout technology**, however, entails inactivating (knocking out) specific genes in an organism. The purpose is to understand the effects of gene inactivation or overexpression in an organism. Researchers can use this data to understand the pathogenic mechanisms of genetic and infectious diseases. Gene knockout technology also provides animals for testing new genetic and drug therapies. Several researchers developed the technology in 1989, including Mario Cappechi of the University of Utah, and Elizabeth Robertson of Harvard Medical School. Knockout models abound in mice, with over 100 genes knocked out. The technology is also applicable to research in plants.

Using the mouse as an example, knockout gene development starts with isolation and cloning of the normal version of the gene to be inactivated, which is subsequently inactivated by inserting a marker (antibiotic resistance) gene. The altered gene is then transferred to cultured embryonic stem (ES) cells, which are special cells derived from early embryos which are totipotent. The altered gene replaces the normal gene in some of the ES cells through a recombinational event. Cells carrying one copy of the normal allele and the recombined gene are then selected by an antibiotic selection system for culturing. The cultured ES cells are injected into the blastocyst stage of a normal mouse embryo, which is then implanted into a surrogate female mouse to complete its development. By incorporating coat-color genes as markers in the procedure, chimeric offspring (those carrying the altered gene) can be selected on a phenotypic basis. These heterozygous mice are bred with wildtype mice to eventually produce mice that are homozygous for the altered gene, called knockout mice (see Figure 14–9). Development of a knockout mouse takes about one year to complete.

Mouse models that have been developed by this technology include the RAG-1- and RAG-2-lacking mice. These genes are associated with the immune system. Others are Gaucher's disease (a progressive and fatal autosomal recessive disorder of lysosomal enzymes), cystic fibrosis, and Lesch-Nyhan syndrome (an X-linked fatal disorder). These models and others are helping scientists understand human disorders, thereby assisting in the development of treatments.

Knockin technology is used to introduce additional DNA to the middle of a gene, often by using transposon mutagenesis.

REVERSE GENETICS

In classical genetics (forward genetics), the researcher starts with a gene product and then tries to identify the gene itself. In reverse genetics, an altered phenotype (mutant) is obtained (e.g., by insertional mutagenesis), then the question is asked, "What is the resulting change in the phenotype?" Another way of stating this concept is that researchers employing reverse genetics start with a gene and set out to determine what its normal function is. Reverse genetics depends on the ability to introduce specific mutations in the genome of an organism. For most genetic diseases, the normal function of the gene involved is not known. Scientists may employ reverse genetics to localize a disease-causing gene without knowing anything about the gene's molecular or biochemical nature. Identifying genes based on their position in the genome with no knowledge of the gene product is called **reverse genetics** or sometimes **positional cloning**. Positional cloning is facilitated by the availability of saturated or dense genetic maps (where many visible and molecular markers exist). This task will be greatly accelerated with the availability of genome sequences for organisms.

There are different ways of accomplishing gene disruption. Site-specific mutations may be made by oligonucleotide-directed mutagenesis, splicing of restriction fragments,

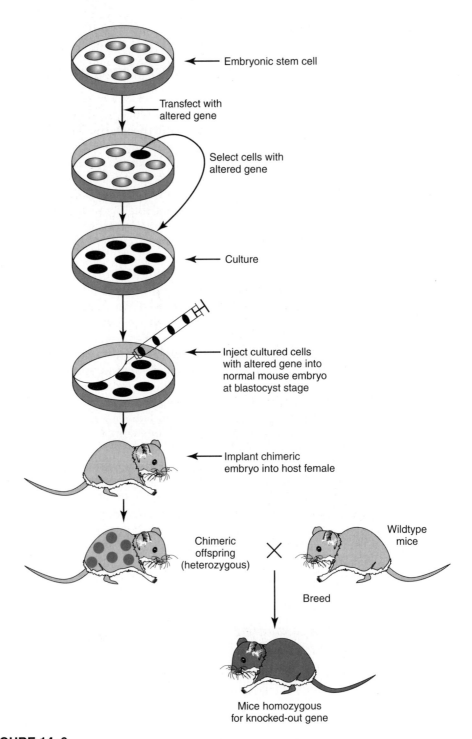

FIGURE 14–9

Steps in the development of a knockout mouse. A cloned normal mouse gene is inactivated by the insertion of an antibiotic-resistant gene, which serves as a marker. Embryonic cells are obtained and transfected with the altered gene. Through recombinational events, the altered gene will replace the normal gene in some part of the embryonic cell. Transfected cells have the antibiotic resistant marker gene and will therefore grow. They are selected and injected into the blastocyst stage of a normal mouse embryo where they are involved in the formation of adult tissues and organs. The chimeric embryo is implanted in a surrogate female to fully develop. The chimeric offspring is bred with wildtype mice to produce mice that are homozygous for the knockout gene.

transposons, or by PCR with mutation-creating primers. More recent technologies include chimeroplasy and triple-helix induced mutation. Gene disruption is introduced into individual organisms to create knockout individuals for research.

Site-directed mutagenesis allows scientists to study subtle differences in genotype and phenotype, thereby enhancing our understanding of the relationships among amino acid sequence, protein structure, and protein function.

INSERTIONAL MUTAGENESIS (TRANSPOSONS/T-DNA)

Genetic elements may be inserted into genes to disrupt their function or to enhance their expression.

◼ LOSS-OF-FUNCTION MUTATIONS

DNA elements that are able to insert within chromosomes at random can be used as mutagens to induce loss-of-function mutations. The DNA sequences of such inserts are known. Consequently, they can be used as a rapid way of cloning genes since the genes in which they insert can be readily recovered. In effect, the transposon, with its recognizable DNA sequence, acts as a "tag" of the affected gene, allowing the simultaneous identification of gene function and isolation of that gene. The commonly used agents are the heterologous transposons (especially those of maize origin like *Ac/Dc, En/Spm,* or *Mu*), and the T-DNA of *Agrobacterium tumifaciens*. Mutant populations of many plants (e.g., *Arabidopsis*, petunia, snapdragon, tomato, maize, and rice) have been generated utilizing this technique. The level of saturation (the probability of having at least one insertion in any gene) is variable.

T-DNA mutagenesis has significant advantages over conventional methods, including the low copy number (1.5 copies per insertion per transformant) and random insertion of T-DNA. The drawbacks to using gene disruption technology are that the transposon insertion renders the gene completely inactive (complete loss of function) in most cases. Consequently, phenotypes that require subtle changes cannot be generated and studied by this method. Furthermore, genes that are organized and function as families, or have redundant genes, cannot be studied this way.

◼ ENHANCER TRAPPING

Another variation in the use of insertional mutagens is to engineer a reporter cassette into the transposon. The purpose of this is to signal or report the expression of the chromosomal gene at the site of insertion. A reporter cassette (e.g., *lacZ* gene) fused to a minimal (weak) promoter close to the end of the insertion element can be dis-activated when inserted close to a transcriptional enhancer that will drive the expression of the reporter gene. The expression pattern of the reporter gene tends to reflect the expression of nearby genes. Consequently, the researcher can identify genes that are active in specific tissues and developmental periods under their environmental conditions. This strategy of using a reporter gene is called **enhancer trapping** (or **gene trapping**).

◼ ACTIVATION TAGGING

The development of an organism is governed by the temporal and special expression patterns of its genes. Therefore, the identification of developmentally regulated genes or of regulatory genes determining the expression pattern of other genes is paramount to the understanding of gene regulation and developmental genes in the organism. In plants, T-DNA mutagenesis is useful for the generation of loss-of-function genes. However, because

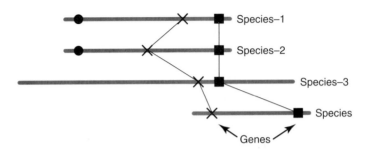

FIGURE 14–10
The concept of synteny. Certain gene sequences occur in different species on different parts of the chromosomes. These genes are not equally spaced on the chromosomes in different species.

T-DNA insertion often leads to complete loss of gene function, another strategy is needed for the study of phenotypes that require subtle mutations. A technique called **activation tagging**, based on the use of an insertion element carrying a strong promoter or enhancer to direct the transcription into the region flanking the insertion, leads to overexpression of genes next to the integration point (gain-of-function). The mutant phenotype is based on the ectopic expression of these genes, and because any tissue-specific or developmental regulation will be overridden, the functions of these genes will become apparent in a dominant phenotype.

SYNTENY

Gene order in chromosomes is conserved over wide evolutionary distances. In some comparative studies, scientists discovered that large segments of chromosomes, or even entire chromosomes in some cases, had the same order of genes. However, the spacing between the mapped genes was not always proportional (see Figure 14–10). The term **colinearity** is used to refer to the conservation of the gene order within a chromosomal segment between different species. The term **synteny** is technically used to refer to the presence of two or more loci on the same chromosome that may or may not be linked. The modern definition of the term has been broadened to include the homeology (i.e., homologous chromosomes are located in different species or in different genomes in polyploid species and originate from a common ancestral chromosome) of originally complete homologous chromosomes. Whole genome comparative maps have been developed for many species, but are most advanced in the *Gramineae* family (*Poaceae*). Some researchers have attempted to clone a gene in one plant species based on the detail and sequence information (microsynteny) in a homologous region of another genus.

KEY CONCEPTS

1. Proteomics (a contraction of protein and genome) is an analysis of protein profiles of tissues.
2. Two generalized categories of analysis are identified in proteomics: (1) expression proteomics (involved in studying global changes in protein expression) and (2) cell-map.
3. To determine gene function, the protein is first isolated. It may be converted to a DNA sequence (or used as is). The appropriate computer database is searched to identify the gene that encodes the protein.

4. Frontline technologies used in proteomics include two-dimensional gel electrophoresis, yeast two-hybrid method, and X-ray crystallography.
5. Mass spectrometry is also used as frontline equipment in proteomics.
6. Other strategies of determining gene function include metanomics, gene knockout, and synteny.

OUTCOMES ASSESSMENT

1. Define the term proteome and contrast it with genome.
2. Describe the roles of two-dimensional electrophoresis and mass spectrometry in proteomics.
3. Discuss the goal of metanomics.
4. Discuss the knockout technology and its importance in understanding the function of proteins.
5. Describe how the yeast two-hybrid method of proteins is conducted.
6. Describe how a mass spectrophotometer is used in proteomics.

INTERNET RESOURCES

1. Intro to proteomics: *http://www.e-proteomics.net/tech.html*
2. NMR: *http://www.e-nmr.com/*
3. Intro to X-ray crystallography and NMR: *http://www.e-nmr.com/index.html#crystal*
4. Proteomics journal: *http://www.mcponline.org/*
5. Proteomics glossary: *http://www.genomicglossaries.com/content/proteomics.asp*
6. Overview of proteomics: *http://www.incyte.com/proteomics/tour/index.shtml*
7. Synteny human-pigs: *http://www.toulouse.inra.fr/lgc/pig/compare/compare.htm*
8. Intro to concept: *http://opbs.okstate.edu/~melcher/MG/MGW1/MG124.html*
9. Mouse-human synteny: *http://www.sanger.ac.uk/HGP/Chr22/Mouse/*
10. Arabidopsis synteny: *http://www.sgn.cornell.edu/maps/tomato_arabidopsis/synteny_map.html*
11. Synteny in grasses: *http://www.sccs.swarthmore.edu/users/00/aphilli1/genetics/synteny.html*
12. Terminologies, description of synteny: *http://exon.tn.nic.in/~vardhini/synteny.htm*

REFERENCES AND SUGGESTED READING

Bassett, D. E. Jr., M. B. Eisen, and M. S. Boguski. 1999. Gene expression informatics—it's all in your mind. *Nature Genet. 21:*51–55.

Fields, S., and O. Song. 1989. A novel genetic system to detect protein-protein interactions. *Nature, 340:*245–246.

Gyuris, J., E. Golemis, H. Chertkov, and R. Brent. 1993. Cdi1, a human G1 and S phase protein phosphatase that associates with Cdk2. *Cell, 75:*791–803.

Palmitter, R. D., and R. L. Brinster. 1985. Transgenic mice. *Cell, 41:*343–345.

Persidis, A. 1998. Proteomics. *Nature Biotechnol., 16:*393–395.

Schena, M. 1999. *DNA microarrays: A practical approach.* New York: Oxford University Press.

15

Modifying Protein Production and Function

PURPOSE AND EXPECTED OUTCOMES

Genes are expressed as proteins, which are polymers assembled from peptide bonding of various combinations of 20 commonly occurring amino acids. In the same way breeders manipulate plants and animals for enhanced performance, proteins can be tweaked for enhanced performance as well.

In this section, you will learn:

1. The concept of protein engineering.
2. Why protein engineering is desirable in certain cases.
3. How protein engineering is done.

WHAT IS PROTEIN ENGINEERING?

Proteins are fundamental to all life processes, as they are involved in numerous functions and structures of organisms. As enzymes, proteins function as catalysts that facilitate biochemical reactions. As antibodies, they are involved in immune reactions, thereby affording protection to the organism. Carrier proteins are involved in intercellular transport of metabolites, while protein hormones are involved in signal transduction, impacting growth and development. Proteins perform structural roles in an organism, while they also function as surface receptors on the cell allowing cells to communicate among themselves. Each protein has a unique amino acid sequence that determines how it folds to produce its functional configuration, which is a 3-D atomic structure.

Spontaneous mutations in the gene encoding a protein may arise in nature to change the structure and function of a protein. These changes are random and may be advantageous or disadvantageous to the organism in terms of its evolution. **Protein engineering** is conceptually like molecular evolution, the key difference being that mutations can be targeted at specific locations in a gene to generate desired variation in the protein in a short time. Hence, protein engineering is concerned with the structural modification of an existing protein to change its properties and functions in a desired and predetermined way. Scientists may also design and construct new proteins. The outcomes of such molecular restructuring may include altered substrate specificity of an enzyme or increased stability of a protein for a specific application (e.g., industrial, therapeutic). Even though such modification in molecular property may be accomplished by chemical or enzymatic techniques, direct DNA manipulation is the preferred way of making either minor or major modifications to proteins.

Our knowledge is incomplete in terms of how structure determines function or how the amino acid sequence of a protein determines its 3-D structure. As such, even though techniques are available for amino acid sequences in an encoded protein to be rearranged or substituted with precision, the final product resulting from this manipulation can be defined only in terms of a desired function.

WHY ENGINEER PROTEINS?

Enzymes have extensive industrial applications. In particular, hydrolytic enzymes are needed to break down carbohydrates, fats, and proteins in various applications. However, such applications frequently involve heating that endangers the enzymes. Thermal stability is a desired property of industrial enzymes. Protein engineering has been used to make proteins thermostable, thereby increasing their stability in solvents, increasing their resistance to oxidative processes, and increasing their stability to protease digestion. The Cetus Corporation, for example, introduced disulfide bonds into the peptide hormone Interlukin II to increase its thermal stability. Sometimes, proteins can be engineered to alter their catalytic activity. The substrate specificity of enzymes may be changed as well as the substrate binding or affinity for a substrate.

Protein engineering has an application in drug design. A drug either interacts with a receptor to produce an effect or interacts to abolish an effect. Consequently, 3-D models of the drug-receptor complex provide researchers an opportunity to study these interactions at the molecular level. Pharmaceutically active proteins can be engineered to enhance their utility, stability, or shelf life. The ability to construct chimeric or hybrid proteins is of value in the development of synthetic vaccines, as well as monoclonal antibodies for diagnostic kits.

STEPS IN PROTEIN ENGINEERING

Protein engineering is a cyclical process involving several steps (see Figure 15–1).

1. **Identify and characterize the target protein**

    The first step in protein engineering is to identify the protein that needs to be modified. This target protein is assayed and biologically characterized as much as possible. Comprehensive characterization is necessary so that the mutant (modified) protein can be compared with the normal type to see if real functional modifications were made.

2. **Biological characterization**

    Biological characterization may be conducted before or after modification of the protein. The advantage of conducting functional analysis prior to purification is that it reduces the work involved in the more tedious structural analysis of proteins by advancing only those proteins that have good potential to be useful. Properties of enzymes measured include rates of the catalyzed reactions as a function of temperature and pH, as well as the stability of enzymes to denaturing agents of interest.

3. **Define the specific modification desired**

    The next step is to define clearly the goal or nature of the outcome desired in terms of changes in the protein function expected. The purpose of protein engineering is to tailor protein properties and activities purposefully and in a predetermined fashion.

4. **Clone or chemically synthesize the coding DNA sequence**

    The coding DNA sequence may be selected from an existing library using an appropriate probe. The gene could also be chemically synthesized. The clone with the coding sequence for the target protein is grown to produce large quantities of the protein, which is then purified.

FIGURE 15–1
A summary of the steps in protein engineering.

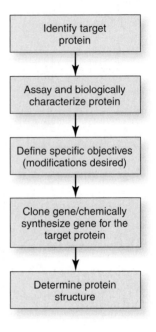

5. **Determination of protein structure**

Knowledge of 3-D protein structure is fundamental to protein design and protein engineering. A 3-D model shows the relative positions of all the atoms. If possible, the positions of substrate or ligand atoms are observable in a 3-D model. The 3-D molecular structure is commonly determined by X-ray crystallography (or protein crystallography, as it is sometimes called).

Protein structure was discussed previously. The secondary structure of protein architecture entailing α-helices, turns, and β-sheet secondary motifs of each polypeptide chain are folded into a compact 3-D tertiary structure. This structure may be divided into smaller functional units called **domains.** These regions contain contiguous amino acid sequences that independently fold into defined structural units.

PROTEIN ENGINEERING ACTIVITIES

The specific activity and methodologies employed in protein engineering are as follows:

1. **Isolation**

Gene isolation may be accomplished by using the gene pattern of expression or its ability to bind a complementary strand of DNA. It may also be isolated based on its ability to code for a specific protein that can be assayed for with antibodies. The isolated gene is analyzed to reveal the amino acid sequence that corresponds to the cloned protein product. The cloned gene is subjected to genetic engineering techniques to edit its information, one base at a time, to substitute one amino acid with another. This strategy enables the researcher to examine the role of individual amino acids in the structure and function of the target protein. The process of altering the information in the nucleotide sequence to modify the target protein is called **site-directed mutagenesis.** The altered or engineered protein may exhibit increased stability or modified catalytic properties.

2. **Purification**

   Protein purification is a challenging task, requiring the isolation of a specific protein from a complex mixture of thousands of proteins, DNA, RNA, lipids, and carbohydrates. The purity of the final product depends on the intended use: it may be more pure for pharmaceutical purposes and less pure for industrial applications (e.g., industrial enzymes). High-performance liquid chromatography (HPLC) produces a high purity of products. Recombinant DNA technology may be used to over-express a gene so that in some cases about 50 percent of a cell's total protein would comprise the target protein, thereby making it easier to isolate and purify. Sometimes, scientists may facilitate the purification process of a protein by engineering modifications into the encoding gene that would alter the protein product in specific ways. For example, to facilitate the purification of **subtilisin,** scientists at Genentech engineered a cysteine residue into this serine protease, which allowed them to use a thiol-specific-binding resin to isolate the product. Researchers may also engineer the protein to be extracellular in location so that it is secreted into the culture medium to help the purification process. This is especially desirable for proteins that contain disulfide bonds.

3. **Functional analysis (biological characterization)**

   Biological characterization of the protein entails the use of specific and sensitive assays of protein function to characterize the protein on the basis of functional properties like catalysis, ligand binding, or stability to denaturing agents like urea.

4. **Structural analysis**

   In order to accomplish the primary goal of protein engineering—the rational design and construction of novel proteins with modified or unique properties—it is critical to have a 3-D model of the target protein. Such a model will show the relative positions of all atoms, including, if possible, the positions of substrates of ligand atoms. X-ray crystallography (or protein crystallography) is the primary technique for determining the molecular structure of proteins. However, protein structure may also be determined in solution, using nuclear magnetic resonance (NMR), thereby avoiding the crystallization step that is an involved process.

   Modifying the amino acid content of the protein by site-directed mutagenesis can have one of two basic effects. It may change the amino acid function without changing protein structure. An example of such an alteration is the replacement of glutamic acid with glutamine. On the other hand, replacing glutamic acid residue with aspartic acid would change the protein structure but not its function.

5. **Knowledge-based structural prediction**

   By knowing how proteins with similar sequences fold, and with the aid of powerful computers and sophisticated software, scientists can make structural predictions of proteins. This approach to modeling the 3-D structure of protein based on strong sequence homologies of other proteins of known structure is called **knowledge-based structural prediction.** This strategy depends on the availability of 3-D structures for the comparison and will therefore become more useful as the database grows. To do this, scientists align the homologous amino acid sequence of a protein of unknown structure with those of known structure. This allows secondary structural motifs, domains, or ligand interactions to be identified. A 3-D structure is put together by borrowing pieces of the structure from the homologous protein closest in sequence to the modeled proteins (e.g., connecting chains at junctures between defined proteins).

6. *De Novo*

   Ultimately, protein engineering aims at designing protein structure with specified activity, correct solubility and stability, as well as other functional properties from the application of basic principles. Currently, this application is in its infancy. Various researchers have constructed 20 to 40 amino acid polypeptides. Efforts are continuing to build binding or catalytic sites into these rudimentary structures.

SELECTED APPLICATIONS

1. **Protein engineering of laundry detergent**

   One of the successful commercial applications of protein engineering is the engineering of the laundry detergent additive called subtilisin. Subtilisin is a protease enzyme (a protein-digesting enzyme) produced by bacteria. It has a broad specificity for proteins that commonly soil clothing. Manufacturers of certain laundry detergents enhance the efficiency of their products by including this enzyme. Such improved detergents often have a label that may read "with biologically active enzymes." Unfortunately, the enzyme is inactivated by bleach through the oxidation of the amino acid methionine at position 22 of the subtilisin molecule. Genetic engineers have manipulated the subtilisin gene in *E. coli* to change methionine to other amino acids and found that substitution of methionine by alanine was the best in terms of activity and stability. Consequently, laundry detergent manufacturers currently utilize the subtilisin from the recombinant source in their product.

2. **Other applications**

   Other applications of protein engineering include the modification of proteins for thermal stability (e.g., lysozyme), improved binding of small molecules to protein receptors on cell surfaces, and altered specificity of DNA-binding proteins and specific metabolic enzymes. Protein engineering can also be utilized in protein design for the production of protein and peptide mimics (e.g., neuropeptidase inhibitors), for enzyme inhibitors for pharmaceutical production, and the synthesis of oligopeptides that can be used in vaccine development. Protein engineering also has potential applications in some of the more radical areas such as bioelectronics.

KEY CONCEPTS

1. Protein engineering is concerned with the structural modification of an existing protein to change its properties and functions in a desired and predetermined way.
2. Protein engineering is conducted according to certain steps that include protein isolation, purification, characterization, and structural analysis.
3. Protein engineering is used in various practical ways including enhancing the stability of enzymes, altering catalytic activity, and in designing drugs.
4. Protein engineering may be likened in some way to biological evolution.
5. X-ray crystallography is a frontline technology of protein engineering.
6. 3-D protein structure is critical to protein design and protein engineering.

OUTCOMES ASSESSMENT

1. Give specific reasons why it is necessary for certain proteins to be modified.
2. Describe the steps in protein engineering.
3. Give specific practical applications of protein engineering.
4. In what way is protein engineering like biological evolution?
5. Explain why a 3-D structure is important to understanding protein function.
6. 3-D protein structure is fundamental to protein design and engineering. Explain.

INTERNET RESOURCES

1. Journal of protein engineering: *http://protein.oupjournals.org/*
2. Direct mutagenesis and protein engineering: *http://photoscience.la.asu.edu/ photosyn/courses/BIO_343/lecture/protein.html*
3. Protein engineering protocols: *http://departments.colgate.edu/chemistry/rsr-protocols.pdf*
4. Genentech site on protein engineering: *http://www.gene.com/gene/research/ biotechnology/proteinengineering.jsp*
5. General information on protein engineering: *http://www.biores-irl.ie/biozone/protein.html*

REFERENCES AND SUGGESTED READING

Cleland, J. L., and C. S. Craik. 1996. *Protein engineering: Principles and practices.* New York: John Wiley and Sons.

Moody, P. C. E., and A. J. Wilkinson. 1990. *Protein engineering.* Oxford, England: Information Press Ltd.

Oxender, D. L., and T. J. Graddis. 1991. *Protein engineering.* In *Biotechnology: The science and business.* V. Moses and R. E. Cape (eds). London, UK: Harwood Academic Publishers.

SECTION 2
Antisense Technology

PURPOSE AND EXPECTED OUTCOMES

The complementary nature of the two DNA strands and the manner in which the genetic message encoded is deciphered offers opportunities for scientists to regulate the expression of genes.

In this section, you will learn:

1. The strategy of gene silencing.
2. The phenomenon of RNA interference.
3. Practical applications of the antisense technology.

WHAT IS THE ANTISENSE TECHNOLOGY?

The DNA is double helix, as previously described, comprising two complementary polynucleotide chains. The genetic information enciphered in the DNA is deciphered through a series of steps, beginning with the transcription of information. Only one of the two DNA strands is transcribed. The transcribed strand is called the **antisense strand** (the other strand is called the **sense strand**), which becomes the template for transcribing the enzymes that assemble the mRNA (see Figure 15–2). The sequence of nucleotides

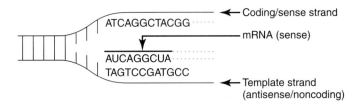

FIGURE 15–2
Sense versus antisense strands. The template strand is the noncoding strand, which is transcribed into mRNA. This strand is identical to the coding strand, except for the exchange of thymine with uracil in the mRNA.

in the resulting mRNA is called "sense" because it results in a gene product (protein). The mRNA is single stranded. However, it can form duplexes like DNA, provided a second strand of RNA with a complementary base sequence is present. If an mRNA forms a duplex with the antisense RNA, it is not difficult to see that it will be inhibited from translation into a gene product. Consequently, the strategy of antisense technology is to employ genetic engineering methodologies to introduce synthetic genes (DNA) that encode antisense RNA molecules into organisms, to interfere with the translation of sense mRNA molecules that are transcribed from genes that encode undesirable products.

A major use of antisense technology in genetic engineering is in changing the expression of a gene. In this regard, genetic engineering employs two basic strategies. A gene can be expressed in a position where it normally would not be expressed. This is called **ectopic expression.** The normal expression of a gene can be suppressed (called **gene silencing**).

Gene silencing may be accomplished by introducing an antisense gene into the organism or by cosuppression. Cosuppression is the silencing of an endogenous gene by introducing a transgene into the organism. In this case, rather than overexpression because of the two genes, the combined presence of the endogenous gene and the transgene leads to a loss in gene function.

SYNTHETIC OLIGONUCLEOTIDES

The antisense technique starts with the introduction of a copy of a construct containing all or part of the target gene in an antisense orientation into the organism (see Figure 15–3).

There are three basic types of oligonucleotides used in antisense technology. The first two operate after transcription, while the third is pre-transcriptional.

1. The classic antisense oligonucleotide is designed to target and bind a specific gene location. The result of such a bond is that the mRNA is inhibited from being translated into its product (protein). In certain cases, the target RNA that has been bound by the antisense oligonucleotide is susceptible to degradation by the enzyme RNase H. This not only accomplishes the purpose of eliminating the target product, but the introduced antisense construct is released to repeat the process.
2. The second group of antisense oligos differs from the first in that the degradation of the target gene is effected by a ribozyme (not RNase H).
3. Not considered a true antisense ("psuedoantisense"), this third group of antisense oligos has a triple-helix DNA structure. The construct binds to the gene (DNA) to prevent its transcription.

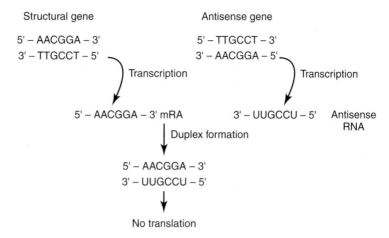

FIGURE 15–3

How antisense technology works. The sequence in the template strand is inserted in the reverse order, the 3′ end becoming the 5′ end. This is the antisense gene. Both genes are inserted into the same cell. The effect is that both coding and non-coding strands are transcribed, thereby creating a condition for complementary bonding (duplex formation), like a Velcro effect. The effect of duplex formation is the inhibition of transcription.

ANTISENSE OLIGOS DELIVERY

The most widely used and efficient method of delivery of antisense oligos into cells is by liposome encapsulation. However, these constructs can be delivered by micro-injection into cells individually. Though very effective, it is a very slow process, thus limiting its application (e.g., as a practical strategy for gene therapy). Oligos may be introduced into the cell via receptor-mediated endocytosis. The last method is least desirable. It is inefficient and unpredictable in results.

NATURALLY OCCURRING ANTISENSE RNA

Double-stranded RNA (dsRNA) corresponding to specific genes are known to be synthesized by a wide variety of organisms (including plants, fungi, mice, zebrafish, *Drosophila,* and worms) that have been studied. The consequence of the presence of these duplex RNA molecules is the suppression of the expression of their corresponding genes, a phenomenon called **RNA interference (RNAi).** This phenomenon is also called **post-transcriptional gene silencing.** In fact, studies have revealed that RNA interference (in animals), post-transcriptional gene silencing or cosuppression (in plants), and quelling (in fungi) all refer to a group of interrelated, sequence-specific, RNA-targeted gene-silencing mechanisms. Furthermore, it has been shown that a family of RNA-dependent RNA polymerases have a genetic role in the mechanism of RNA silencing. Also, even though RNA silencing phenomena share a common biochemical machinery, a diversity of silencing triggers exist, one of which is the double-stranded DNA. It has been shown that in both *Drosophila* and human cells, synthetic or purified small interfering RNAs (siRNAs) are capable of replacing dsRNA as an RNAi trigger.

The RNA molecules found in the cytoplasm of a cell are normally single-stranded RNA (ssRNA). When a cell encounters dsRNA, it uses an enzyme to cut them into fragments of 21 to 25 base pairs long. In the *Drosophila,* this degradative enzyme is called the

Dicer. The two strands of the short fragments subsequently separate enough, exposing the antisense strand, which in turn hybridizes to the complementary sequence on the mRNA transcribed from the endogenous sense target gene. The binding of the mRNA creates a dsRNA, making it susceptible to degradation and consequently triggering the degradative enzyme. The RNA transcript is therefore not translated into a product. Researchers have discovered that the gene for insulin-like growth factor 2 receptor (*Igf2r*) inherited from humans and mice synthesizes an antisense RNA that is implicated in the suppression of the synthesis of the mRNA for *Igf2r*. Consequently, the expression of the *Igf2r* gene in one individual gene depends upon the source of the gene, whether inherited from the male (father) or female (mother). This phenomenon of variation in gene expression in humans is called **genomic** or **parental imprinting.**

DOWN-REGULATION OF GENE EXPRESSION

One of the applications of the techniques of genetic engineering is the manipulation of gene expression. Specifically, transformation technology is often used to down-regulate the expression of endogenous genes that encode undesirable products that affect quality traits, health, or the general development of an organism. The so-called "knockout" technique is used for targeted silencing or inactivation of genes to study gene function. Down-regulation of gene expression via antisense is a powerful tool for research, with enormous practical (realized and potential) applications in product development and therapeutics. Antisense experimentation has indicated that sometimes both the transgene and the homologous endogenous gene were simultaneously silenced, a phenomenon called **cosuppression.** Because the down-regulation of gene expression via antisense and cosuppression involves interactions between homologous and complementary nucleic acid sequences, their effect is termed **homology-dependent gene silencing.** Homology-dependent gene silencing may occur at two levels—DNA level (termed **transcriptional gene silencing**) or after transcription (called **post-transcriptional gene silencing**). These phenomena are not completely understood.

■ TRANSCRIPTIONAL GENE SILENCING

Transcriptional gene silencing (TGS) prevents gene transcription into mRNA. This normally involves an interaction of genes that share homology in promoter regions. Transcriptional gene silencing is associated with increased promoter methylation in both plant and animal systems that can be meiotically heritable. In *Petunia hybrida*, such methylation was observed to cause alterations in chromatin structure in transgenic lines. It is suggested that methylation does not directly prevent gene transcription, but rather DNA methylation is associated with the targeting of transcriptional repressive protein complexes on the transgene region. These proteins are believed to bind to the methylated DNA, causing chromatin condensation and thereby preventing gene transcription. In TGS events where cosuppression occurs, the mechanism must involve *trans*-acting methylation signals.

■ POST-TRANSCRIPTIONAL GENE SILENCING

In post-transcriptional gene silencing (PTGS), gene transcription is not affected. However, the RNA transcript is subsequently degraded, and hence there is no accumulation of RNA. As previously discussed, the presence of double-stranded RNA (dsRNA) molecules appears to trigger PTGS by the mechanism of RNA interference. However, this model does not explain the event of cosuppression. Alternative sources of dsRNA must occur to trigger PTGS. Suggested sources include the production of dsRNA from sense transgenes via the activity of an RNA-dependent polymerase, and the production of aberrant RNA, possibly from DNA methylation that might cause premature termination of transcription.

SELECTED APPLICATION IN PLANTS: THE MAKING OF THE FLAVR SAVR TOMATO

Tomatoes are widely grown throughout the world. They are eaten fresh or cooked. They are sold fresh or processed into a variety of products (e.g., juice, ketchup, sauce, or sold as puree or crushed). Each of these products has certain qualities that are desired for optimal product quality. Many of the quality factors are impacted by the process of ripening. These include soluble solids, acids, color pigment, flavor, fruit firmness, and others.

Flavor is determined in part by sugars and acids, while the modifications of cell wall structure and composition affects the fruit texture and processing characteristics. Scientists have isolated specific genes for several tomato fruit quality parameters. This includes genes impacting ethylene (ACC synthase and ACC oxidase), color (e.g., phytoene synthase), cell wall (polygalacturonase), and others (e.g., proteinase inhibitor, storage protein).

A quality factor of great importance to the tomato industry is texture. It is important to the processing of both fresh and processed products. Texture depends on cell-wall characteristics such as cellulose, hemicellulose, and pectins. An enzyme, **polygalacturonase (PG),** which hydrolyses α-1,4 linkages in the polygalacturonic acid component of the pericarp cells of tomato, has been isolated, purified, and the gene encoding it has been cloned. It is synthesized only during tomato ripening. Three isoforms produced from post-translational modification of a single polypeptide from a single gene are known. Another enzyme, pectinesterase (PE), present in two isoforms is also associated with fruit ripening in tomatoes.

The conventional practice is to pick tomatoes from the vine before they ripen. These fruits that are green have firm texture and are able to withstand transportation without bruising. At their destination, these unripe fruits are exposed to ethylene to induce forced ripening (which actually is just a cosmetic change in color). Vine-ripened tomatoes have good color and are tastier. However, they are prone to damage during transportation.

Scientists at Calgene took up the challenge to develop tomatoes that will vine-ripen, be transportable, and have a longer shelf life. Having identified PG as the enzyme responsible for cell-wall degradation that results in fruit softening, they mapped out a strategy to down-regulate the gene that encodes this enzyme by the antisense technology.

They created a cDNA of the entire *PG* gene. Then they excised a 730 bp region, including a 50 bp non-coding region called the *Hin*f1 fragment, from the sequence. This fragment was cloned into a plasmid. The cloned fragment was excised and ligated into a second plasmid, inserting it in reverse orientation, just after the CaMV 35S promoter. The antisense gene was cloned into a plasmid pBIN19, and subsequently inserted into another plasmid pJR16A containing the *KanR* gene (for kanamycin resistance marker) and delivered into cells in tissue culture via *Agrobacterium*-mediated gene transfer (see Figure 15–4). Because CaMV 35S is a constitutive promoter, the *PG* gene is constantly expressed, making the antisense mRNA abundantly present.

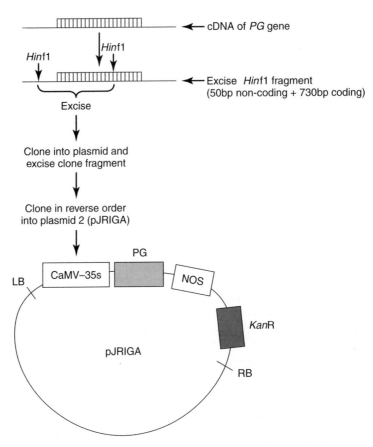

FIGURE 15–4
A summary of the steps in the creation of the Flavr Savr tomato using antisense technology. A cDNA of the entire PG gene is created. A *Hin*f1 fragment comprising a 730 bp region (including a 50 bp non-coding region) is excised and cloned into a plasmid. The fragment is cloned in reverse order into a second plasmid and used to transform tomato cells. As both the antisense gene and the gene in the normal orientation are transcribed, there is an opportunity for the duplex formation, thereby reducing the number of mRNAs available for translation into PG.

SELECTED APPLICATION IN INDUSTRY

■ ANTISENSE THERAPEUTICS

The use of antisense technology in drug discovery is a relatively new application for biotechnology. The rationale of this strategy is that nearly all human diseases are the result of inappropriate protein production or improper protein performance. Host diseases (e.g., cancer) and infectious diseases (e.g., HIV-AIDS) are protein based. These diseases can be treated at the root by preventing the undesirable proteins from forming in the first place.

Conventional Drugs versus Antisense Drugs

Conventional drugs are designed to act on the disease-causing proteins while an antisense drug is designed to inhibit translation of specific mRNA targets into disease-causing proteins (see Figure 15–5). Essentially, an antisense drug will be comprised of oligonucleotides designed to bind to specific sequences of nucleotides in the target mRNA. To design an antisense drug, the offending protein is first identified and its 3-D form characterized. This

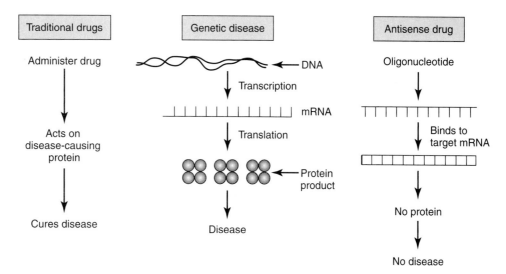

FIGURE 15–5

A summary of how conventional drugs and antisense drugs work. Gene products (proteins) cause disease. Conventional drugs act on disease-causing proteins. Antisense drugs act on mRNAs to prevent their translation into disease-causing proteins.

enables researchers to develop an appropriate prototype compound that can effectively interact with the target protein. The challenge is the correct prediction of structures of complex molecules like proteins. To design an antisense drug, the 3-D structure of the receptor site of the mRNA also needs to be identified, but this is a less challenging task. Because antisense drugs can lead to the mRNA at multiple points of interaction at a single receptor site (as opposed to binding at only two points of interaction in the case of conventional drugs), they have the potential to be more selective and more effective.

Challenges of Antisense Therapeutics

In spite of the potential of antisense drugs, certain hurdles need to be overcome before these drugs become practical reality. The Isis Pharmaceutical Company of Carlsbad, California, is at the forefront of antisense drug discovery and development. Several antisense compounds are being evaluated in clinical trials. Antisense drug development has challenges that have been addressed to varying extents. The possibility of the body mistaking the antisense oligonucleotide (the antisense drug) as a foreign invader and consequently using the immune system to attack it is realistic. It was shown that antisense oligos fragments containing the 2-base CpG (cytosine-phosphate-guanine) sequence triggered an immune response, activating mammalian B cells and natural killer cells in culture. However, methylated CpG sequence did not trigger this response. Whereas unmethylation is common in bacterial DNA, it is less common in mammalian DNA. Another major concern is the failure of antisense oligos to correctly bind to target cells or tissue, resulting in unintended damage (e.g., reduction in the production of critical proteins near target sites). This may possibly be the cause of the observed side effect in test subjects, including a drop in white blood cell counts and elevated blood pressure. Because the enzyme DNAse occurs throughout the body, antisense oligos are designed with terminally inverted polarity to resist exonuclease degradation. First synthesized in 1969, phosphorothiotate oligos are more stable to nucleases. Modifying the phosphodiester bonds between nucleotide bases to form phosphorothiotate bonds extends the half-life of antisense. However, phosphorothiotate oligos have limitations with implications in various aspects of drug therapy, including pharmacodynamic (e.g., low affinity per nucleotide unit), pharmacokinetic (e.g., limited bioavailability, limited blood-brain penetration), and toxicologic (e.g., release of cytokines, clotting effects).

The ability of antisense oligos to bind targeted cells and not neighboring proteins may be enhanced by attaching oligos to DNA-protein complexes on cationic liposomes. Receptor ligands or cell-specific antibodies are then incorporated into the oligo sequence to help navigate them to the desired target cells or tissue region. Side effects of antisense drug therapy may be minimized by the method of delivery into the patient. A slow and continuous delivery via intravenous injection of low doses of the drug is preferred to an acute dose (high dose over a short period). The use of metallotexaphyrin-oligos promises to address most of the shortcomings of antisense drugs as discussed previously. The lanthanide (III) texaphyrin complex being developed by Pharmacyclics, Inc. acts like a synthetic ribozyme capable of cleaving DNA. Because ribozymes are very large and very complex molecules, there is the need to find single oligos that can bind to a specific location in order to catalyze the hydrolysis of RNA. The lanthanide (III) metal texaphyrin complex is effective in this task.

KEY CONCEPTS

1. The strategy of antisense technology is to employ genetic engineering methodologies to introduce synthetic genes (DNA) that encode antisense RNA molecules into organisms in order to interfere with the translation of sense mRNA molecules that are transcribed from genes that encode undesirable products.
2. Gene silencing may be accomplished by introducing an antisense gene into the organism or by cosuppression.
3. There are three basic types of oligonucleotides used in antisense technology. The first two operate after transcription, while the third is pre-transcriptional.
4. Double-stranded RNA (dsRNA) corresponding to specific genes are known to be synthesized by a wide variety of organisms. The consequence of the presence of these duplex RNA molecules is the suppression of the expression of their corresponding genes, a phenomenon called RNA interference (RNAi).
5. Studies have revealed that RNA interference (in animals), post-transcriptional gene silencing or cosuppression (in plants), and quelling (in fungi) all refer to a group of interrelated, sequence-specific, RNA-targeted gene-silencing mechanisms.
6. Antisense technology may be used to down-regulate gene expression.
7. One of the high-profile applications of antisense technology is the development of the Flavr Savr tomato by Calgene.

OUTCOMES ASSESSMENT

1. Describe how the antisense technology works.
2. Describe how RNA interference works. Contrast it with antisense.
3. Describe how the Flavr Savr tomato was developed.
4. Discuss the challenges in the application of antisense technology.
5. Discuss the types of synthetic antisense oligos used in antisense technology.
6. Distinguish between transcriptional gene silencing and post-transcriptional gene silencing.
7. Distinguish between how conventional drugs and antisense drugs work.

INTERNET RESOURCES

1. Intro: *http://www.isip.com/antisens.htm*
2. Antisense and RNA interference: *http://www.ultranet.com/~jkimball/BiologyPages/A/AntisenseRNA.html*
3. Journal of antisense: *http://www.liebertpub.com/ARD/default1.asp*
4. Antisense therapeutics: *http://www.antisense.com.au/default.htm*
5. Prospects of antisense therapy and hurdles: *http://www.hosppract.com/genetics/9909mmc.htm*
6. How antisense works: *http://opbs.okstate.edu/~melcher/MG/MGW1/MG1234.html*

REFERENCES AND SUGGESTED READING

Ahlquist, P. 2002. RNA-dependent RNA polymerases, viruses, and RNA silencing. *Science, 296*:1270–1273.

Crooke, S. T. 1998. An overview of progress in antisense therapeutics. *Antisense Nucleic Acid Drug Dev, 8:*115.

Gewirtz, A. M., D. L. Sokol, and M. Z. Ratajczak. 1998. Nucleic acid therapeutics: State of the art and future prospects. *Blood, 92:*712.

Helene, C., and J. Toulme. 1990. Specific regulation of gene expression by antisense, sense and antigene nucleic acids. *Biochem. Biophys. Acta., 1049:*99–125.

Li, H., W. X. Li, and S. W. Ding. 2002. Induction and suppression of RNA silencing by an animal virus. *Science, 296:*1319–1321.

Milligan, J. F., M. D. Matteucci, and J. C. Martin. 1993. Current concepts in antisense drug design. *J. Medicinal Chem., 36:*1923–1937.

Plaster, R. H. A. 2002. RNA silencing: The genome's immune system. *Science, 296:*1263–1265.

Stein, C. A., and M. Arthur. 1998. *Applied antisense oligonucleotide technology.* New York: John Wiley and Sons.

Weintraub, H. M. 1990. Antisense RNA and DNA. *Sci. Amer., 262:*40–46.

Zamore, P. D. 2002. Ancient pathways programmed by small RNAs. *Science, 296:*1265–1269.

PART IV

Specific Applications

Selected applications of biotechnology will be discussed in this part. The purpose of the discussion is not to do the impossible task of listing every known application of biotechnology but rather to sample the applications in areas that have already benefited society or have great potential to do so. Whenever possible, the applications that made headlines have been included.

16 Food Biotechnology

SECTION 1
Plants

PURPOSE AND EXPECTED OUTCOMES

Areas of application of genetic engineering in plants are varied. However, the key applications that have made the most economic impact include the production of *Bacillus thuringiensis* (*Bt*) crops and herbicide-tolerant crops. Early headliners were the introduction of "ice minus" and "Flavr Savr" tomatoes.

In this section, you will learn:

1. Selected landmark applications of biotechnology in plant agriculture.
2. Selected major applications of biotechnology in plant agriculture.

These two applications will also be discussed for their historic roles in the application of biotechnology to agriculture. Other selected applications will be introduced very briefly.

OVERVIEW OF PLANT FOOD BIOTECHNOLOGY

One of the major areas of biotechnology applications is food biotechnology. This application is one that the public watches with very close attention because everybody needs to eat food to survive. It is not surprising that food biotechnology generates considerable public debate, as this technology has both critics and supporters. The supporters (who come primarily from academia, industry, and trade organizations, among others) promote food biotechnology for benefits such as increased crop yields, improved nutritional content of foods, reduced pesticide usage, and general agricultural efficiency. Critics are quick to point out the potential risks to humans and the environment, as well as adverse socioeconomic impacts on the disadvantaged producers. The expanding market power of a few multinational companies, the souring international trade over genetically modified organism (GMOs), and the role of the federal government in regulation of the biotechnology industry are contentious issues. The specific issues are discussed in detail in Chapters 20 and 22. This chapter will focus on the scientific strategies employed to accomplish the enhancement of food crops.

The first generation of agricultural biotechnology food products has directly benefited primarily agricultural producers through increased yields and reduced production costs. The most dominant products are the *Bt* products (crops with resistance to lepidopteran pests) and herbicide-resistant products (crops with resistance to major herbicides like those which are glyphosate based). The crops that have been modified in this fashion are generally field crops (e.g., corn, potato, soybean, cotton). The second generation GMOs are beginning to address consumer needs (for example, flavorful produce, increased nutritional value, and extended shelf life of perishable produce).

Some of the GMOs that made headlines in plant biotechnology include ice minus and the Flavr Savr tomato. These products were the trailblazers in plant food biotechnology. They have since been superseded by the pest protection products (*Bt* and herbicide resistance). Nonetheless, it is instructive to trace the path of plant food biotechnology by discussing these early products.

ENGINEERING ICE PROTECTION: THE CASE OF "ICE MINUS"

Crop producers in areas that are subject to subzero temperatures during the growing season are prone to frost damage, especially if the crops being produced are frost sensitive. Plants are intolerant of ice crystals forming in their tissue. It has been known that water may remain supercooled without forming ice crystals. However, supercold water can be seeded (just like seeding clouds for rain) to induce ice formation. Furthermore, certain soil bacteria (e.g., *Pseudomonas syringae*) are capable of promoting ice nucleation by producing a specific protein that catalyses the formation of ice crystals at temperatures of between -1.5 to $15°$ C. Normally, most compounds that are capable of ice nucleation (called **ice nucleating agents–INA**) are active at lower temperatures of $-10°$ C.

Studies have shown that bacteria could limit frost damage of frost-sensitive plants at temperatures above $-5°$ C. Furthermore, reducing the numbers of ice nucleation bacteria or their activity also reduced injury to plants. *P. syringae* is widely distributed in nature. Scientists thought it would be an environmentally safe strategy to reduce frost damage by reducing the population of this bacterium.

Advanced Genetic Science (AGS) initiated commercialization of a microbial frost inhibition system. The company proposed the creation of a mutant without the capacity to produce the ice nucleation proteins. Through chemical mutagenesis, researchers developed a mutant strain of *P. syringae* without ice nucleation capability. The mutant (called **INA⁻** or "**ice minus**") lacked the gene for producing the ice nucleation protein. Because the mutant strain was not produced by rDNA technology, no special permit was required for field-testing of the product. Test results showed a decrease in frost damage as a result of the application of a spray of the mutant bacterium. However, the company, seeking to improve the effectiveness of the product, proceeded to use rDNA technology to precisely delete the ice nucleation gene, arguing that the technique would not only be precise but that the mutants produced would be more stable and better genetically characterized. Their tests showed that the rDNA product was more effective in frost control without any additional ecological risk than the product of chemical mutagenesis. Their proposal to field-test their product was submitted to NIH in 1984, and then later to the EPA for its Experimental Use Permit, intended for a test in Monterey County, California. Unfortunately, this drew considerable public protest that ended in the withdrawal of the permit. Part of the controversy with this field-test arose because AGS was found to have illegally tested their bioengineered bacteria prior to receiving EPA approval.

The University of Berkeley team decided to push forward with field-testing of their own ice minus bacteria. Similarly, their efforts were greeted with organized public protests. However, the Berkeley team pursued a legal recourse, which eventually paved the way for them to successfully test their product. AGS also succeeded the second time around in 1986. The field-test of ice minus marked the first time a genetically engineered product was field-tested. It is significant to note that the controversy generated by ice minus prompted the development and implementation of the current U.S. regulatory system for biotech products.

ENGINEERING FRUIT RIPENING

Certain fruits exhibit elevated respiration during ripening with concomitant evolution of high levels of ethylene. Called **climacteric fruits** (e.g., apples, bananas, tomatoes), the ripening process of these fruits involves a series of biochemical changes leading to fruit softening. Chlorophyll, starch, and the cell wall are degraded. There is an accumulation of lycopene (red pigment in tomato), sugars, and various organic acids. Ripening is a complex process that includes fruit color change and softening. Ripening in tomatoes has received great attention because it is one of the most widely grown and eaten fruits in the world. Ethylene plays a key role in tomato ripening. When biosynthesis of ethylene is inhibited, fruits fail to ripen, indicating that ethylene regulates fruit ripening in tomatoes. The biosynthesis of ethylene is a two-step process in which s-adenosyl methionine is metabolized into aminocyclopropane-1 carboxylic acid, which in turn is converted to ethylene. By knowing the pathway of ethylene biosynthesis, scientists may manipulate the ripening process by either reducing the synthesis of ethylene or reducing the effects of ethylene (i.e., plant response).

In reducing ethylene biosynthesis, one successful strategy has been the cloning of a gene that hydrolyzes s-adenosyl methionine (SAM), called SAM hydrolase, from a bacterial virus, by Agritope of Oregon. After bioengineering the gene to include, among other factors, a promoter that initiates expression of the gene in mature green fruits, *Agrobacterium*-mediated transformation was used to produce transgenic plants. The effect of the chimeric gene was to remove (divert) SAM from the metabolic pathway of ethylene biosynthesis. The approach adopted by researchers was to prevent the aminocyclopropane-1-carboxylic acid (ACC) from being converted to ethylene. A gene for ACC synthase was isolated from a bacterium and used to create a chimeric gene as in the Agritope case.

The technology of antisense has been successfully used to develop a commercial tomato that expresses the antisense RNA for ACC synthase and ACC oxidase. USDA scientists pioneered the ACC synthase work, while scientists from England in collaboration with Zeneca pioneered the ACC oxidase work. Because transgenic tomatoes with incapacitated ethylene biosynthetic pathway produced no ethylene, they failed to ripen on their own, unless exposed to artificial ethylene sources in ripening chambers. The technology needs to be perfected so that fruits can produce some minimum amount of ethylene for autocatalytic production for ripening over a protracted period.

■ THE FLAVR SAVR TOMATO

Another application of antisense technology is in preventing an associated event in the ripening process, fruit softening, from occurring rapidly. Vine-ripened fruits are tastier than green-harvested and forced-ripened fruits. However, when fruits vine-ripen before harvesting, they are prone to rotting during shipping or have a short shelf life in the store. It is desirable to have fruits ripen slowly. In this regard, the target for genetic engineering is the enzyme polygalacturonase (PG). This enzyme accumulates as the fruit softens, along with cellulases that break cell-wall cellulose, and pectin methylesterase that together with PG break the pectic crosslinking molecules in the cell wall. Two pleiotropic mutants of tomato were isolated and studied. One mutant, never ripe (*Nr*), was observed to soften slowly and had reduced accumulation of PG, while the second mutant, ripening inhibitor (*rin*), had very little accumulation of PG throughout the ripening process. This and other research evidence strongly suggested an association between PG and fruit ripening. PG is biosynthesized in the plant and has three isoenzymes (PG1, PG2, and PG3).

This technology was first successfully used by Calgene to produce the **Flavr Savr tomato**, the first bioengineered food crop, in 1985. The protocol was previously described (see antisense). As previously noted, this pioneering effort by Calgene flopped because of the poor decision to market a product intended for tomato processing as a fresh market variety.

ENGINEERING INSECT RESISTANCE

About 80 percent of animal life consists of insects. Insects are very widely adapted and may cause damage directly to crops or be carriers (vectors) of pathogens. An estimated 600 species of insects are crop pests. Insect pests damage crops in the field at all stages of growth, and also affect stored products. There are two basic approaches to genetic engineering of insect resistance in plants:

1. The use of protein toxins of bacterial origin.
2. The use of insecticidal proteins of plant origin.

■ PROTEIN TOXINS FROM BACILLUS THURINGIENSIS (BT)

The *Bacillus thuringiensis* (*Bt*) endotoxin is a crystalline protein. It was first identified in 1911 when it was observed to kill the larvae of flour moths. It was registered as a biopesticide in the United States in 1961. *Bt* is very selective in action; that is, one strain of the bacterium kills only certain insects. Formulations of whole sporulated bacteria are widely used as biopesticide sprays for biological pest control in organic farming. There are several major varieties of the species that produce spores for certain target pests: *B. thuringiensis* var. *kurstaki* (for controlling lepidopteran pests of forests and agriculture), var. *brliner* (wax moth), and var. *israelensis* (dipteran vectors of human disease). The most commercially important type of the crystalline proteinaceous inclusion bodies are called **δ-endotoxins**. To become toxic, these endotoxins, which are predominantly protoxins, need to be proteolytically activated in the midgut of the susceptible insect to become toxic to the insect. These endotoxins act by collapsing the cells of the lining of the gut regions.

Bt resistance development has been targeted especially at the European corn borer that causes significant losses to corn in production. Previous efforts developed resistance in tobacco, cotton, tomato, and other crops. The effort in corn was more challenging because it required the use of synthetic versions of the gene (rather than microbial *Bt per se*) to be created.

Two genes, *cryB1* and *cryB2*, were isolated from *B. thuringiensis* subsp. *kurstaki* HD-1. These genes, which were cloned and sequenced, differed in toxin specificities, with *cryB1* gene product being toxic to both dipteran (*Aedes aegypti*) and lepidoteran (*Manduca sexta*) larvae, while *cryB2* affected only the latter. The *Bt* toxin is believed to be environmentally safe as an insecticide. In engineering *Bt* resistance in plants, scientists basically link the toxin to a constitutive (unregulated) promoter that will express the toxin systemically (i.e., in all tissues).

Transgenic plants expressing the δ-endotoxin gene have been developed. The first attempt involved the fusion of the *Bt* endotoxin to a gene for kanamycin resistance to aid in the selecting of plants (conducted by a Belgian biotechnology company, Plant Genetic Systems, in 1987). Later, Monsanto Company researchers expressed a truncated *Bt* gene in tomato directly by using the CaMV 35S promoter. Agracetus Company followed with the expression of the *Bt* endotoxin in tobacco with the CaMV 35S promoter linked to an alfalfa mosaic virus (AMV) leader sequence. Since these initial attempts were made, modifications to the protocols have increased expression of the toxin in transgenic plants by hundreds of fold. Transformation for expressing the chimeric *Bt* genes was *Agrobacterium*-mediated, using the TR2′ promoter. This promoter directs the expression of manopine synthase in plant cells transformed with the TR DNA of plasmid pTiA6.

The original *Bt* coding sequence has since been modified to achieve insecticidal efficacy. The complete genes failed to be fully expressed. Consequently, truncated (comprising the toxic parts) genes of *Bt* var. *kurstaki* HD-1 (*cry1A*[b]) and HD-73 (*cry1A*[c]) were expressed in cotton against lepidopteran pests. In truncating the gene, the *N*-terminal

TABLE 16–1
Types of commercially available *Bt* corn. Some companies have licensed their
Bt inventions to other companies for development of various *Bt* products.

Trade Name	*Bt* Gene	Transformation Event	Company
Knockout	*crylA (b)*	1st/2nd	Ciba (Novartis)
NatureGard		Generation 176	Mycogen
YieldGard	*crylA (b)*	*Bt* 11	Northrup King (Novartis)
YieldGard	*crylA (b)*	MON 810	Monsanto
Bt-Xtra	*crylA (b)*	DBT 418	Dekalb

Note: The *Bt* product produced by Aventis (called StarLink™) used the *cry9 (c)* gene, and was approved for feed only.

half of the protein was kept intact. For improved expression, various promoters, fusion proteins, and leader sequences have been used. The toxin protein usually accounts for about 0.1 percent of the total protein of any tissue, but this concentration is all that is needed to confer resistance against the insect pest.

Genetically engineered *Bt* resistance for field application is variable. For example, Ciba Seed Company has developed three versions of synthetic *Bt* genes capable of selective expression in plants. One is expressed only in pollen, another in green tissue, and the third in other parts of the plant. This selectivity is desirable for several reasons. The European corn borer infestation is unpredictable from year to year. The life cycle of the insect impacts the specific control tactic used. The insect attack occurs in broods or generations. The *Bt* genes with specific switches (pollen and green tissue) produce the *Bt* endotoxin in the parts of the plants that are targets of attack at specific times (i.e., first and second broods). This way, the expression of the endotoxin in seed and other parts of the plant where protection is not critical is minimized. Monsanto's YieldGard corn produces *Bt* endotoxin throughout the plant, and protects against both first and second broods of the pest. The commercially available *Bt* corn cultivars were developed by different transformation events, each with a different promoter (see Table 16–1).

Bt cotton is another widely grown bioengineered crop. The pest resistance conferred by the *Bt* gene has led to dramatic reduction in pesticide use and consequently a reduced adverse impact on the environment from agropesticides. As indicated previously, *Bt* sprays are widely used in organic farming for pest control. However, such application is ineffective if the insect bores into the plant. Furthermore, *Bt* sprays have a short duration activity.

■ *THE STARLINK*™ *CONTROVERSY*

A discussion of *Bt* would be incomplete without a mention of the StarLink™ controversy of 2000. StarLink™ is a corn variety that was genetically modified to express the *Bt* endotoxin. This product differed from the ones previously discussed in that the pesticide protein is *(cry 9(c))* derived from the subspecies *tolworthi* of *Bacillus thuringiensis*. Another unique characteristic of this product is that it contains, in addition to the *Bt* gene, a gene from the bacterium *Stroptomyces hygroscopicus,* which confers broad-spectrum herbicide resistance to the plant. This strategy of inserting multiple genes of interest into one crop is called **gene stacking**. StarLink™ hybrids are resistant to the European corn borer as well as the southwestern corn borer.

Having determined through tests that the *(cry 9(c))* protein could survive cooking or processing as well as being hard to digest, the company requested approval from the EPA to certify the yellow corn for livestock feed and industrial use. This approval exempted this product (a plant-pesticide) from the requirement for tolerance in animal feed. This kind of approval is called a split approval, a format that is common in approvals for conventional chemicals (i.e., approved for use on specific plants only). The EPA cautioned the company against allowing the product to contaminate the food chain. Unfortunately, in September 2000, StarLink™ was detected in taco shells, leading to the first test case of contamination of the food supply by a GMO. The current owner of the patent for the product, Aventis (who purchased it from AgrEvo USA), was compelled to purchase all of the 2000-crop year StarLink™ corn.

■ *ENGINEERING VIRAL RESISTANCE*

Even though viruses may utilize DNA or RNA as hereditary material, most of the viruses that infect plants are RNA viruses. One of the most important plant viruses in biotechnology is the Cauliflower Mosaic Virus (CaMV) from which the widely used 35S promoter was derived (CaMV 35S promoter). As previously described, a virus is essentially nucleic acid encased in a protein coat. The primary method of control of viral infections is through breeding of resistance cultivars. Also, plants can be protected against viral infection by a strategy that works like inoculation in animals. Plants may be protected against certain viral infections upon being infected with a mild strain of that virus. This strategy, called **cross-protection**, provides protection to the plant against future, more severe infections.

Engineering transgenic plants with resistance to viral pathogens is accomplished by a method called **coat protein-mediated resistance**. First, the viral gene is reverse transcribed (being RNA) into DNA from which a double-stranded DNA is then produced. The product is cloned into a plasmid and sequenced to identify the genes in the viral genome. A chimeric gene is constructed to consist of the open reading frame for the coat protein, to which a strong promoter is attached for high level of expression in the host. This gene construct is transferred into plants to produce transgenic plants.

Successes with this strategy have been reported in summer squash (the first product developed by this approach), and for resistance to papaya ringspot virus (a lethal disease of papaya), among others.

ENGINEERING HERBICIDE RESISTANCE

Herbicides constitute one of the most widely used agrochemicals in crop production. GMOs engineered for herbicide resistance are among the major applications of biotechnology in plant food biotechnology.

■ *WHY ENGINEER HERBICIDE-RESISTANT CROPS?*

A successful herbicide should destroy weeds only, leaving the economic plant unharmed. Broad-spectrum herbicides (nonselective) are attractive but their use in crop production can be problematic, especially in the production of broadleaf crops like soybean and cotton. There is a general lack of herbicides that will discriminate between dicot weeds and crop plants. Preplant applications may be practical to implement. However, once the crop is established and too tall for the safe use of machinery, chemical pest management becomes impractical. Grass crops (e.g., wheat, corn) may tolerate broadleaf herbicides better than the reverse situation. Consequently, when cereal crops and broadleaf crops are grown in rotation or adjacent fields, the latter are prone to damage from residual herbicides in the soil, or drift from herbicides applied to grasses. When a crop field is infested by weed species that are closely related to the crop (e.g., red rice in rice crop or nightshade in potato crop), herbicides lack sensitivity enough to distinguish between the plants.

To address these problems, one of two approaches may be pursued: (1) the development of new selective post-emergent herbicides or (2) genetic development of herbicide resistance in crops to existing broad-spectrum herbicides. The latter strategy would be advantageous to the agrochemical industry (increased market) and farmers (safer alternative to pesticides that are already in use). New herbicides are expensive to develop and take time.

■ MODES OF ACTION AND HERBICIDE-RESISTANCE MECHANISMS

Most herbicides are designed to kill target plants by interrupting a metabolic stage in photosynthesis. Because all higher plants photosynthesize, most herbicides will kill both weeds and desirable plants. Plants resist phytotoxic compounds via one of several mechanisms.

1. The plant or cell does not take up toxic molecules because of external barriers like cuticles.
2. Toxic molecules are taken but sequestered in a subcellular compartment away from the target (e.g., protein) compounds the toxin was designed to attack.
3. The plant or cell detoxifies the toxic compound by enzymatic processes into harmless compounds.
4. The plant or cell equipped with resistance genes against the toxin may produce a modified target compound that is insensitive to the herbicide.
5. The plant or cell overproduces the target compound for the phytotoxin in large amounts such that it would take a high concentration of the herbicide to overcome it.

MOLECULAR METHODS OF WEED CONTROL

The molecular genetic strategies for engineering herbicide resistance in plants can be grouped into two broad categories:

1. Modification of target enzyme of the herbicide.
2. Development of herbicide-tolerant genes.

■ MODIFICATION OF THE TARGET OF THE HERBICIDE

The common mechanisms by which genetically engineered plants resist herbicides are by:

1. Inhibiting a pathway in photosynthesis.
2. Inhibiting amino acid biosynthesis.

Genetic analysis of weeds that developed resistance to triazine herbicides, and their progeny, revealed that genetic resistance was cytoplasmic in origin. It was determined that a mutant gene, *psbA*, located in the chloroplast, encoded a 32-kDa protein in the thylakoid membrane that is involved in the photosystem-II electron transport system. Triazine herbicides (and others including triazinones, urea derivatives, and uracils) bind to this protein in susceptible plants. This binding blocks quinone/plastoquinone oxidoreductase activity and incapacitates or interferes with electron transport. It was also discovered that resistant weeds had about a 1,000-fold decrease in affinity for atrazine. [Further analysis showed that the amino acid composition of the 32-kDa chloroplast proteins had minor modifications in.] Gene sequencing has revealed that more than 60 percent of the mutations were the result of a substitution of Ser-264 (the number indicates amino acid position in the polypeptide) by Gly or Ala.

Using the *psbA* modification as a genetic engineering strategy for developing herbicide resistance is problematic. In the first place, techniques for transforming chloroplasts are lacking. However, selectable markers (streptomycin phosphototransferase gene) have been developed to facilitate chloroplast transformation. Other researchers have converted the *psbA* mutant gene from plastid to nuclear gene. In *Amaranthus hybridus,* this transfer was achieved by fusing the coding region of the mutant gene to the transcription regulation and peptide–encoding sequences of a nuclear gene for a small subunit of rubisco. The chimeric construct was transferred to herbicide-sensitive tobacco plants via *Agrobacterium* transformation protocols. Some transgenic plants exhibited atrazine resistance and produced the protein product of the nuclear *psbA* gene in the chloroplasts.

◼ *HERBICIDE-TOLERANT GENES*

Certain herbicides inhibit amino acid biosynthesis (e.g., Glyphosate (or Roundup®)) while others are photosynthetic inhibitors (e.g., S-triazines like atrazine).

Inhibition of Amino Acid Biosynthesis

Acetolactate Synthase (ALS) Genes
Only plants and microorganisms can synthesize about 50 percent of amino acids found in proteins. Examples of these are phenylalanine, tyrosine, tryptophan, lysine, and methionine. Enzymes that are involved in the biosynthesis of these amino acids by pathways that are exclusive to plants provide a unique opportunity for targeting herbicides. Somatic selection techniques have been utilized to isolate mutants with resistance to sulphonylureas. Genetic analysis indicated that resistance was conferred by a single dominant or semidominant nuclear gene. Further studies with the bacterium *Salmonella typhimurium* showed that growth inhibition by sulphonylureas on minimal media could be prevented by inclusion of certain branched-chain amino acids (leucine, isoleucine, and valine) in the medium. Consequently, the enzyme **acetolactate synthase (ALS)**, also called **acetohydroxyacid synthase (AHAS)**, was implicated as the target for the herbicide (the enzyme is required for the biosynthesis of these amino acids). This has been confirmed by other tests. ALS/AHAS is also the protein target for the **imidazolinones** (e.g., Impazapyr) and the **triazopyrimidines** (or **sulphonanilides**) herbicides. These classes of herbicides are structurally distinct. This makes ALS a particularly susceptible target for herbicides.

Attempts to genetically engineer herbicide resistance in plants by targeting ALS is also problematic. Like the *psbA* gene, ALS occurs in the chloroplast. The metabolic pathways for amino acid biosynthesis in plants and microorganisms are identical, and therefore attract opportunities for microbial information to be applied to the understanding of plant processes. However, because bacterial ALS comprises two different subunits, the generation of herbicide resistance in plants by using microbial genes is complicated. It would require the expression of two different protein subunits of the enzyme. Furthermore, like *psbA*, chloroplasts would have to be transformed. ALS genes have been isolated from several species including *Nicotiana tabaccum* and *Arabidopsis thaliana.* Site-directed mutagenesis was used to induce mutations in plants with ALS genes. Tobacco plants were successfully transformed with the mutant genes.

Engineering Glyphosate Resistance—the aro A gene
Glyphosate (*N*-[phosphonomethyl-glycine]) is the active ingredient in the broad-spectrum herbicide developed by Monsanto called "Roundup®." This herbicide is toxic to plants, fungi, and bacteria. It is phloem-mobile and has little residual effect as well as being less toxic to all major crops. Because of these desirable properties, engineering glyphosate resistance has received tremendous attention in the scientific community.

The primary target of glyphosate is an enzyme in the aromatic amino acid biosynthetic pathway called 5-enolpyruvyl-shikimate-3-phosphate synthase (EPSPS or EPSP synthase). It is also called 3-phosphoshikimate-1-carboxyvinyl transferase. The role of EPSPS in aromatic amino acid biosynthesis is the catalysis of the condensation of phosphoenolpyruvate and shikimate-3-phosphate to form 5-enolpyruvylshikimate-3-phosphate, a precursor of several amino acids (tryptophan, phenylalanine) as well as a number of aromatic secondary metabolites. Glyphosate binds to prevent the binding of phosphoenolpyruvate to EPSP synthase, thereby incapacitating the enzyme.

It has been discovered that the gene *aro A* encodes EPSP synthase in bacteria. Mutants of this gene have conferred glyphosate resistance upon *Salmonella typhimurium* and *E. coli.* The mechanism of resistance is by overproduction of the enzyme or overexpression of the gene. Mutant *aro A* genes have been transferred to plants under the control of a variety of gene promoters (e.g., the CaMV 35S RNA gene). The mutant bacterial genes have also been fused to plant EPSPS chloroplast transit peptide-encoding sequences of the rubisco small subunit gene and transferred to plants resulting in a 1,000-fold higher level of resistance.

Whereas the previously discussed strategies produced significant "laboratory resistance" to glyphosate, "field resistance" that enabled commercial application of the technology came as a result of isolating a bacterial strain, *Agrobacterium* CP4, with a suitable EPSP synthase. Likewise, the gene for this EPSP was fused to the coding sequence for the plant EPSPS chloroplast transit peptide and coupled to a powerful form of CaMV 35S. The Monsanto Company used this strategy to develop its Roundup Ready® soybean.

Engineering Herbicide Resistance Using Herbicide-Detoxifying Genes

Herbicide-detoxifying genes may be derived from plants or bacteria.

Plant Origin

A strategy for engineering herbicide resistance in plants is introducing herbicide-detoxifying enzymes into plants. This strategy is considered superior to those that modify the target for the herbicide in plants. Target modification may alter the enzyme to a degree that may adversely affect its physiological function to the detriment of the plant. On the other hand, suitable herbicide-detoxifying enzymes are limited. Those enzymes occur in both plants and microorganisms. Plant herbicide detoxifying enzymes include conjugation enzymes and mixed function oxidases. Researchers have isolated genes for the detoxifying enzyme **glutathione-S-transferase (GST)** from maize. These genes encode detoxifying enzymes for alachlor and atrazine. In tomato, another conjugation enzyme *N*-glucosyltransferase provides resistance to metribuzin through increased activity of the enzyme. Mixed function oxidases have been found to detoxify 2,4-D in pea and dicamba in barley.

Bacterial Origin

Soil bacteria have the capacity to detoxify many herbicides. Consequently, soil bacteria are a potential source for herbicide-tolerant genes. Several bacterial genes have been isolated and expressed in plants. For example, the plasmid-encoded nitralase gene, *bxn*, converts the herbicide 3, 5-dibromo-4-hydroxy benzonitrite to its inactive metabolite, 3,5,-dibromo-4-hydroxy benzoic acid (bromoxynill). BXN cotton is available on the market, produced by using a gene obtained from a strain of *Klebsiella ozaenae*. This bacterium was isolated from a contaminated field.

L-phosphinothricin (PPT) is an antibiotic produced by *Streptomyces* spp. It is a competitive inhibitor of glutamine synthetase, the enzyme that catalyzes the conversion of glutamate to glutamine. This is the only enzyme in plants capable of detoxifying ammonia that is generated as a product of nitrogen metabolism (e.g., amino acid degradation or nitrate reduction). In the absence of this enzyme, ammonia accumulates to lethal levels in the cell. Another antibiotic used as an herbicide is bialaphos, a tripeptide produced by *Streptomyces hygroscopicus*. Resistance to bialaphos is conditioned by a gene, *bar*, which codes for the enzyme phosphinothricin acetyltransferase (PAT). This enzyme acetylates the free amino acid group of PPT. The *bar* gene has been transferred into plants including tobacco, tomato, and potato, in which the gene is under the transcriptional control of the CaMV 35S promoter. Resistant cultivars have been produced.

Derived from the *Streptomycete* fungus, **glufosinate** blocks the synthesis of glutamine, causing an accumulation of toxic levels of ammonia in the plant. The fungus produces an enzyme that detoxifies the antibiotic it produces. The gene encoding the detoxifying enzyme has been isolated, cloned, and genetically engineered for optimal expression in plants. The technology was used in the development of resistance in corn to the commercial herbicide Liberty® (also called Ignite or Basta). The transgenic corn hybrid is marketed as LibertyLink®.

ENGINEERING NUTRITIONAL QUALITY

Nutritional quality augmentation through addition of new quality traits, removing or reducing undesirable traits, or other manipulations is an important goal in bioengineering food crops. Crops that feed the world are primarily cereals, roots and tubers, and legumes. Unfortunately, they are nutritionally inadequate in providing certain amino acids required for proper growth and development of humans and monogastric animals. For example, cereals are generally deficient in lysine and threonine, while legumes are generally deficient in sulfur amino acids. In some species (e.g., rice) where the amino acid balance is relatively appropriate, the overall protein quantities are low.

Molecular genetic approaches are being adopted to genetically engineer seed protein. They may be categorized as follows:

1. Altering the amino acid profile of the seed.
2. Selective enhancement of expression of existing genes.
3. Designing and producing biomolecules for nutritional quality.

■ THE MAKING OF "GOLDEN RICE"

"Golden rice" is so called because it has been genetically engineered to produce β-carotene (responsible for the yellow color in certain plant parts like carrot roots) in its endosperm. This rice produces β-carotene or pro-vitamin A, the precursor of vitamin A, which does not occur in the endosperm of rice.

An estimated 3 billion people of the world depend on rice as a staple food. Of this number, about 10 percent are at risk for vitamin A deficiency and the associated health problems that include blindness, as well as deficiency of other micronutrients like iron and iodine. The effort to create golden rice was led by Dr. Ingo Potrykus, a professor of plant science at the Swiss Federal Institute of Technology. In 1990, Garry Toenniessen, the director of food security for the Rockefeller Foundation, recommended the use of the sophisticated tools of biotechnology to address the problem of lack of vitamin A in rice. Later, at a Rockefeller-sponsored meeting, Potrykus met Peter Beyer of the University of Freiburg in Germany, an expert on the β-carotene pathway in daffodils. In 1993, with seed money of $100,000 from the Rockefeller Foundation, the two embarked upon an ambitious project to create a transgenic plant in a manner unlike any before. After seven years, the duo announced to the world their outstanding achievement, the golden rice, at a cost of $2.6 million. The bill was partly footed by the Swiss government and the European Union.

The Science Behind Golden Rice

The scientific feat accomplished in engineering β-carotene into rice is that it marks the first time a metabolic pathway has been engineered into an organism. Rice lacks the metabolic pathway to make β-carotene in its endosperm. Potrykus and Beyer had to engineer a metabolic pathway consisting of four enzymes into rice (see Figure 16–1). Immature rice endosperm produces **geranylgeranyl-diphosphate (GGPP)**, an early precursor of β-carotene. The first enzyme engineered was **phytoene synthase**, which converts GGPP to phytoene (a colorless product). Enzyme number 2, called **phytoene desaturase**, and enzyme number 3, called **β-carotene desaturase**, each catalyzes the introduction of two double-bonds into the phytoene molecule to make lycopene (has red

FIGURE 16–1

A summary of the key steps in the biosynthesis of beta-carotene. Starting from geranylgeranyl-diphosphate, three enzymes are involved in the conversion of intermediary products to beta-carotene, the precursor of vitamin A.

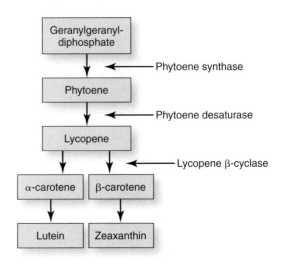

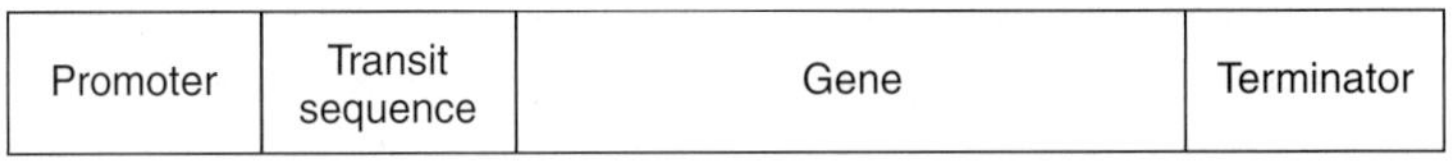

Promoter	Transit sequence	Gene	Terminator

FIGURE 16–2

The general characteristics of an expression cassette. An expression unit consists of the transgene flanked by the transit sequence and a terminating sequence. The transit sequence directs the location of expression of the transgene in the host. The promoter is the engine that drives the expression of the transgene.

color). Enzyme number 4, called **lycopene β-cyclase**, converts lycopene into β-carotene. A unit of transgenic construct (called an **expression cassette**) was designed for each gene for each enzyme. These expression cassettes were linked in series or "stacked" in the final construct (see Figure 16–2).

The source of genes for enzymes 1 and 4 was the daffodil, while genes for enzymes 2 and 3 were derived from the bacterium *Erwinia uredovora*. Three different gene constructs were created, the first and most complex combining enzyme 1 with the enzymes 2–3 combo, together with an antibiotic resistance marker gene that encodes hygromycin resistance, along with its promoter, CaMV 35S. The second gene construct was like the first, except that it lacked any antibiotic resistance marker gene. The third gene construct contained the expression cassette for enzyme 4, plus the antibiotic marker. By separating the genes for the enzymes and antibiotic resistance marker into two different constructs, the scientists reduced the chance of structural instability following transformation (the more cassettes that are stacked, the more unstable the construct).

These gene constructs were transformed into rice via *Agrobacterium*-mediated gene transfer in two transformation experiments. In experiment 1, the scientists inoculated 800 immature rice embryos in tissue culture with *Agrobacterium* containing the first transgenic system. They isolated 50 transgenic plants following selection by hygromycin marker. In the second experiment, they used 500 immature rice embryos, inoculating them with a mixture of *Agrobacterium* (T-DNA) vectors carrying both the second and third constructs. This experiment yielded 60 transgenic plants. The second experiment was the one expected to yield the anticipated results of a golden endosperm. This was so because it had all four enzymes required for the newly created metabolic pathway. However, the scientists also recovered transgenic plants with yellow endosperm from experiment 1. Subsequent chemical analysis confirmed the presence of β-carotene, but no lycopene. This finding suggests that enzyme 4 may be present in rice endosperm naturally, or it could be induced by lycopene to turn lycopene into β-carotene. Analysis also showed the presence of lutein and zeaxanthin, both products derived from lycopene. None of the above was found in the control (nonengineered) plants.

How Much Vitamin A?

The initial golden rice lines produced 1.6 to 2.0 micrograms of β-carotene per gram of grain. The recommended daily allowance (RDA) set by health agencies for children is 0.3 mg/day. Estimates of bioavailability of β-carotene has been put at less than 10 percent in some cases. Activists like Greenpeace continue to strongly protest what they believe to be a hype and overstatement of the benefits of this biological invention. The organization estimates that an adult would have to eat about 9 kg of cooked rice daily to satisfy his or her daily need of vitamin A. However, the scientists intend to refine their invention to make it produce three to five times its present level of β-carotene.

Problems with Intellectual Property

Upon an international intellectual property rights (IPR) audit commissioned by the Rockefeller Foundation through the International Service for the Acquisition of Agri-Biotech Applications, Potrykus and his team realized their invention utilized 70 IPRs and TPRs (technical property rights) owned by 32 different companies and universities. Because of the humanitarian goal of the project, the development team negotiated with owners of these patents to allow the use of their inventions under the "freedom to operate" clause. Furthermore, because of public pressure and the need for big business to tone down their profit-oriented public image, the key companies (e.g., Monsanto) offered free licenses for their IPRs involved with golden rice.

What Next?

It might take an estimated additional 5 to 10 years before the ordinary person meant to benefit from golden rice can produce it. The characteristics of the present genotype must be bred into as many locally adapted varieties and ecotypes in as many rice-growing countries as quickly as possible. It is also important that such breeding efforts be organized such that all rules and regulations concerning the handling and use of genetically modified organisms be strictly followed to avoid stirring up additional controversies. Agronomic and other studies will have to be conducted to determine how well golden rice yields, its palatability and digestibility, and public acceptance.

The next phase requires additional funds. Because of public protests against the product for a variety of reasons, public funding began to quickly dwindle. In January of 2001, a new effort was launched in the Philippines for a comprehensive set of tests to determine the efficiency, safety, and usefulness of golden rice for people in the developing world. The joint effort includes the Philippines-based International Rice Research Institute (IRRI), Syngenta, and the Rockefeller Foundation. Furthermore, IRRI has set up a Humanitarian Board to oversee this project and to ensure that the highest standards for testing, safety, and support are achieved. The Board includes several public and private organizations, like the World Bank, Cornell University, the Indo-Swiss Collaboration in Biotechnology, and the Rockefeller Foundation.

OTHER SELECTED APPLICATIONS

Other applications of biotechnology in crop plants include the following:

1. **Cultivar identification** technologies have been developed. In the event of patent infringement, it is incumbent upon the plaintiff to demonstrate that the cultivar in litigation is identical to the protected product. Plant cultivar identification is beneficial to consumers, as well as helpful in protecting plant breeders' rights. It will help

to guarantee the authenticity of cultivars that farmers and end users purchase. A customer can check to see if the vendor supplied the correct cultivar with the desirable qualities purchased.

2. **Detection of biotechnology content** of grains and processed foods is needed. In the face of mounting pressure from activists for biotechnology commodities to be labeled as such, there is the need to conduct tests that will detect genetically engineered products in the commodity supply chain. Secondly, with the emergence of value-enhanced products (e.g., high-oil corn) that may or may not be bioengineered, there is the need to institute rapid, accurate, and economical tests to preserve specific end-use characteristics.

3. **Genetic engineering of flower color** is accomplished by two main approaches: (1) production of novel pigments (previously nonexistent pigments) by direct gene transfer, thereby expressing pigment biosynthesis genes from a heterologous species in a target plant; and (2) selective inhibition of particular flavonoid gene(s), thereby obstructing the pigment biosynthesis pathway, leading to accumulation of certain previously limited flavonoid pigments.

4. **Genetic engineering stress resistance** is being pursued by various research groups. Some of the major objectives in engineering stress resistance are (1) to protect plants from water deficit or drought and osmotic stress through modification of the synthesis of compatible solutes (e.g., mannitol and proline), (2) to protect plants from freezing injury by constitutively activating cold responses, (3) to protect plants from aluminum toxicity (aluminum is the most abundant metal ion in the rhizosphere but is not required by plants), and (4) protection of plants against oxidative damage (e.g., by engineering plants to overexpress glutathione-S-transferases (enzymes involved in the removal of toxic compounds that accumulate in plant cells)).

COMMERCIAL PLANT PRODUCTS

Several major field crops have been bioengineered to express certain important genes. Some of the currently available commercial products are presented in Table 16–2. There are many other products in the pipeline. These products are expected to be commercially available within the next decade. They include products that have been engineered for either input traits (e.g., incorporating genes for disease resistance, frost tolerance, and stacking of traits, that is, expressing more than one gene in the plant) as well as output traits (affecting food and feed quality, value-added traits, and chemical production). Some of these upcoming products are presented in Table 16–3.

KEY CONCEPTS

1. Many applications of biotechnology to crop production have been made, with most of the first generation efforts being directed at solving problems that benefit the producer directly (e.g., herbicide resistance, insect resistance).

2. The microbial frost inhibition system (ice minus) was the first genetically modified organism to be field-tested.

3. Goals in engineering fruit ripening by reducing ethylene biosynthesis using antisense RNA technology for ACC synthase and ACC oxidase have been embarked upon with varying success.

TABLE 16–2
Selected genetically modified products approved for commercial production.

Company	Product
AgrEvo	LibertyLink® corn—resistant to Liberty® herbicide
AgrEvo	StarLink™ corn—resistant to a variety of lepidopteran pests
American Cyanamid	SMART® canola seed—tolerant to imidazolinone herbicides
Calgene	Laurical®—rapeseed with high laurate
Monsanto	Bollgard® insect-protected cotton—protection against cotton bollworms and others
Monsanto	Roundup® Ready cotton—resistant to Roundup® herbicide
Monsanto	YieldGard® insect-protected corn—protection against European corn borer and others
Mycogen	NatureGard® hybrid seed corn—protection against European corn borer
Mycogen	High oleic sunflower
Novartis	NK Knockout™ corn—protection against various pests
Novartis	Attribute™ *Bt* sweet corn—protection against the European corn borer
Zeneca	Increased pectin tomato—remain firm longer and retain pectin during processing into paste
Hansen	Chymogen®—biotech version of chymosin for cheese production
Monsanto	Prosilac® bovine somatotropin—rBST for improved milk production

TABLE 16–3
Some genetically modified products in the pipeline. Some products are awaiting approval from the regulatory agencies, whereas others are in various stages of development and testing.

Company	Product
Agrecetus	Cotton fiber with enhanced performance
LibertyLink®	Soybean, cotton, and sugar beet resistant to Liberty® herbicide
American Cyanamid	CLEARFIELD® wheat, rice, and sugar beet tolerant to imidazoline herbicides
Calgene	High sweetness tomato, high-stearate oil
AquaAdvantage®	Salmon, tilapia, trout, and flounder with enhanced growth rate
EDEN Bioscience	Messanger™ with harpin protein against various diseases
Monsanto	High solids potato, Roundup® Ready canola, sugarbeet
Zeneca Plant Sciences	Resistance of banana to Black Sigatoka, modifying ripening of banana
Mycogen Corp.	*Bt* sunflower, wheat, canola

4. The Flavr Savr tomato was the first bioengineered food crop to be marketed. It contained a gene for repressing the expression of polygalacturonase by antisense RNA technology.
5. *Bt* products are genetically modified crops designed to express the *Bt* endotoxin conditioned by a gene obtained from the bacterium *Bacillus thuringiensis* for protection against lepidopteran pests.
6. Common *Bt* products on the market are corn, soybean, and cotton.
7. Herbicide-resistant GM crops with resistance against major herbicides (e.g., Monsanto's glyphosate) are available on the market.
8. The second generation of GM products appear to address consumer benefits directly (e.g., nutritional quality as exemplified by the golden rice).

OUTCOMES ASSESSMENT

1. Discuss the rationale of engineering microbial frost inhibition for commercial application.
2. Discuss the two basic approaches to genetic engineering of insect resistance.
3. Discuss the various types of *Bt* endotoxins and their applications in developing GM crops.
4. Discuss the strategies of molecular genetic engineering of herbicide resistance.
5. Discuss the development of golden rice.

INTERNET RESOURCES

1. CIMMYT report—application of biotechnology to maize: *http://www.cimmyt.org/Resources/archive/What_is_CIMMYT/Annual_Reports/AR95-96/htm/AR95-96biotech.htm*
2. Devoted to discussion on application of biotech: *http://www.biotech-info.net/rating-biotech.html*
3. Case for biotech application: *http://www.biotech-info.net/battling_hunger.html*
4. Global review of biotech adoption of transgenic crops: *http://www.isaaa.org/publications/briefs/Brief_12.htm*

REFERENCES AND SUGGESTED READING

Krimsky, S. 1995. The genetics and politics of frost control. In: *The biotechnology revolution?* M. Fransman, G. June, and A. Roobeek (eds). Oxford, UK: Blackwell.
Lindow, S. E. 1983. Methods of preventing frost injury caused by epiphytic ice-nucleation-active bacteria. *Plant Disease, 67(3):*327–333.
Lindow, S. E. 1985. Ecology of *Pseudomonas syringae* relevant to the field use of ice-deletion mutants constructed *in vitro* for plant frost control. In: *Engineered organism in the environment: Scientific issues.* H. O. Halvorson, D. Pramer, and M. Rogul (eds). Washington, DC: American Society for Microbiology, 23–35.

SECTION 2
Animals and Microbes

PURPOSE AND EXPECTED OUTCOMES

Biotechnology applications in animal agriculture have had their fair share of the headlines. The development and use of growth hormones, as well as the case of "Dolly" the sheep, drew considerable press. The application of bacteria in various food technologies has its origins in antiquity.

In this section, you will learn:

1. Selected major applications of biotechnology in food animal production.
2. Selected major applications of microbes in food product development.

OVERVIEW OF ANIMAL FOOD BIOTECHNOLOGY

Genetic manipulation of animals has certain unique challenges. For example, animal cells generally lack the totipotency of plant cells, that is, they are not capable of regenerating from a single somatic cell into a fully differentiated animal. Consequently, researchers are unable to take advantage of low-frequency transformation methods that rely on chemical selection and regeneration. Instead, they often utilize the technically challenging and low-throughput method of direct injection of DNA into the nuclei of the cells of the target organism at a very early stage of development. However, current advances in the technology of animal cell manipulations such as somatic cell nuclear transfer are gradually overcoming the limitations of pronuclear microinjection.

Other challenges include the fact that reproductive technologies in large animals are less developed and less efficient as compared to plants. Unlike plants in which populations are rapidly replaced, populations in animals are slow to replace (except in fish and poultry), as they are more localized and diffuse. Consequently, it is more challenging to have rapid and large-scale market penetration by GM animal genotypes as has been the case with plants. This constraint adversely impacts investments in GM animals for farm production.

Animal food biotechnology generally is focused on either altering the genome of animals to achieve higher yield of specific products (e.g., milk, meat), or enhancing certain characteristics of products (e.g., reduced fat, reduced cholesterol content). The application of reproductive biotechnology to generate clones of animals without genetic modification is a slow and much more expensive proposition than the use of micropropagation to clonally produce plants on a large scale. Except for transgenic fish that appear to be close to becoming commercially marketable, most of the efforts at developing transgenic animals are still in the laboratory phase. One of the widely commercialized applications of animal food biotechnology is the growth hormone. Other applications that are being pursued include altered carcass composition in meat producing and compositional modification of milk and eggs.

ENGINEERING GROWTH HORMONES

One of the earliest applications of biotechnology that drew headlines was the development and use of a growth hormone called **bovine somatotropin (bST). Somatotropins** are protein (not steroid) growth hormones that are species-specific in their effect. Cows, humans, pigs, mice, and other organisms produce their own unique somatotropins, which are not cross active. However, the somatotropin of a "higher" order species (e.g., cow) may be active in a "lower" order species (e.g., mice) but not humans (which is of a higher order than cows). Bovine somatotropin is found in cows, in which it plays a significant role in growth and development. As early as the 1930s, scientists discovered that milk production in lactating cows could be significantly boosted by supplementing the natural production of the bST in cows with an artificial injection of additional amounts of bST. Between 1954 and 1987, the use of bST in the dairy industry dramatically boosted milk production from 5,400 pounds per cow to 13,700 pounds per cow per year. The supplemental dose of bST was obtained from the pituitary glands of slaughtered cattle. This source of growth hormones yielded limited amounts of bST, consequently making artificial supplementation expensive to farmers.

Enter biotechnology. The gene that encodes bST production was isolated and inserted into *E. coli*. The transgenic bacteria are cultured under controlled environments in fermentation chambers to produce large quantities of bST. The protein is isolated and purified for use in the dairy industry. The Food and Drug Administration has approved it for this purpose. It is administered to cattle intravenously at periodic intervals, the response being higher when cows have been lactating for about 100 days rather than soon after calving. Furthermore, it takes about six days after injection to realize the maximum effect of the hormone. To meet this increased production, the cows increase their feed intake by about 10 to 20 percent. Recorded increase in milk yield as a result of bST supplementation ranged from 8.3 to 21.8 percent in research studies.

In terms of milk composition, bST supplementation is not known to cause significant deviation from milk from untreated cows, regarding common constituents such as free fatty acids, cholesterol, and flavor. Minor variations in milk fat and protein in the early parts of treatment are eliminated after about a month. Humans break down both natural and biotech-produced bST into inactive products upon digestion. Administration of bST directly to humans as a growth hormone to promote growth in children was ineffective. Humans and cows differ in about 30 percent of their amino acids in their somatotropins.

The USDA estimates about a 2 to 5 percent increase in national milk production as a result of the bST technology. Producers with better-managed operations would benefit more from this technique than others. The FDA gave approval to the Monsanto Company of St. Louis, Missouri, the company that developed the biotechnology growth hormone, to market it in 1993 as a drug under the trade name Prosilac®.

Like all new technologies, the use of bST attracts some opposition from the dairy industry as well as special interest groups. Some dairy producers are concerned about possible adverse economic impacts from overproduction of milk and competitive disadvantage from producers who are able to afford the technology (economies of scale work against the small producer). Other activists feel biotechnology products are unsafe for human consumption.

Porcine somatotropin (pST) is the pig equivalent of bST. It is produced and administered just like bST. The goal of the use of pST in the pig industry is to reduce back fat. Some research data show nearly 25 percent reduction in back fat and about 10 percent gain in muscle. The amount of feed consumed per unit of weight gained decreased by about 20 percent, while the average growth rate was about 15 percent faster than in untreated pigs. This means the use of pST requires nutrient supplementation in terms of proteins and amino acids in the diet. Unlike bST, FDA approval was still pending by 2001.

CLONING ANIMALS

Cloning of plants is now scientifically a routine process. Scientists working with animals have long sought to clone mammals. Asexual production in animals is not common (a few known examples include the starfish and some invertebrates). Attempts to clone animals have been made in frogs with varying success. Embryonic tissues are split (called **embryo twinning**) to provide clones, commonly in cattle. So what was unique about "Dolly" the sheep? The answer is simply the starting material. A mammal was cloned from nonembryonic cells (i.e., somatic cells)—in other words, adult cell cloning. The challenge was deprogramming a differentiated cell, then reprogramming it to develop into a full organism.

Such a feat was accomplished in 1997 by researchers at the Roslin Institute in Scotland, led by Dr. Ian Wilmut. The target cells to be cloned were obtained from the udder of a pregnant Finn Dorset ewe. These cells were cultured on a medium that was low in concentration of nutrients. Consequently, the cells were induced into a state of inactivity (ceasing to divide). Next, an unfertilized egg (oocyte) was obtained from a Scottish Blackface ewe. The nucleus of the egg was removed completely, leaving the surrounding cytoplasm with all the capabilities to produce an embryo. The two cells were placed side by side (see Figure 16–3). Fusion was induced by the application of an electric pulse, which also induced cell division. The fused cells began to divide and, after about six days, the embryo was implanted into the uterus of another Blackface ewe, the surrogate mother. Out of 277 attempts, only 29 embryos were able to survive longer than six days.

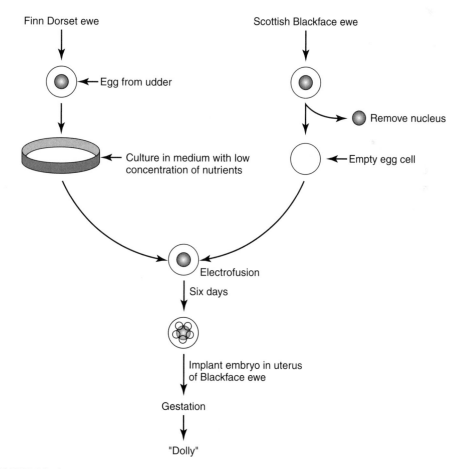

FIGURE 16–3
A summary of the steps in the making of "Dolly" the sheep. The procedure is also described as nuclear transfer. An unfertilized egg from the Blackface ewe is extracted and the nucleus removed to create an empty egg cell. A cell from the udder of the Finn Dorset ewe is removed and cultured. The empty egg cell and the cultured cell are united through electrofusion. The resulting embryo is implanted in a surrogate female to full gestation.

"Dolly" was born after the normal gestation, the first cloned mammal from nonembryonic cells. All the other embryos failed to reach birth.

Since this historic event, other researchers have announced duplication of the feat in other mammals. Furthermore, some have developed clones from cloned animals. Yet, other scientists are contemplating doing the ultimate, cloning humans!! This has stirred up a storm of controversy, prompting various governments to preempt such research by imposing a ban on human cloning.

APPLICATIONS IN AQUACULTURE

One of the major areas of emphasis in the development of transgenic fish has so far been the development of enhanced phenotypes for the aquaculture industry. The first report on transgenic development in fish occurred in 1985. Since then, transgenic fish development has been reported to occur in several dozen fish species including rainbow trout (*Oncorhynchus mykiss*), Atlantic salmon (*Salmo salar*), coho salmon (*O. kisutch*), channel catfish (*Ictalurus punctatus*), zebrafish (*Danio rerio*), common carp (*Cyprinus carpo*), tilapia (*Oreochromis niloticus*), goldfish (*Carasius auratus*), and striped bass (*Micropterus salmoides*). The first commercial transgenic fish in North America was the Atlantic salmon, which was engineered for enhanced growth.

The primary method of transgene delivery in fish is microinjection into the cytoplasm of the developing embryo. Growth enhancement often occurs at a rate averaging between 200 and 600 percent, depending on the species. However, the transgenic fish do not attain final sizes that are significantly larger than the nontransgenic fish. The increased growth rates decrease the time required to rear the fish to market sizes. Another application being pursued is the transfer of the so-called antifreeze gene from marine fish such as winter flounder (*Pseudopleuronectes americanus*) to commercially viable fish for aquaculture production.

ENHANCING THE USE OF FEED BY ANIMALS

Certain parts of the digestive tract of animals are known to harbor microflora that aid in the digestion of food. This microbial community is subject to the harsh environment of the gut that is sometimes exacerbated by the introduction of orally administered therapeutics. One of the applications of biotechnology is to engineer resistance to various compounds into the major gut bacteria. The transfer of a gene into the rumen bacteria was achieved in 1988 and involved the transfer of the tetracycline-resistant Tc^R gene into *Prevotella ruminicola*. Other recent genetic modifications include the engineering of acid tolerance into cellulolytic rumen bacteria. In addition to these applications, gene transfer for cellulase activity has been made into hind-gut bacteria, while rumen bacteria have received genes to improve the protein yield (essential amino acid) and hydrogen scavenging (to reduce methanogenesis in rumen bacteria). The major limitation to such applications of biotechnology is being able to establish these engineered microbes in the appropriate regions of the digestive tract.

PROBIOTICS

Probiotics is the supplementation of animal feed with beneficial, live microbes to improve the intestinal microbial balance for better utilization of feed and for good health. The added bacteria may improve digestion of feed and absorption of nutrients, stimulate immunity to diseases, or inhibit the growth of harmful microbes. Research in this area is focused, among other goals, on bioengineering bacteria that can detoxify naturally occurring toxins that threaten livestock and also bacteria that are more efficient at digesting fiber in the digestive tract.

MICROBIAL APPLICATIONS IN DEVELOPMENT OF FOOD PRODUCTS

Microbial biotechnology has evolved through the ages from a period in which microbes were used without any understanding of the underlying science. The Babylonians and Sumerians were noted to have brewed alcoholic beverages using the technology of fermentation. Primitive applications of biotechnology were product-oriented, primarily for making fermented foods (alcoholic beverages, leavened bakery products, cheese, yoghurt, and others). Through the years, the methods have been significantly improved to become more efficient and versatile, based on scientific information. Nonetheless, primitive techniques continue to be used, especially in the developing world. The state-of-the-art technology entails the deliberate engineering of the microbial genome for specific goals.

■ TRADITIONAL MICROBIOLOGICAL INDUSTRIES

Selected major traditional microbiological industries include the production of alcoholic beverages, baking, and cheese preparation.

Alcoholic Beverages

Alcoholic beverages represent, perhaps, the single most important group of biotechnology products in terms of world value. Ethanol is the alcohol in alcoholic beverages. These beverages may either derive directly from fermentation (e.g., wines and beers), or may be products of the distillation of fermented products (called spirits). Spirits have higher alcoholic content than wines and beers. The primary ingredient in an alcoholic beverage industry is a carbohydrate source, which predominantly is of plant origin. Major plant sources of fermentable sugars (mainly glucose, fructose, and sucrose) are fruits (e.g., grapes) and plant storage organs or modified plant parts like stems (e.g., sugar cane) and roots (e.g., sugarbeet). Cereal crops (e.g., barley, wheat, rice, sorghum, corn, millet) and roots (e.g., potatoes) are sources of polysaccharides (mainly starch) in various cultures in the world.

Alcoholic beverages are produced using yeast fermentation, involving *Saccharomyces cerevisiae* or *S. uvarum* as the principal yeast species. The fermentation reactions may be summarized as follows:

$$C_6H_{12}O_6 + 2ADP + 2Pi - \rightarrow 2C_2H_5OH + 2CO_2 + 2ATP + 2H_2O$$

Since the reaction represents anaerobic metabolism by the microbe, the less efficient this energy generating process, the higher the production of alcohol (which is a by-product of yeast fermentation). It should be pointed out that yeast is not a true anaerobic organism. It requires oxygen for the synthesis of unsaturated fatty acids and sterols that are key components of the membranes of yeast cells. Because of the oxygen-deficient conditions of fermentation, yeast is not able to synthesize these cellular components. Producers may include sources of sterols such as ergosterol and unsaturated fatty acids in the fermentation tank when the duration of the fermentation process is protracted. In cases where a continuous fermentation production is being undertaken, the medium may be aerated with small amounts of oxygen. Similarly, the fermentation medium may be enriched with amino acids and peptides to compensate for the low proteolytic activity of yeasts.

Apart from ethanol, yeasts produce a variety of compounds that either enhance the flavor of the product or taint it. Yeasts can form sulfur compounds (hydrogen sulfide and sulfur dioxide) that contribute to foul aromas. The production process allows yeast to be collected and reused as inoculum in subsequent fermentation reactions. *S. cerevisiae* tends to rise to the top of the fermentation tank and is called a top fermenter. *S. uvarum* is a bottom fermenter because it sinks to the bottom of the fermenting vessel in beer production.

Microbial involvement in wine production is similar to its role in beer production as previously described. However, there are certain key differences. Wine is produced from the berries of grapes (*Vitis* spp.). It is important that the berries be crushed immediately after picking to avoid infestation by microbes introduced by the fruit fly. These microbes can start undesirable fermentation reactions. Crushed grapes contain numerous yeasts and bacteria, some of which are undesirable (e.g., those of the genera *Brettanomyces, Candida, Hansenula, Kloeckera,* and *Pichia*). These undesirable microbes produce acetic acid and lactic acid. Their growth can be suppressed by introducing sulfur dioxide into the crushed grapes. The strain of yeast used to inoculate the fermentation tank is critical to the flavor produced in the final product. Commonly, wine producers use pure cultures of *S. cerevisiae* var. *ellipsoids* in batch fermentation systems. Batch fermentation yields higher quality wine.

Grape juice contains two major natural acids—D-tartrate and L-malate. Small amounts of succinate acid are introduced from the yeast used in fermentation. Contaminating lactic acid bacteria that decarboxylate malate to lactate by the process called **malolactic fermentation** can reduce the natural acidity of wine. However, this reaction is critical to the production of red wine in certain cultures where cooler environments prevail during production. Malolactic fermentation is difficult to control. Any strategy employed to control it should be done with caution to avoid destroying the desirable microbe, *Leuconostoc oenos*. The major undesirable lactic acid bacteria include *Lactobacillus brevis, L. plantarum,* and *Pediococcus cerevisiae*.

A key objective in the production of spirits is to maximize the yield of ethanol from fermentation. The production environment is intolerable to most contaminating microbes. Heavy contamination by lactic acid bacteria may reduce yeast growth and consequently ethanol yield. A distinction is made between brewer's yeast, distilling yeast, and baker's yeast.

Baker's Yeast

A wide variety of bakery products require the use of baker's yeast to leaven the dough to make the product light. This modern practice has its origins in antiquity. The leavening of dough is the result of CO_2 accumulation from the fermentation of sugar. Baker's yeast is commercially produced by culturing under highly aerobic conditions that maximize its growth. It is started on agar slants, then transferred into a liquid medium. The yeast is concentrated by centrifugation and used to inoculate a larger vessel. The yeast from this vessel is in turn used to seed other vessels.

Cheese

The art of cheese making, like the use of yeast, has ancient roots, where milk was converted to clots (curds) by exposure to warm conditions. The science underlying cheese making is the conversion of the protein and fat of the milk to a form that resists the rapid spoilage (putrification) associated with fresh unrefrigerated milk. Lactic acid bacteria (undesirable in alcoholic beverage production) are encouraged to grow to convert lactose (milk sugar) to lactic acid. This lowers the pH of the product and thereby prevents the growth of putrefying bacteria and other pathogens for a long period. The formation of curd (which is separated from the fluid or whey) is the result of an enzymatic process. A protease called chymosin is added to the milk to disrupt the micellar structure of casein (the major milk protein). The products of this activity are insoluble at acidic conditions resulting from bacterial metabolism, causing the individual polypeptides to precipitate to form curds. The curds are usually further processed to yield various products with varying tastes and aromas, as well as physical characteristics (especially hardness). The fresh curds are subjected to a period of maturation in which a secondary set of bacteria or fungi operate to develop characteristic aromas and appearances.

Many microorganisms are involved in cheese making. Those included at the start (called starter bacteria) include *Streptococcus lactis, S. cremoris, S. thermophilus, Lactobacillus acidophilus, L. casei,* and *L. lactis*. The primary role of these organisms is to ferment lactose to produce lactic acid, and thereby create acidic conditions (pH of about 4.6) for curd formation. The secondary microbial flora (called **finishing organisms**) that function to produce various types of cheese include *Propionibacterium freundrichii* spp., and *Shermanii,* which is involved in Swiss cheese making. This bacterium produces propionic acid for flavoring the cheese, and the CO_2 gas that forms the bubbles in the cheese, resulting in the characteristic holes in Swiss cheese. Roquefort cheese depends on the fungus *Penicillium roquefortii* that is responsible for the blue-green veining.

The starter bacteria (lactic acid bacteria) are susceptible to phage attack, which can be devastating to the fermentation process. Because bacteriophages are highly host-specific, the buildup of one strain can be discouraged by adopting a system of starter culture rotation.

Yoghurt

Yoghurt is a fermented and acidified milk product that originated in the Middle East. Its introduction into Western culture has resulted in a variety of modifications, especially in flavors and color. The starter culture consists of an equal mixture of two bacteria with symbiotic relationships, *Lactobacillus bulgaricus* and *Streptococcus thermophilus*. The milk is heated to kill unwanted bacteria as well as promote protein aggregation.

Food Fermented with Salt

The addition of salt to the fermentation medium discourages the growth of pathogenic microbes like *Clostridium botulinum* and *Staphylococcus aureus*. Fermented foods, with and without the addition of salt, are common in tropical regions where food spoilage is rapid due to high temperatures and the lack of refrigeration because of pervasive poverty. Some of the widely known foods fermented with salt and the principal microbe is used include the following: kenkey, a corn product from Ghana (*Saccharomyces* spp.); gari, a cassava product common to West Africa (*Candida* spp.); soy sauce, a soybean product (*Aspergillus oryzae, Pediococcus soyae*); pickles, a cucumber product (*Lactobacillus* spp.); and sauerkraut, a cabbage product (*Lactobacillus* spp., *Leuconostoc* spp.).

Citric Acid

Citric acid may be derived from fruits like lemon and pineapple wastes. However, about 99 percent of commercially available citric acid is a fermentation product. Citric acid is used primarily in the food, beverage, and confectionery industries, and also in the pharmaceutical industry. It is used to flavor jams, confections, and soft drinks. It is used as an antioxidant in foods. Citric acid is used in the detergent industry among other applications. The microbe used in the fermentation process is the *Aspergillus niger*. The process involves both glycolysis and the tricarboxylic acid cycle (TCA). The key is to accumulate citrate and minimize its oxidation in the TCA cycle. For accumulation to occur, the culture medium must be free from certain metallic ions (Cu, Fe, Mg, Mn, Mo, Zn), or have a low concentration of these ions.

KEY CONCEPTS

1. Genetic manipulation of animals has certain unique challenges. For example, animal cells generally lack the totipotency of plant cells, that is, they are not capable of regenerating from a single somatic cell into a fully differentiated animal.

2. Other challenges of genetic manipulation include the fact that reproductive technologies in large animals are less developed and less efficient as compared to plants.

3. Animal food biotechnology generally is focused on either altering the genome of animals to achieve higher yield of specific products (e.g., milk, meat), or enhancing certain characteristics of products (e.g., reduced fat, reduced cholesterol content).

4. The application of reproductive biotechnology to generate clones of animals without genetic modification is a slow and much more expensive proposition than the use of micropropagation to clonally produce plants on a large scale.

5. One of the widely commercialized applications of animal food biotechnology is the growth hormone. Other applications that are being pursued include altered carcass composition in meat producing and compositional modification of milk and eggs.

6. One of the earliest applications of biotechnology that drew headlines was the development and use of a growth hormone called bovine somatotropin (bST).

7. Growth enhancement of fish often occurs at a rate averaging between 200 and 600 percent, depending on the species. However, the transgenic fish do not attain final sizes that are significantly larger than the nontransgenic fish.

8. Another application in aquaculture being pursued is the transfer of the so-called antifreeze gene from marine fish such as winter flounder (*Pseudopleuronectes americanus*) to commercially viable fish for aquaculture production.

9. One of the applications of biotechnology is to engineer resistance to various compounds into the major gut bacteria.

10. Selected major traditional microbiological industries include the production of alcoholic beverages, baking, and cheese preparation.

OUTCOMES ASSESSMENT

1. Discuss the use of genetically engineered growth hormones in animal production.
2. Give three specific major challenges that face scientists in the application of biotechnology in the animal production industry.
3. Discuss the potential of commercial application of cloning in the animal production industry.
4. Describe one major commercial application of biotechnology in aquaculture.
5. Give an overview of the use of microbes to make food products for both humans and livestock.

INTERNET RESOURCES

1. Application of biotech to animal health: *http://www.accessexcellence.org/AB/BA/Animal_Health_Update.html*
2. Transgenic fish: *http://pewagbiotech.org/buzz/index.php3?IssueID= 10*
3. Roslin Institute biotech: *http://www.ri.bbsrc.ac.uk/molbiol/mcwhir/applicat.htm*
4. FDA writer on animal biotech: *http://www.fda.gov/fdac/features/2001/101_fish.html*
5. bST issues: *http://www.nal.usda.gov/bic/BST/Other/other.html*

REFERENCES AND SUGGESTED READING

Hammer, R. E., V. G. Pursel, C. E. Rexroad, Jr., R. J. Wall, D. J. Bolt, K. M. Ebert, R. D. Palmiter, and R. L. Brinster. 1985. Production of transgenic rabbits, sheep and pigs by microinjection. *Nature, 315:*680–683.

Wilmut, I., A. E. Schnieke, J. McWhir, A. J. Kind, and K. H. S. Campbell. 1997. Viable offspring derived from fetal and adult mammalian cells. *Nature, 385:*810–813.

Human Health and Diagnostics

SECTION 1
Applications in Medicine

PURPOSE AND EXPECTED OUTCOMES

Biotechnology applications with bearing on human health occur in the areas of therapeutics, diagnosis, and counseling. Some recent applications seek to directly manipulate humans. It is these latter types of applications that have stirred up considerable public concern.

In this section, you will learn:

1. Selected classic applications of biotechnology in human health issues.
2. Selected major applications of biotechnology in human health issues.

GENE THERAPY

Gene therapy is the molecular biotechnological technique of correcting genetic disorders by replacing defective genes with functional or normal ones. This procedure in effect is equivalent to the reverse of gene knockout technology. There are two basic approaches to correcting genetic disorders:

1. **Replacement therapy**

    The replacement therapy strategy involves an addition of an intact fully functional gene to the patient's genome to make up for the defective one. Gene replacement does not necessarily replace all the normal expression signals.

2. **Targeted gene repair**

    In this strategy, attempts are made to use molecular toolkits to correct mutations in the genome. The procedures employed do not disturb the stretches of DNA flanking the faulty gene that are involved in the regulation of the gene. Consequently, after the repair is completed, the cell is able to function as it was originally intended to. So far, successes have been realized in animals for gene defects caused by mutations involving insertions, deletions, and substitutions of a few nucleotides.

The cells that multiply throughout the patient's life (e.g., bone marrow cells) are the ones that are subjected to manipulation in this technique. The technique is most readily applicable to disorders that are conditioned by single defective genes. Even though the technology of gene therapy originally targeted hereditary diseases, gene therapies for the prevention and treatment of infectious diseases, cancer, and heart disease are being developed. The strategies are the same—reliance on the gene's ability to produce a key protein when and where it is needed.

■ FORMS OF GENE THERAPY

There are two forms of the gene therapy:

1. **Somatic cell gene therapy**

   Genetic manipulation in this form of gene therapy is limited to the somatic (body) cells and is therefore not transmissible from the patient to his or her offspring. This is currently the form of research that can be supported with federal funds.

2. **Germline gene therapy**

   This form of gene therapy is heritable and conducted on germline or sex cells (sperm and egg). If performed early in the individual's life (e.g., at the embryonic stage), it is possible that the individual could be treated in addition to his or her offspring.

■ METHODS OF GENE THERAPY

There are two general methods of correcting defective genes in organisms.

1. *Ex vivo*

   The *ex vivo* approach entails the manipulation of the patient's cells in laboratory containers and transplanting the amended cells back into the patient. The general steps in this approach are as follows:

   a. The normal gene to be transferred is cloned in a vector. One of the earliest and most common vectors for gene transfer is the genetically engineered Maloney murine leukemia virus in which the *gag, pol,* and *env* genes are deleted from the viral genome and replaced by human genes. These missing genes prevent viral replication. The cloned target gene is packaged into the viral protein coat.

   b. This vector is mixed with a suspension of target cells (e.g., bone marrow cells). The vector enters by a cell-surface receptor and then integrates the human gene into one of the chromosomes in the target cell.

   c. The engineered cells are injected into the patient. The cell with the incorporated normal gene will multiply throughout the life of the patient, expressing the missing protein and effecting a cure of the disorder.

2. *In vivo*

   In vivo gene therapy entails the use of gene delivery systems (e.g., vectors or liposomes) to deliver corrected genes into the individual directly. Liposomes are artificially produced in the laboratory. DNA is mixed with liposomes so that synthetic lipid membranes encapsulate the DNA. The DNA is released into the cell when this artificial membrane fuses with the cell plasma membrane.

TOOLS FOR GENE THERAPY

The tools for gene therapy are basically the delivery systems and the sequences delivered.

■ VIRAL DELIVERY SYSTEMS

One of the oldest and most dominant methods in the molecular toolkit for gene therapy is the viral vector gene delivery system that is used in about 75 percent of all gene therapy protocols. There are four common types of viral vector systems in use.

1. **Retroviruses**

   One of the commonly used retroviruses is the Moloney retrovirus. The virus inserts its genes into the host's genome and hence any genes deliberately inserted into this vector are expressed for a long period. Retroviral systems are efficient and generally do not produce a strong immune response. However, they are active only in dividing cells and also integrate their DNA randomly. This property of random location gives the system the potential to promote disease (e.g., cancer), should it locate in a wrong place in the genome.

2. **Adenoviruses**

Adenoviruses are easy to grow. They also readily infect both dividing and nondividing cells. Furthermore, they can express their genes without necessarily inserting themselves into the genome of the host or posing a cancer risk. However, adenoviruses produce proteins that stimulate strong immune reactions, which sooner than later eliminate the vector from the host body. Consequently, the genes they carry express only for a short time, making it necessary for high doses to transfer enough genes to be of benefit to the patient. Unfortunately, such high doses administered intravenously can have very toxic consequences. Nonetheless, the use of adenoviruses in gene therapy is increasing as the use of retroviruses decrease.

Such an event is suspected to have occurred in 2002, according to a revelation by a French gene-therapy team. The scientists, Alain Fischer and Marina Cavazzana-Calvo are noted for their breakthrough curing of children with a lethal immune deficiency, in 2000. Unfortunately, one of 10 patients is suspected to have developed a blood disorder resembling leukemia, following gene therapy with the Moloney retrovirus as vector. They speculated that the vector triggered an insertional mutagenesis event by splicing itself into a gene on Chromosome 11 known to be aberrantly expressed in a form of childhood acute lymphoblastic leukemia, stepping up its production.

3. **Adeno-associated viruses**

Adeno-associated viruses infect both dividing and nondividing cells and are nearly invisible to the human immune system. They require an associated (helper) adenovirus to replicate. When they integrate into the host genome, they do so at a known location that is apparently safe. Because they have a small genome, adeno-associated viruses are restricted in the amount of therapeutic DNA they can carry. Furthermore, it is difficult to grow them to high concentrations.

4. **Lentiviruses**

The AIDS virus belongs to this class of slow-growing retroviruses. Lentiviruses are able to infect both dividing and nondividing cells without producing a strong immune reaction in the host.

■ TOOLS FOR TARGETED GENE REPAIRS

The tools for targeted gene repairs are in their infancy stage of development and application in gene therapy. There are several techniques being developed.

Triplex-forming Oligonucleotides

Triplex-forming oligonucleotides are single-stranded DNA molecules that, upon recognizing double-stranded DNA with identical or nearly identical sequences, engage them into forming a triplex structure. There are two variations of this technique. In one method, the oligonucleotide is linked to a snippet of double-stranded DNA that has the correct sequence of the defective gene. The double-stranded DNA inserts itself near where the oligonucleotides bind, thereby replacing the faulty gene. In another method, the oligonucleotide is used alone. The formation of the triplex eventually leads to the disruption in the activity of the mutant gene.

Homologous Recombination

This method is also called the small fragment homologous replacement and exploits the cell's ability to have the exchange of small stretches of DNA between chromosomes. Researchers introduce a fragment (400 to 800 bases long) of DNA that is complementary to the section of the gene in which the error is located, except at the site of the error. At this site, a nucleotide that is complementary to that which is supposed to be in the DNA sequence of the normal gene is inserted. Upon engaging its complement, a bulge is created in the synthetic oligomer, triggering the cell's DNA repair mechanism to replace that corresponding nucleotide in the defective gene with the correct nucleotide. Success with this methodology is very low.

Viral Gene Targeting

This is a variation of the adeno-associated virus system. Part of the normal gene is inserted into the single-stranded DNA of the virus and transferred into the host, where it is inserted into the genome. The host cell then utilizes this insert to repair errors in its genome.

STATUS OF GENE THERAPY

One of the earliest successes in gene therapy was recorded by French Anderson, who treated two girls with ADA (adenosine deaminase) deficiency. A defect in the gene that encodes this enzyme is the cause of an autosomal recessive form of SCID (severe combined immunodeficiency). SCID-affected individuals have no functioning immune system and consequently normally succumb to even minor infections. Individuals with ADA deficiency lack functional T and B cells. A subpopulation of T cells from the patient is isolated and mixed with vectors carrying the ADA gene. Following the viral infection, the T cells are cultured in the lab to ensure that the target gene is active, prior to injecting into the patient.

Other single-gene disorders being treated with gene therapy include cystic fibrosis, Duchene muscular dystrophy, and familial hypercholesterolemia. The last disease afflicts 1 in 500 individuals. Victims are unable to metabolize dietary fat, resulting in elevated blood cholesterol and eventually premature death from coronary heart disease. The condition arises because the defect produces a lack of cell-surface receptors that normally remove cholesterol from the bloodstream. The recombinant vector with the target gene is mixed with cells obtained by dissolving a lobe of the liver from an affected individual into cells. More efficient vectors are being developed that will have a large cloning capacity. These include adenoviruses and herpes viruses. Also, scientists are considering the development of a 47th chromosome as a large vector to coexist with the basic 46. Gyula Hadlaczky of Hungary pioneered the concept of the human artificial chromosome.

Gene therapy has so far been marginally successful. The development and application of new vectors is likely to increase the success rate. New target-cell strategies are also needed. At the moment, the technique is not applicable to cells that do not multiply through life (e.g., nerve cells). Other ethical issues hang over gene therapy. Some of these have a genetic basis. For example, it is feared that deleting unwanted alleles from a population is risky. A gene that is not important to survival of the species at this point in time might be beneficial under a new set of conditions. There is also the potential for abuse, whereby athletes, for example, might desire enhancements in physique, intellect, and ability in certain areas. The procedure is very expensive, raising questions as to who will have access to it and who will not. The age-old question of "playing God" can be asked here. Should we attempt to use cells in the germline as target cells to ensure heritable genetic alteration?

Gene therapy is more readily applicable to diseases conditioned by single genes. Most genetic disorders involve more than one gene, plus an interaction with other environmental factors.

CHALLENGES OF GENE THERAPY

The slow progress made with gene therapy is due to several reasons.

1. **Understanding gene function**

    Before genes can be manipulated, the researchers need to understand their function. With the current emphasis on functional genomics, the functions of genes are going to be elucidated to facilitate gene manipulation.

2. **Gene delivery**

    Scientists using viral delivery systems are aware of the possibility that surprises could arise anytime whereby the virus becomes pathogenic. There is the need to search for safer and more effective and efficient gene delivery systems for gene therapy.

3. **Multigenic traits**

    So far, success with gene therapy has been recorded for diseases conditioned by single genes. Many of the major diseases of interest are conditioned by multiple genes, not to mention the role of the environment and the interaction between genes and their environment. Methodologies need to be developed for handling these complex traits.

4. **High cost**

    The cost of gene therapy trials average over $50,000 per patient. This places the therapy out of the reach of many people.

DNA/RNA VACCINES

Vaccines are designed to protect an individual against invasion by certain pathogens. They are administered prior to infection and provide varying degrees of protection for varying durations of time. Several major debilitating diseases have been eradicated (e.g., smallpox) while the spread of several others has been significantly abated (e.g., hepatitis A, hepatitis B, measles, typhus, and tetanus) through vaccination.

■ IMMUNE RESPONSE

In order to understand how vaccines are genetically engineered, it is important to understand how the immune system works. An **antigen** (a contraction of antibody-generating) by definition is a molecule that elicits an immune response. These molecules are usually proteins, protein fragments, or polysaccharides that are unique to pathogens (viruses, bacteria) and also found on the surfaces of spores, pollen, house dust, cancer cells, and transplanted organs. The immune response is adaptive in the sense that it can "remember" (develop memory cells for) antigens it has encountered before and react against these foreign agents more promptly and vigorously in subsequent encounters. The host body may acquire immunity (resistance to specific invaders) naturally through natural infection or artificially through vaccination (injection of a vaccine). Immunity acquired by one of these modes is called **active immunity**. However, immunity may be described as **passive immunity** if antibodies (rather than antigens) are used. A fetus may be passively immunized by receiving antibodies from the mother's bloodstream. Travelers may also receive a short duration (few weeks or months) immunity from injections of antibodies against certain pathogens endemic in the countries of their destination.

The immune response is provided by lymphocytes (white blood cells) that originate from the bone marrow. Some immature lymphocytes develop into specialized cells called **B-lymphocytes** or **B cells**. Others are carried by the blood to the thymus gland where they develop into **T-lymphocytes** or **T cells** (also called **cytotoxic or killer T-lymphocytes**). The B and T cells both unite in the lymph nodes and other lymphatic organs (e.g., spleen). Both cells are involved in immune defense by two different modes. The B cells secrete antibodies into the blood, which is called **humoral immunity.** This immunity defends against pathogens in the body fluids (blood, lymph, intestinal fluids). This immunity can be passively transferred through blood transfusion. The T cells produce cell-mediated immunity. T cells also move in the blood and lymph and attack body cells that have been invaded by pathogens, killing the invading pathogens. They also fight infection caused by fungi and protozoa. They promote phagocytosis, but other lymphocytes also help B cells in producing antibodies. Cell-mediated immunity can be transferred only if a nonimmune person receives actual T cells. As both B and T cells develop, certain genes are turned on, causing the cell to synthesize specific protein molecules attached to the plasma membrane. The molecules are called antigen receptors. Each receptor can recognize a specific antigen and subsequently can provide an immune response against it. Antibodies identify and bind to localized regions on the antigen (called **antigene determinants**) that are recognized by specific regions on the antibody molecule (called **antigen building sites**). An antigen usually has several determinants so that different antibodies may bind to it. This way an antigene can stimulate the immune system to produce different antibodies against it.

■ *CONVENTIONAL VACCINES*

Conventional vaccines are prepared from killed or weakened versions or parts of pathogens. These materials act as antigens that provoke the host immune system to produce antibodies to fight what is perceived to be an invasion. By "priming" the immune system in this fashion, the host is ready to fight real live pathogens when they attack. Conventional vaccines offer varying degrees of protection for varying durations of time. Those prepared from killed pathogens (e.g., hepatitis A and injected polio vaccines) or from antigens isolated from disease-causing agents (e.g., hepatitis B subunit vaccine) provide humoral immunity. This is because they are unable to penetrate cells and cannot activate killer T cells. Consequently, conventional vaccines of this type are ineffective against many microbial pathogens that enter cells. Furthermore, their protection is not indefinite; that is why people need "booster shots" periodically.

On the other hand, attenuated (weakened) live vaccines have dual activity and are capable of penetrating cells to induce killer T cells to attack as well. Common vaccines manufactured this way include those for use against measles, mumps, rubella, oral polio, and smallpoxes. These vaccines often have persistent activity that lasts a lifetime. However, attenuated live vaccine has significant shortcomings. Apart from being capable of failing to protect the vaccinated individual against certain diseases, people with compromised immune systems from debilitating diseases like cancer and AIDS, as well as the elderly, can be overwhelmed and end up suffering full-blown diseases against which they were immunized. There is always the possibility that a weakened virus can mutate back to the virulent form. Another drawback of conventional vaccines using full organisms is that molecules from the organism that are not involved directly in the immune process are retained in the host. These may pose other unanticipated health risks.

■ *GENETIC VACCINES*

Genetic vaccines are prepared by inserting genes for one or two antigenic proteins in a plasmid that has been genetically reconstructed to lack the ability to reconstitute itself and cause disease. The recombinant plasmids are delivered into the muscle cells of the host by a syringe or gene gun. The recombinant plasmids enter the nucleus of the cell from where the host cell is compelled to express the enclosed antigenic protein. The antigen-encoding genes are transcribed into mobile mRNA and subsequently translated into antigen proteins. The antigen (protein) may be fragmented into pieces and attached to special molecules in the cell called **major histocompatibility** (MHC) class I molecules. The two structures form an antigenic complex that partially protrudes to the surface of the cell (see Figure 17–1). Alternatively, the intact protein molecules may exist in the cell as free antigens. These proteins can produce humoral immunity or can also elicit cell-mediated immunity upon escaping from the cell and breaking down and being displayed on the cell surface.

Other strategies are being adopted to produce DNA vaccines with higher efficiency. It has been discovered that plasmid DNA yields the most potent immune response when CG sequences are flanked by two purines (A or G) to the C side and two pyrimidines (TC) to their G side. These are described as "immunostimulatory sequences." Increasing these sequences in plasmids might increase the immunogenecity of the antigenic code in a DNA vaccine. It is known that cells of the immune system release cytokines to regulate their own and others' activities. One proposed strategy for amplifying the immuno-genecity of the DNA vaccine is to incorporate cytokine genes into the plasmids carrying the antigen or other plasmids.

DNA vaccines have certain advantages in addition to those of the conventional ones. They are capable of eliciting both humoral and cell-mediated immunology. They can be engineered to carry genes from different strains of a pathogen and thereby provide immunity against several strains at once. This capability would be advantageous, especially in cases where the pathogen is highly variable (influenza and HIV). Using modern rDNA technology, it should be possible to produce DNA vaccines rapidly and cheaply.

Many DNA vaccines have been developed and are in the early stages of trials at this point in time. These include those for hepatitis B, herpes, HIV, influenza, and malaria prevention. Others have been developed as therapy for HIV, B cell lymphoma, adenocarcinomas of the breast and colon, prostrate cancer, and cutaneous T cell lymphoma. There are many questions that need to be answered and more fine-tuning required to make DNA viruses commonplace. Generally, DNA viruses are known to stop yielding much protein after about one month. There is the need to find out if people vary in their response to these new generation vaccines. The dosage and delivery schedules need to be worked out. It might be advantageous to utilize both conventional and DNA vaccines together for an overall superior effect.

THE HUMAN GENOME PROJECT

The Human Genome Project was discussed previously as an example of how a genomics project is conducted. One major application of biotechnology in human health is genomics—structural genomics (to discover genes) and functional genomics (to decipher the functions of genes). With the first phase completed, the more challenging second phase of understanding gene function is underway. The project will greatly impact clinical medicine, genetic counseling, and disease treatment. Genes for genetic disorders will be identified. By understanding how they directly affect biological activity, drug manufacturers will be better equipped to design more effective and efficient drugs to treat these disorders.

DNA PROFILING

A DNA profile of an individual reveals the detailed genetic pattern that is unique to this individual, enabling it to be used for a variety of diagnostic and identification purposes.

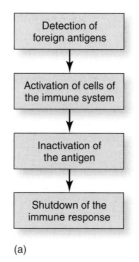

(a)

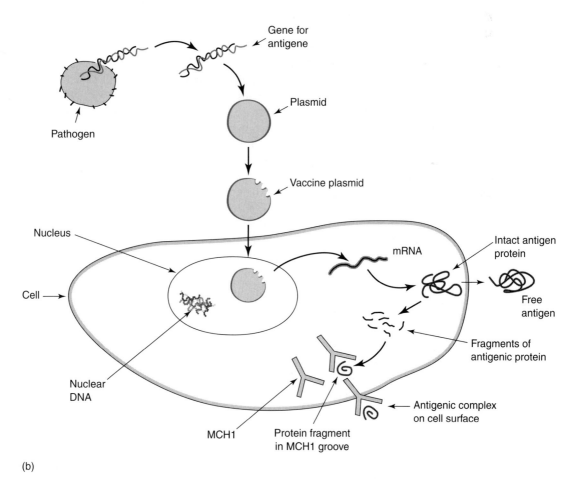

(b)

FIGURE 17–1

(a) A summary of the steps in how the immune system works. (b) A summary of how a genetic vaccine would work. The gene for antigen is transferred into the host cell. The transgenic protein forms a complex with the major histocompatibility complex (MHC) on the cell surface, or may be free in the bloodstream.

■ TESTING FOR GENETIC DISORDERS

Biotechnology may be used for testing for genetic disorders at two stages: prenatal and presymptomatic. Prenatal testing is conducted to determine the presence of genetic disorders in an embryo or fetus. Such tests have traditionally been conducted by **amniocentesis** or **chronic villus sampling (CVS)**. Amniocentesis uses cells collected from the amniotic fluid surrounding the fetus (see Figure 17–2). This test is usually not performed before the 15th or 16th week of pregnancy. The cells are cultured for karyotype analysis or assayed for biochemical defects. The fluid can also be tested for biochemical disorders. Over 100 such disorders have been detected by this technique. CVS can be conducted sooner, at 8 to 10 weeks of gestation (i.e., first trimester of pregnancy). This test is quick, yielding results in only a few hours or days. Fetal tissue is removed by inserting a catheter through the vagina (or abdomen) to sample the chorionic villi (fetal tissue that forms part of the placenta). Biochemical tests are performed directly on the sample, eliminating the need for tissue culture as is the case in amniocentesis.

The role of biotechnology procedures in the detection of genetic disorders at the prenatal level is an extension of the two traditional techniques, providing more information sooner than traditional techniques. For example, in sickle cell anemia, the defective gene product is not produced before birth and therefore cannot be tested by conventional techniques. RFLP analysis can be performed on fetal cells obtained by amniocentesis or CVS. In the case of *in vitro* fertilization procedures, a fertilized human embryo can be tested before implantation (called **preimplantation testing**) into the womb. This is done by removing a single cell from the blastomere (6-to 8-cell embryonic stage) and extracting its DNA. The DNA is increased by PCR and then used for various tests to determine the presence of the mutated sequence in the genome (e.g., muscular dystrophy, Lesch-Nyhan syndrome, hemophilia, and cystic fibrosis have been successfully diagnosed). The embryos without genetic defects are then implanted in the womb and nurtured for full development.

Genetic disorders may also be screened at the adult stage (called **presymptomatic screening**) for adult-onset disorders like polycystic kidney disease (a dominant trait). This diagnosis can be performed at any time before the disease symptoms appear. The Human Genome Project will assist in diagnosis. DNA-based gene tests are available for many inherited conditions. Additional tests are provided later.

FIGURE 17–2

The technique of amniocentesis involves drawing amniotic fluid containing fetal cells for cytogenetic analysis.

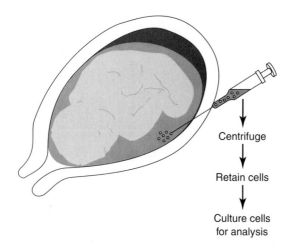

Centrifuge

Retain cells

Culture cells for analysis

■ GENETIC COUNSELING

Genetic counselors use results from genetic testing to provide information for practitioners to assist their clients in making informed choices when confronted with a diagnosis of genetic disorder. Genetic counseling is not a recent application of genetics. As previously indicated, biotechnology tools enable such information to be obtained more readily.

PHARMACOGENETICS

Drug therapy is widely used against a variety of diseases ranging from those of pathogenic origin to physiological disorders. These therapies are not always curative. Individuals do not respond to drugs in the same way; some have adverse effects to certain drugs. Some drugs are therapeutic when administered to some people but have no effect on others. Traditionally, drug manufacturers test drugs on a broad population basis and use statistical analysis to make predictions on their impact on individuals. Drug dosages are prescribed on the basis of average response obtained from such population studies. In effect, the traditional approach to drug therapy is a one-size-fits-all strategy. A study has reported that adverse drug reactions are the fifth-leading cause of deaths in hospitalized patients.

It is clear from the foregoing that individualizing drug therapy is a much safer alternative to the current approach. Pharmacogenomics and **pharmacogenetics** deal with the genetic basis of variable drug response in individuals. Pharmacogenetics examines the impact of sequence variations in genes suspected of affecting drug response. Pharmacogenomics looks at the total genome's impact, recognizing the fact that numerous genes and other factors interact to impact drug response in individuals. Such factors include age, nutrition, and the general health of the individual. Pharmacogenomics, in short, is a study of drug response as a function of genetic differences among individuals.

The availability of advanced biotechnology techniques is making pharmacogenomics more attainable. The strategy of pharmagenomics is to group people with the same phenotypic disease profile into smaller and more homogenous subpopulations according to genetic variations associated with disease, drug response, or both. A drug regimen is then tailored to such small groups with the hope of increasing efficacy of drugs and reducing toxicity. Anticipated benefits of pharmacogenomics include improved doctor prescriptions, allele-specific nucleotides, and drug development.

■ IMPROVED DOCTOR PRESCRIPTIONS

Doctors sometimes resort to trial-and-error in their attempts to find the best drug for a patient. Some patients may have their medications changed several times over the course of a treatment. Doctors will also be able to adopt a "precision prescription" approach in their drug prescriptions to patients and tailor drugs and dosages to individual needs to avoid overprescription.

■ ALLELE-SPECIFIC NUCLEOTIDES

Allele-specific nucleotides are suitable for diagnosing genetic disorders caused by point mutations. Allele-specific oligonucleotides (ASO) are probes that are so specific that they will hybridize with sequences that differ by only one nucleotide. In disorders where the nucleotide sequences for both the normal and mutant genes are known, ASOs can be used to screen heterozygote carriers. For example, in the autosomal recessive disorder associated with the protein cystic fibrosis transmembrane conductance regulator (CFTR), the cause of cystic fibrosis, about 70 percent of mutant alleles have a three-nucleotide deletion. This mutation causes the deletion of phenylalanine at position 508 in the gene product. Allele-specific oligos are created from nucleotide primers to detect a heterozygote.

FIGURE 17–3

A summary of the techniques of allele-specific oligonucleotides (ASOs). Denatured DNA from the sample is spotted on strips of DNA-binding filters. Each strip is hybridized to a specific ASO and visualized on X-ray film. The genotype is read directly from the filters. A homozygous normal (AA) genotype would produce a dark spot because it carries two copies of the normal allele. A heterozygous genotype has one copy of the normal allele and would hence produce a light spot, whereas the homozygous recessive genotype would show no spot for lack of a normal allele of the gene.

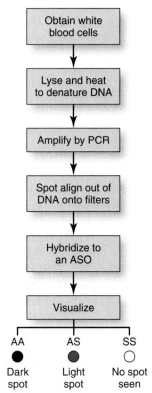

(Example: screening sickle cell anemia)

DNA samples from white blood cells are amplified by PCR and spotted on filters (nylon, nitrocellulose) and then hybridized to each of the ASO probes. Only ASO prepared from the mutant allele will hybridize. This technique is also used in the diagnosis of sickle cell anemia. When an ASO for the normal sequence is used, homozygotes produce a very dark spot while heterozygotes (one allele) produce a lighter spot. Homozygous recessives produce no color (see Figure 17–3).

■ DRUG DEVELOPMENT

Drug development will help pharmaceutical companies to develop more powerful drugs targeted to individual needs. By understanding the nature and functions of biomolecules, scientists are better able to design drugs to be more effective. For examples, the **cytochrome P450 monoxygenase** system of enzymes is known to be involved in metabolism of drugs in humans. Numerous P450 subtypes have been studied. Individuals that are homozygous for CYPD26 null alleles are known to be poor metabolizers of many drugs and consequently more prone to exhibiting adverse reactions to drugs.

Drug manufacturers will be able to discover drugs that are more likely to be successful at their first introduction. This will reduce the time taken to get drugs to the market and reduce the cost of drugs. Bioinformatics will allow scientists to discover biomolecules that are candidates for drug manufacture.

STEM CELLS

Animal stem cells are cells with the capacity to divide for infinite periods of time to give rise to specialized cells. They are somewhat the equivalent of callus in plants. Upon fertilization, the zygote produced is totipotent (has unlimited capacity to specialize into all the cells, tissues, and organs of the organism). The zygote undergoes cycles of cell division to form a hollow sphere of cells called the **blastocyst.** This occurs within approximately four days. A cluster of cells, called the **cell inner mass,** forms inside the hollow sphere surrounded by an outer layer. The outer layer of cells develops into the placenta and supporting tissue. The remainder of the fetus is formed from the inner mass of cells. Because these cells cannot form the placenta (but can form all other tissues), the inner mass of cells is described as pluripotent (as opposed to totipotent). Furthermore, these pluripotent cells become specialized into a class of cells called **stem cells** that are more specialized in function (e.g., blood stem cells and skin stem cells). Because stem cells can specialize into only a selected number of structures (e.g., blood stem cells produce blood-related components like platelets, red blood cells, white blood cells, and so on), they are described as multipotent. Blood stem cells are found in the bone marrow.

■ *PRODUCING PLURIPOTENT STEM CELLS*

Human embryonic stem cells were discovered in 1998. Scientists obtain human stem cells from several basic sources:

1. *In vitro* **fertilization embryos**

   Stem cells from this source may be isolated from the inner cell mass of human embryos obtained from embryos from *in vitro* fertility clinics. Because couples contract these embryos for reproductive purposes, they are obtained with the consent of the donor couples.
2. **Aborted fetuses**

   Stem cells may be obtained from fetuses derived from terminated pregnancies.
3. **Somatic cell nuclear transfer**

   An egg cell is obtained and its nucleus removed. Another somatic cell is then fused with this "egg shell," the product of which is totipotent. Under proper culture, it will form a blastocyst from which stem cells may be harvested from the inner cell mass.

■ *ADULT STEM CELLS*

Adult stem cells may be retrieved from some adult tissue (e.g., the blood cells, nervous system cells). Stem cells from some tissues of interest still have not been successfully isolated. Because stem cells are pluripotent, it is important to isolate them from all tissues of interest. Another factor that limits the use of adult stem cells is that they are usually present in minute quantities and are difficult to isolate and purify. To be used for therapeutic purposes, adult stem cells first have to be isolated from the patient and cultured to obtain sufficient quantities. However, in emergency cases, there may not be enough time to wait for these cells to multiply. Furthermore, in cases where the disease is genetic in origin, the adult cells would also be defective and therefore not useful. There is a possibility that, even in normal individuals, adult stem cells may have accumulated some mutations with age. There is no clear evidence that adult stem cells are pluripotent, thereby limiting their usefulness in scientific research.

Adult stem cells are less controversial to use than the more versatile pluripotent cells from embryonic sources. The use of these cells is heavily criticized and protested by various interest groups for being unethical. Scientists are therefore working to find less controversial sources of pluripotent stem cells. Furthermore, adult stem cells lose their ability to divide and differentiate after a period of time in culture, making them appear unsuitable for medical uses.

■ POTENTIAL APPLICATIONS OF PLURIPOTENT STEM CELLS

Scientists are interested in stem cell research for a variety of reasons:

1. It would help in understanding the events that occur during development. Scientists need to understand more completely how cellular decisions are made to direct the specialization of cells. This will help in understanding how birth defects and diseases associated with abnormal cell division and cell specialization occur. This could also shed light on how diseases like cancer occur and how to cure them.
2. Pluripotent stem cells could provide materials for the initial testing of new drugs in more cell types prior to clinical testing. Whole animals and humans would be used for testing after an initial screening of the product, thereby limiting the risk associated with clinical trials.
3. Pluripotent cells could be manipulated to develop into specialized cells needed to replace specific damaged cells and tissues, thereby reducing the need for organ and tissue transplants. Because organ donation lags behind the need, pluripotent cells could be a renewable source of replacement cells. Diseases that could benefit from "cell therapy" include Parkinson's, Alzheimer's, heart disease, and diabetes.

■ CHALLENGES WITH CELL THERAPY USING PLURIPOTENT STEM CELLS

1. Scientists need to understand how specialization occurs so pluripotent cells can be effectively directed to produce specific cells or tissues needed for transplantation.
2. Because pluripotent cells derived from embryos or fetal tissue would be genetically different from the genotype of the recipient, the question of immune rejection from tissue incompatibility needs to be overcome. The use of somatic cell nuclear transfer may be used to overcome this problem.

Because of the controversy surrounding the use of embryonic stem cells of human origin, scientists continue to search for new sources of less controversial materials. In early 2002, an announcement was made to the effect that scientists have discovered a new source of adult stem cells—unfertilized eggs.

■ STEM CELL BUSINESS

Since James Thomson's announcement that he had successfully cultured human embryonic stem cells *in vitro*, the race has been on for how to commercialize its potential. Geron Corporation of Menlo Park, California, appears to have the early lead, having funded Thomson's work and secured first rights to exploit the discovery. The company also funded research by John Gearhart that resulted in the discovery of very early primordial cells from humans. Other companies are hot in pursuit, developing adult stem-based therapies. Whereas Geron controls the uses of the cell line produced by Gearhart through an exclusive license from John Hopkins, the University of Wisconsin is holding on to Thomson's discovery and has promised to keep the cell lines pure (not mixed with human embryonic stem cells) and to make it accessible to the public research community. The university has gone a step further to create a nonprofit subsidiary called WiCell to handle stem cell issues.

ANIMAL FARMING FOR CELLS AND ORGANS

Biotechnology is creating yet another "new kind of farming" in which animals are not reared for food, but for parts—cells, tissues, and organs—for transplanting into humans.

Again, the question arises—how new is this practice? It depends on what is involved. Heart valves from pigs have been surgically implanted in humans for some time. What is new is breeding animals such that their major organs (e.g., heart and lungs) can be transplanted into humans, a practice dubbed **xenotransplantation** (cross-species transplantation).

Early successes have been reported with transplanting porcine cells into humans. A 19-year-old stroke victim dramatically improved after a pig-cell brain graft. A pig liver was used outside the body of a 17-year-old diagnosed with fulminant hepatic failure as a stop-gap measure for about six hours while the boy awaited a human donor. Early clinical studies involving Parkinson's patients showed promising results, but certain hurdles need to be overcome before the technology becomes routine.

So why pigs? The pig is the farm animal of choice for harvesting parts for xenotransplantation for several reasons. Pig and human organs are similar in size and structure. The physiology of the two species are similar. Pigs are easy to breed and care for. Being non-primate animals, pigs do not attract the negative publicity from activists who oppose the use of chimpanzees and baboons, which are most closely related to humans, in scientific research. Even though it would appear then that these primates would be more likely candidates for xenotransplantation, there are significant problems. Baboons' organs are too small in size to be useful for long-term support of humans. Chimpanzees are not only scarce, but as they are closely related to humans, there is an increased risk of transmission of diseases from these primates to humans. In fact, the human AIDS virus is known to have originated in chimpanzees and monkeys. Nonetheless, scientists have successfully sustained humans on baboon livers for 71 days and chimpanzee kidneys for nine months.

However, using pigs for parts does have its unique problems. Pigs carry virus-like genetic materials called **porcine endogenous retroviruses (PERV)** that have been demonstrated to be transmissible to human cells in the lab, as well as lab mice. More importantly, pig organs carry certain sugar molecules that trigger immunological attacks against the human immune system, leading to rejection of organs. The pig antigen called *Gal* speeds immune rejection. Scientists are working on producing "*Gal*-negative" pigs. This new breed of pigs will also carry human genes that will help pig organs become acceptable to the human immune system, the so-called "humanized pigs" (or better still, "pigman," a contraction of pig and human).

KEY CONCEPTS

1. Gene therapy is the molecular biotechnological technique of correcting genetic disorders by replacing defective genes with functional or normal ones.
2. There are two basic approaches to correcting genetic disorders: gene replacement and targeted gene repair.
3. There are two forms of gene therapy: somatic cell therapy and germline therapy.
4. Correcting genetic disorders by gene therapy may be done *in vivo* or *ex vivo*.
5. The most widely used method of gene delivery in gene therapy is by using viruses.
6. The tools for targeted gene repairs are in their infancy stage of development and application in gene therapy.
7. One of the earliest successes of gene therapy involved the treatment of two girls with ADA (adenosine deaminase) deficiency.
8. Conventional vaccines are prepared from killed or weakened versions or parts of pathogens.
9. Genetic vaccines are prepared by inserting genes for one or two antigenic proteins in a plasmid that has been genetically reconstructed to lack the ability to reconstitute itself and cause disease.

10. One major application of biotechnology in human health is genomics—structural genomics (to discover genes) and functional genomics (to decipher the functions of genes).

11. A DNA profile of an individual reveals the detailed genetic pattern that is unique to this individual, enabling it to be used for a variety of diagnostic and identification purposes.

12. Pharmacogenomics and pharmacogenetics deal with the genetic basis of variable drug response in individuals. Pharmacogenetics examines the impact of sequence variations in genes suspected of affecting drug response. Pharmacogenomics looks at the total genome's impact, recognizing the fact that numerous genes and other factors interact to impact drug response in individuals.

13. Adult stem cells may be retrieved from some adult tissue (e.g., the blood cells, nervous system cells). Stem cells from some tissues of interest still have not been successfully isolated. Because stem cells are pluripotent, it is important to isolate them from all tissues of interest.

14. Biotechnology is creating yet another "new kind of farming" in which animals are not reared for food, but for parts—cells, tissues, and organs—for transplanting into humans.

OUTCOMES ASSESSMENT

1. Give the two basic approaches of gene therapy and discuss their pros and cons.
2. Distinguish between somatic cell gene therapy and germline gene therapy.
3. Briefly describe the steps employed in *ex vivo* gene therapy.
4. Compare and contrast the various viral gene delivery systems used in gene therapy.
5. Give the challenges of gene therapy.
6. Compare and contrast conventional vaccines and genetic vaccines.
7. Discuss the application of gene profiling in humans.
8. Discuss the concept of "personalized medicine or therapy."
9. Discuss the potential of stems for therapeutic use in humans.
10. What is xenotransplantation and what are the challenges in this application of biotechnology?

INTERNET RESOURCES

1. Biotech and health: *http://www.bio.org/aboutbio/guide1.html#health*
2. Gene therapy: *http://www.ornl.gov/hgmis/medicine/genetherapy.html*
3. Biotech in medicine: *http://www.jic.bbsrc.ac.uk/exhibitions/bio-future/medbiotech.htm*
4. Fundamentals of gene therapy: *http://www.fda.gov/fdac/features/2000/gene.html*
5. Questions and answers about gene therapy: *http://cis.nci.nih.gov/fact/7_18.htm*
6. Genetic counseling: *http://www.accessexcellence.org/AE/AEC/CC/counseling_background.html*
7. Medical genetics information for physicians: *http://www.genetests.org/*

REFERENCES AND SUGGESTED READING

Alcamo, E. I. 2000. *DNA technology: The awesome skill.* Academic Press.

Holland, S., K. Lebacqz, and L. Zoloth (eds). 2001. *The human embryonic stem cell debate: Science, ethics, and public policy.* Boston, MA: MIT Press.

SECTION 2
Forensic Applications

PURPOSE AND EXPECTED OUTCOMES

Except identical twins, no two individuals have identical genotypes. Because of this uniqueness or individuality, the DNA profile or characteristic pattern of nucleotide distribution is used as a powerful basis of identification of individuals in a population. The use of DNA to provide evidence in the judiciary system is now widely acceptable in many parts of the free world.

In this section, you will learn:

1. The application of DNA in the law courts.
2. The science behind DNA profiling.

WHAT IS DNA PROFILING?

DNA profiling is the application of DNA for diagnostic purposes whereby the profile of DNA from the sample of interest is compared with the pattern from another sample of known characteristics. This application is used in the health care and judiciary systems, among others. In the health care system, as previously discussed, it is used for diagnosis of hereditary diseases to predict the chance of an individual inheriting a disease from afflicted parent(s). It can also be used to detect the predisposition of an individual to cancer, or chromosomal aberrations.

In the judicial system, forensic experts use DNA profiling to identify suspects in criminal cases, especially where body fluids (e.g., in rape, murder) and other particles like hair and skin samples can be retrieved. Forensic experts can use minute specimens that contain DNA to associate a person with a crime. DNA profiling is also used in disputed family relations (e.g., paternity suits) and in immigration cases (e.g., where a child is seeking citizenship based on family relationships; the profile of the applicant and the established citizen are compared for determination of family ties).

Alec Jeffreys and his colleagues at the University of Leicester in England pioneered the technique of RFLP analysis and its application in forensics. He also helped found the company called Cellmark Diagnostics, a private DNA lab operating in both the United Kingdom and the United States, which was involved in the O. J. Simpson murder trial in 1994. The first case to be solved by DNA diagnostics evidence occurred in 1986. Jeffreys and company used DNA fingerprinting to free an innocent man in a case involving the rape and murder of two 14-year-old girls in the English Midlands. The technique also led to the conviction of the perpetrator of the crime.

SOURCES OF DNA FOR FORENSICS

DNA may be obtained from any body part including fluids in which cells are likely to be found (blood, saliva, urine). Both nuclear and mitochondrial DNA are used. A mitochondrion contains 5 to 10 mtDNA molecules. A cell can have hundreds to thousands of mitochondria. Because of this high copy number, mtDNA more than nuclear DNA is the most likely to be recovered from an ancient specimen of biological stains. Being cytoplasmic in origin, mtDNA can be inherited maternally (i.e., a woman, her children, mother, maternal grandmother, and so on have identical mtDNA sequences). Mitochondrial DNA is therefore important in solving cases involving biological relationships, provided there is an unbroken female line.

Even though the DNA in a species is stable, it is not static. Mutations occur to alter the topology of the genome. If they occur in the non-coding sequences, they may remain undetected and with no consequence to the individual. Such neutral mutations are responsible for the differences between individual genomes.

METHODS OF ANALYSIS

Both the RFLP and PCR techniques can be used for DNA profiling. Classical DNA profiling entails the use of **variable number tandem repeats** (**VNTRs**).

■ VARIABLE NUMBER TANDEM REPEATS

The success of DNA technology rests in identifying the segments of the genomes that vary the most and can be used to discriminate among individuals in a population. Markers like blood groups are ineffective because too many individuals fall into the four categories (A, B, AB, O). The most variable regions of the genome occur in clusters of sequences called **minisatellites**. Minisatellites consist of 2 to 100 nucleotides in length. These sequences characteristically are tandemly repeated DNA sequences that occur between two restriction enzyme sites. An example is GGATGGATGGATGGATGGAT, which is four tandem repeats of the basic sequence GGAT. The complementary strand will have CCTA. The unit or basic sequence typically comprises 14 to 100 nucleotides. Furthermore, the number of repeats in a cluster at each locus is about 2 to over 100. The repeated nucleotide patterns are called variable number tandem repeats (VNTRs). VNTRs were discovered by Alec Jeffreys and his colleagues while studying the gene encoding myoglobulin (the red oxygen-binding protein in muscles). The number of pattern repeats at a given locus is variable. Furthermore, each variation constitutes a VNTR allele. The VNTR region is a polymorphic locus. Dozens of alleles have been identified at many loci and, consequently, heterozygosity is common regarding VNTRs. In an extended family, one pattern may be repeated 10 times at a specific locus in the DNA of one individual, 15 times in a sibling, and 50 times in an uncle (see Figure 17–4). It is possible that, for the same locus, two individuals in a population can have an identical number of repeats. Consequently, in the application of VNTRs in forensics, the probability of matching repeats at five (instead of one) loci is used to increase the odds dramatically. For example, if the frequency of a VNTR locus is calculated to be 1 in 525 and the frequency at a different locus (two) is 1 in 80, then the frequency of the two different loci occurring simultaneously is 1 in 42,000. If two additional loci (three, four) are included with frequencies of 1 in 75 and 1 in 128, respectively, the combined frequency of these four loci becomes 1 in over 400 million! The odds against matching the number of repeats at five loci are most astonishing.

To apply the technique of RFLP, restriction enzymes with recognition sequences flanking VNTRs are used to digest the DNA extracted from the specimens. The digest is electrophoresed, Southern blotted, and visualized as is done in standard RFLP protocols (see Figure 17–5). Because of the individual enormous variation in the VNTR pattern from this source, each person's pattern theoretically is unique. The pattern is the same for an individual, irrespective of the tissue from which DNA is obtained. This analysis requires only a small amount of specimen (e.g., less than 60 µl of blood).

In using DNA fingerprinting to decide a case, the argument that is made is that these VNTRs are randomly distributed across the population such that the odds against match-

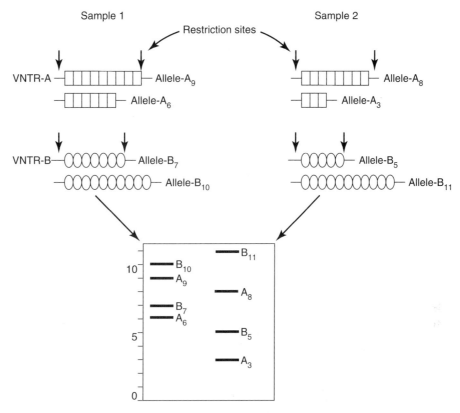

FIGURE 17–4

Distribution of VNTR loci in different samples. Each variation in the number of repeats is a VNTR allele. Following restriction digest that cleaves the DNA at the points marked by arrows, and subsequent electrophoresis and Southern blotting, a pattern of fragment separation may be produced that is characteristic of each individual. Each pattern for a sample is called a DNA fingerprint.

ing five or more loci through chance occurrence alone is astronomical. But what are high enough odds to help decide a case? In other words, what is the chance that a particular VNTR pattern is unique enough to be derived from this source alone in a given population? This chance depends on the frequency of the allele in question in other populations (ethnic groups).

In a highly diverse population like the United States, questions are often raised about the reference population used to calculate the gene frequencies. To remove ethnic biases, certain ceiling standards are proposed. Databases of sample genetic readings should be taken from 100 people each from a variety of races and categories. Matches at any locus that spreads across 5 percent (or less) of the general population would become the ceilings or standards for comparison. A large collection of population data on VNTR frequency in unrelated individuals from numerous human races and ethnic groups is available.

DNA evidence is widely accepted as valid by most courts. However, defense attorneys often attempt to discredit such evidence based on improper procedures used to collect, handle, and process crime scene specimens. A classic example of the use of DNA in forensics was the high-profile murder trial of O. J. Simpson. Defense attorneys, dubbed the "Dream Team," successfully defended the suspect in spite of mounting genetic evidence, citing police misconduct in evidence gathering and processing. In the 1994 U.S. World Trade Center bombing incident, investigators successfully linked a suspect to the crime by DNA

FIGURE 17–5

Steps in the application of RFLP technology in forensics. An appropriate sample in good condition is obtained and DNA is extracted from it. It is restriction digested, electrophoresed, and Southern blotted. The blot is submitted to probing, visualization, and then statistical analysis.

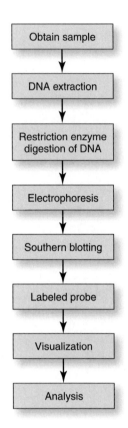

evidence obtained from saliva deposited on an envelope while being sealed by the suspect. Body fluids frequently contain cells from which DNA can be readily extracted. Researchers from the University of Minnesota at Duluth isolated DNA from the lung tissue of a 1,000-year-old mummy of the Chiribaya Indian extraction. They found an exact match of the DNA of the tuberculosis bacterium, conclusively exonerating Christopher Columbus of the charge by some that his crew was the source of the "White Plague" in the New World. In 1992, the former Mayor of Detroit, Coleman Young, settled a paternity suit and agreed to pay child support when DNA evidence proved he was the father of a child in a disputed suit.

DNA evidence has been used by the Armed Forces Institute of Pathology in Gathersburg, Maryland, to identify a number of Vietnam War MIAs. The Innocence Project founded by Barry Scheck, a member of the O. J. Simpson legal team, is devoted to using DNA evidence to exonerate prisoners who have been wrongfully convicted of murder or rape.

■ PCR-based DNA Profiling

Modern DNA profiling is PCR-based and uses short tandem repeats (STRs), which are the short (two to four) VNTRs. This technology debuted in 1994. The advantage of PCR is that it requires minute samples and is quick to perform. STR loci have fewer alleles than VNTRs. PCR data are less compelling in criminal cases, except for excluding a suspect from the pool. However, a variation introduced in 1999 in Britain uses 10 STR loci. This guarantees that the odds of someone sharing the same results are less than one in a billion. In the United States, the FBI used 13 STR loci to raise the odds of two unrelated individuals having the same profile to one in a million billion.

■ *RFLP- VERSUS PCR-BASED DNA PROFILING*

Each technology has its advantages and disadvantages. RFLP-based DNA profiling uses VNTRs in regions that naturally have a high degree of variability (i.e., a large number of alleles for each locus). This makes it highly unlikely that the DNA profile of two unrelated individuals will be identical. Another limitation of this RFLP-based DNA profiling lies in the amount and quality of DNA required. It requires at least 20 nanograms of purified and fairly intact DNA. Consequently, in cases where the sample is minute or partially degraded, the technique is not useful. Another disadvantage is associated with the protocol itself. Because the restriction fragments of VNTRs can be large, their migration in electrophoresis is very slow, leading to improper separation of fragments that are truly different but happen to be similar in size. Different sets of electrophoretic conditions (e.g., different equipment, buffer characteristics) can produce different results for the same gel. The electrophoretic conditions may also introduce another source of error called **band-shifting** in which bands that are produced by identical fragments do not migrate to identical distances on the gel. This occurrence may be verified by using a monomorphic probe to determine if the bands in question are bound by the probe. Reading gel results should be done very carefully to be accurate.

Samples obtained from clothing may incorporate dyes from the fabric that may influence the migration of DNA restriction fragments in electrophoresis. Consequently, when results from such a sample are compared with those from a fresh sample obtained from the same individual, the results could be different and therefore misleading.

PCR-based DNA profiling is a rapid and less expensive technique, but also is less definitive. Furthermore, PCR requires only a minuscule amount of specimen (50 white cells versus 5,000 to 50,000 needed for RFLP). A single cell is less than 10 picograms but is sufficient for PCR analysis. However, the use of very minute samples may cause **allele dropout** during amplification (i.e., some alleles may fail). Because it can utilize a minute sample, the PCR-based diagnostics are very important in criminal investigations where traces of samples are retrieved as evidence. It is the only choice in a situation where the DNA is so degraded that only STR can be analyzed. Many steps in the procedure may be automated. For example, computer-controlled laser equipment is often used to analyze the migration of the PCR products and compare them with DNA standards included in the electrophoresis.

PCR is a very sensitive procedure. Consequently, traces of unknown substances in extracts from stains can hamper the amplification process. DNA may have to be purified in criminal cases to give the purity of product that would not yield ambiguous results.

■ *THE ROLE OF ETHNIC GROUP DATA IN DNA PROFILING*

Population data on the pattern of distribution of alleles (allele frequencies) used in profiling must be obtained. This information is critical for calculations that are used to answer the basic question in, for example, a criminal proceeding: "What is the likelihood that a suspect with a matching DNA profile is not connected with the crime but has the incriminating DNA profile just by chance only?" Population studies have been conducted in various ethnic groups to obtain the frequencies of alleles that have been selected for either PCR- or RFLP-based profiling. One such allele used is the HUMTH01 (human tyrosine hydrogenase), an STR locus, used in PCR-based DNA profiling. This allele is located on the short arm of chromosome 11 on the first intron (hence the designation 01).

Some have criticized population data for being inaccurate because of the wrong assumption that the ethnic groups are genetically homogeneous when in fact some of them are the opposite. For example, the ethnic group designated as "Hispanic" actually includes people of diverse origins (including Mexicans, Cubans, Puerto Ricans, and people of Spanish descent).

PITFALLS IN INTERPRETATION OF DNA PROFILING RESULTS

To have quality and reliable results from DNA profiling analysis, the laboratory protocols should be appropriately conducted. The samples should be properly prepared (excluding contaminants) and the electrophoresis environment properly controlled. As was previously indicated, the electrophoresis process *per se* can be a source of error (e.g., band-shifting). Apart from procedural sources of error, the mistakes may occur during the interpretation of DNA profiles. Lack of a positive match of a suspect's profile with the victim's may be grounds to exclude the suspect. However, a positive match does not necessarily prove the suspect is the perpetrator of the crime. It is possible another individual with the same DNA profile could be responsible for the crime. Identical twins have identical DNA profiles. Similarly, DNA profiles from sibs could be identical if only a few loci were considered in the analysis. Consequently, DNA profiling is not a very powerful tool for resolving issues involving closely related individuals.

In interpreting results from DNA profiles, it is important to have data on the frequency of occurrence of the alleles in the appropriate population used in the analysis, which is used to calculate the probability that an individual picked at random from the population would have the same DNA profile obtained from the crime scene. RFLP-based VNTR loci are more numerous than STR loci. It is customary to analyze more STR regions than VNTR regions in DNA profiling. Then the question is, what odds are deemed adequate to rule in a particular case, 1 in 100,000, 1 in 1,000,000, or better?

One area of common application of DNA profiling is paternity disputes or kinship analysis. Because an individual has two alleles per autosomal locus (humans are diploids), many different DNA profiles may share combinations of alleles. Results from RFLP-based analysis have shown that the probability of a randomly selected male from the population having a DNA profile that is consistent with being the father of a child in a kinship dispute is less than 1 in 10,000 (this could be less for disputes involving closely related individuals). Consequently, if an analysis produces odds that do not exclude a man from fatherhood, the usual interpretation is to declare the person the father with a likelihood of higher than 10,000 in 1 (i.e., the likelihood of the man in question being the father of the child is higher than 99.9%).

It is possible for a child to inherit a VNTR allele that is not present in either parent. This may arise from occasional mutations (e.g., unequal crossing over during meiosis). This will change the number of repeats in an allele from one generation to the next. Because of the possibility of mutations that occur at an estimated rate of 0.1 to 0.5 percent (i.e., 1 in every 1,000 to 200 gametes shows a mutation in the VNTR in question), it is not possible to exclude a man from a paternity dispute with absolute (100 percent) certainty simply on the grounds that a mismatch occurred in a single one of the regions considered in the DNA profiling. To overcome such an argument, the test should be redone with two to three additional loci. The chance of mutations occurring in two out of eight loci is estimated at less than 1 in 100,000.

SELECTED APPLICATIONS OF DNA PROFILING

DNA profiling works best if suspects are not closely related. The probability of identical alleles among sibs is 25 percent for each locus. If four loci are used in an investigation, the probability of encountering identical alleles between two sibs is at least 0.4 percent. As already indicated, using this technique to resolve cases involving members from the same family is more difficult.

Mitochondrial DNA (mtDNA) is also used in DNA profiling. It is valuable in cases involving archeological and anthropological materials. MtDNA is maternally inherited.

Consequently, two individuals who are not related through links in an uninterrupted maternal lineage will most likely differ in one or more mtDNA sequences. Sequence data offer the most specific identity of individuals. The U.S. Armed Forces Pathology laboratory uses this technique to identify the remains of soldiers.

DNA profiling was used to conclusively identify the remains of Czar Nicholas Romanov II, his wife, and their five children, who were executed during the Bolshevick Revolution of 1918. The investigation was a three-prong approach, using DNA extracted from bone fragments. The first analysis was performed to determine the sex of the individuals, the second established family relationships, and the third, involving mtDNA, was used to trace the maternal relationship. The last analysis included samples from HMH Prince Philip, son of a daughter of the Czarina's sisters, to provide an unbroken maternal link.

KEY CONCEPTS

1. DNA profiling is the application of DNA for diagnostic purposes whereby the profile of DNA from the sample of interest is compared with the pattern from another sample of known characteristics. This application is used in the health care and judiciary systems, among others.
2. Alec Jeffreys and his colleagues at the University of Leicester in England pioneered the technique of RFLP analysis and its application in forensics.
3. DNA for profiling may be obtained from any body part including fluids in which cells are likely to be found (blood, saliva, urine). Both nuclear and mitochondrial DNA are used.
4. Mitochondrial DNA is important in solving cases involving biological relationships, provided there is an unbroken female line.
5. The success of DNA technology rests in identifying the segments of the genomes that vary the most and can be used to discriminate among individuals in a population.
6. The most variable regions of the genome occur in clusters of sequences called minisatellites.
7. In using DNA fingerprinting to decide a case, the argument that is made is that these VNTRs are randomly distributed across the population such that the odds against matching five or more loci through chance occurrence alone is astronomical.
8. Modern DNA profiling is PCR-based and uses short tandem repeats (STRs), which are the short (two to four) VNTRs.
9. Population data on the pattern of distribution of alleles (allele frequencies) used in profiling must be obtained. This information is critical for calculations that are used to answer the basic question in, for example, a criminal proceeding: "What is the likelihood that a suspect with a matching DNA profile is not connected with the crime but has the incriminating DNA profile just by chance only?"
10. To have quality and reliable results from DNA profiling analysis, the laboratory protocols should be appropriately conducted. The samples should be properly prepared (excluding contaminants) and the electrophoresis environment properly controlled.

OUTCOMES ASSESSMENT

1. Give the rationale of gene profiling.
2. Discuss the importance of mtDNA in forensic analysis.

3. What are VNTRs, and how are they used in DNA profiling?
4. Compare and contrast PCR-based and RFLP-based DNA profiling.
5. Explain the importance of population data on gene frequency in DNA analysis.
6. Discuss the pitfalls of DNA profiling and how to minimize errors.

INTERNET RESOURCES

1. Consultant for forensic testing: *http://www.execpc.com/~helix/forensic.html*
2. Application of RAPD: *http://www.rvc.ac.uk/review/DNA_1/5_RAPD.cfm*
3. Variety of resources on forensics: *http://www.gwu.edu/gelman/guides/sciences/forensics.html*

REFERENCES AND SUGGESTED READING

Findlay, I., A. Taylor, P. Quircke, R. Frazier, and A. Urquhart. 1997. DNA fingerprinting from single cells. *Nature, 389:*555–556.

Gill, P., P. L. Ivanov, C. Kimpton, et al. 1994. Identification of the remains of the Romanov family by DNA analysis. *Nature Genetics, 6:*130–135.

Hochmeister, M. N., B. Budowle, A. Eisenberg, U. V. Borer, and R. Dirnhofer. 1996. Using multiplex PCR amplification and typing kits for the analysis of DNA evidence in a serial killer case. *J. Forensic Sci., 41:*155–162.

Krings, M., A. Stone, R. W. Schmitz, et al. 1997. Neanderthal DNA sequences and the origin of modern humans. *Cell, 90:*19–30.

Reynolds, R., and J. Varlaro. 1996. Gender determination of forensic samples using PCR amplification of ZFX/ZFY gene sequences. *J. Forensic Sci., 41:*279–286.

Industrial Applications

PURPOSE AND EXPECTED OUTCOMES

The purpose of this section is to discuss the science behind bioprocessing, the technology that enables scientists to use bacteria as bioreactors to produce products like industrial enzymes and pharmaceuticals.

In this section, you will learn:

1. The principles of bioprocessing.
2. The steps in a bioprocess.

WHAT IS A BIOPROCESS?

A **bioprocess** may be broadly defined as an industrial operation in which living systems (microorganisms, animal or plant cell cultures or their constituents; enzymes) are used to transform raw materials (biological or nonbiological) into products. The scientific discipline of bioprocess engineering (more appropriately, a subdiscipline of biotechnology) is described as being responsible for translating life-science discoveries into practical products, processes, or systems for the benefit of society. The key engineering discipline that supports the commercializing of biotechnology is biochemical engineering, which combines chemical or process engineering with biological systems. Bioprocess engineering is multidisciplinary by nature.

PUTTING BIOTECHNOLOGY TO WORK

In order to justify the commercializing of scientific discoveries, there has to be a clear indication that the goods and services to be produced are beneficial to society, and that they can be marketed profitably. Converting science into products entails:

1. The presence of an engineering innovation.
2. The development of appropriate enabling technologies.
3. An economic opportunity.

◼ *ENGINEERING INNOVATION*

Engineering innovation starts with the discovery of a phenomenon. One of the earliest phenomena that has been successfully commercialized was discovered in 1928. The fungus *Penicillin* was found to effectively control the bacterium *Staphylococcus aureus*. In genetic engineering, discovery of the phenomenon of site-specific enzyme cleavage of DNA by restriction endonucleases in bacteria is key to DNA manipulation (recombinant DNA technology).

By itself, an engineering innovation will remain an academic achievement, unless an economic opportunity presents itself for its translation into a product or service beneficial to society. Such an opportunity may arise because of a social emergency for a product or because of a governmental action. For example, the extensive need for antibodies as a result of the consequences of World War II provided a strong impetus for the development of penicillin as a dominant antibody. Similarly, the high sugar prices and the abundance of corn between 1965 and 1980, as well as a government policy to implement a minimum sugar price, provided the needed market forces for the conversion of the 1957 technology (ability of xylose isomers to catalyze the formation of fructose from glucose) to the commercializing of high-fructose corn syrup.

◼ *ENABLING TECHNOLOGIES*

In order to translate the phenomenon into a product, enabling technologies are needed. The key technology that drives biotechnology innovations is the recombinant DNA technology (rDNA). This technology utilizes the phenomenon of the ability of enzymes to cleave specific sites on a DNA strand. Other allied technologies that facilitate rDNA include the polymerase chain reaction (PCR) technology, blotting technologies, electrophoresis, gene transfer technologies, fermenter technology, and others.

◼ *ECONOMIC OPPORTUNITY*

Without an economic opportunity, a scientific innovation will remain at the bench. Economic opportunities arise from identifiable potential benefit(s) that may accrue to society from the development of a product. There is no justification to invest a large sum of money into undertakings that serve no societal purpose. Economic opportunities may be preexisting, but they can be boosted or made more attractive by events or circumstances. These events can generate an unusual demand for a product for a period of time at least. After the discovery that penicillin was most effective in treating infection, the event of World War II created a spike in the demand for such a product.

BIOPROCESS ENGINEERING

There are three key steps in a bioprocess: pretreatment, bioreaction, and downstream processing.

◼ *PRETREATMENT*

The raw material (or feedstock) to be processed must first be converted into a form that is suitable for the process equipment. The design of a bioprocess and the engineering of the process equipment must take into account the physical and chemical properties of the raw material. The process conditions should ensure that the integrity of the organism being used and the products being made is not compromised. The key environmental factors of concern are temperature, pH, shear forces, pressure, and contamination. The ideal levels of these process factors vary among bioreactors. The key is for these factors to be maintained at levels with the physiological tolerance of the organism and enzymes. Bioreactors are generally operated at near neutral pH (about 7), and equal to or less than 37° C and 2Mpa pressure. The increased solubility of CO_2 under high pressure may cause CO_2 toxicity.

The susceptibility of microorganisms to shear forces in a bioreactor is variable, since bioengineered species are usually more prone to mechanical damage than native microbes. Mutants (wall-less mutants) and bioengineered microbes tend to have weaker cell walls or membranes. Generally, bacteria and yeast that grow as small individual cells tend to be relatively shear-resistant. Plant and mammalian cells, however, are sensitive to shearing forces, and so are filamentous bacteria and mycelia fungi. On the other hand, enzymes (except multienzyme complexes and membrane-associated enzymes) are resistant to damage by shearing forces in the absence of gas–liquid interfaces.

■ BIOREACTOR FLUIDS

The medium in which a bioprocess occurs is usually a liquid or slurry. The dynamics of the process medium core is critical to the operation. There are two main kinds of biofluids and slurries used in bioreactors.

1. **Newtonian fluids**

   Newtonian fluids maintain a constant viscosity regardless of the shear rate (the shear stress and shear rates are linearly related). Examples of Newtonian fluids commonly used in bioreactors are water, honey, and most bacterial and yeast fermentation broths.

2. **Non-Newtonian fluids**

   Certain fluids do not flow until a minimum shear (called yield stress) is applied. Non-Newtonian fluids exhibit varying viscosities with varying rates of shear. There are two basic kinds of outcome: one in which the fluid becomes increasingly viscous with shear (dilettante; shear thickening), or one in which the fluid is less viscous (pseudoplastic; shear thinning). Many biofluids exhibit pseudoplastic behavior under shearing.

■ MEDIA STERILIZATION

Contamination is a major enemy of a successful bioprocess. A bioprocess is initiated with specific microorganisms for a specific purpose. It is critical that only this microbial species that was used to inoculate the bioreactor be maintained throughout the process. Contamination may occur through the gaseous and liquid feeds to the fermenter. Sterilization by filtration and heat treatment is the most common technology employed to exclude biological contaminants from the fermentation. Filter sterilization should be used if the fluid (e.g., blood serum) is heat labile, or the liquid is free of suspended materials and of low initial microbial contamination. However, when the fluid is highly concentrated and highly contaminated (e.g., molasses), it is best to heat sterilize.

Heat sterilization may be accomplished in one of two basic ways: **batch sterilization** or **continuous sterilization.** The goal of sterilization is to raise the temperature of the environment to thermally denature one or more enzymes to ensure the destruction of the most resistant microbial contaminant. The most heat resistant bacterial spores are those of *Bacillus stearothermophilus* and *Clostridium botulism.* Unfortunately, elevating the temperature to destroy contaminants may also destroy heat-sensitive essential nutrients in the feed. To minimize this event, the strategy of high-temperature–short-time sterilization may be employed. This entails administering heat for a short duration or a matter of seconds, followed by cooling. This technique is most amenable to injecting steam or indirectly applying heat through a heating jacket to the fluid. Batch sterilization uses longer heating and cooling cycles.

Aerobic fermentation requires the use of large volumes of air that of necessity is sterilized prior to injection. However, it is also important, for the safety of plant personnel, to sterilize the exhaust gas. Air sterilization is achieved mainly by filtration, either depth filtration or absolute filtration. In depth filtration, the inlet directs the air through a bed of packing (e.g.; glass wool). Absolute filters decontaminate the air by sieving it through porous membranes (e.g., polymers, ceramic, metal). To reduce the failing of membranes, the gas may be prefiltered to remove larger particles. After a period of operation, the air filter unit itself needs to be sterilized.

■ GROWTH MEDIA

The growth medium is formulated to maintain microbial growth for economic production of the target product. Its design should be made based on a good understanding of the relationship between microbial growth kinetics and product formation. It should not focus only on needs during fermentation where the need is most crucial, but should consider the entire bioprocess. The choice of medium affects activities at both upstream and downstream stages in the process.

The basic components of a growth medium are a carbon source, nitrogen source, minerals, vitamins, micronutrients, hormones, and oxygen. These components must be supplied in adequate amounts to support the desired cell mass and product yield. The purity of materials (e.g., glucose versus molasses) affects the purification protocol, the byproducts, and their disposal issues. Apart from nutrients, a growth medium may contain additives that facilitate the process (e.g., antifoams).

■ MICROBIAL GROWTH

Microbial growth typically progresses through four basic phases—lag, exponential, stationary, and death phases (see Figure 18–1). Two additional phases (acceleration and deceleration phases) may be identified to mark the beginning and end of the exponential growth phase. The fermentation process begins with preparing (sterilizing) the fermenter or bioreactor. An appropriate volume of an appropriate fermentation medium is placed in the fermenter. The medium is then inoculated with an appropriate concentration of microbial suspension. A concentration of 5 to 10 percent by volume of microbial suspension in a rapid exponential growth phase may be used. Organisms differ in growth rate. It is recommended to use a larger volume of inoculum in the case of slower-growing organisms. This strategy avoids having protracted fermentation times and consequent increases in the cost of the process.

FIGURE 18–1
A typical microbial growth cycle is characterized by four phases. Activities in the lag phase include the adjustment of microbes to their cultural environment. This is followed by a phase of rapid exponential growth.

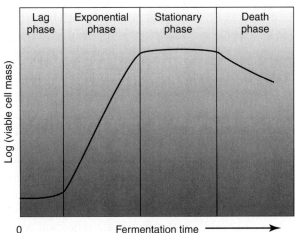

Lag Phase

The **lag phase** of growth is the period immediately following inoculation during which there is no increase in the number of cells. During this period, the organism adapts to its new environment and prepares for the next vigorous and rapid growth. The duration of the lag phase is influenced by factors including the size of the inoculum, the composition of the medium, and the history of the inoculum (e.g., the source of the culture, its phase of growth). If an inoculum is derived from a culture in the stationary phase, it would take a longer time to adjust its metabolism. An inoculum of cells in the late exponential growth phase adjusts more rapidly. Because the lag phase is unproductive in terms of the target product, it should be as minimized as possible.

Exponential Phase

Also called the **log phase**, the **exponential growth phase** is characterized by a number of doublings of the cell population. During the period, the specific growth rate of the culture remains constant and is independent of substrate concentration, provided there are excess nutrients or substrates and no inhibitory factors in the medium. Furthermore, the cell mass and cell number growth rates are not necessarily equal. The rate of increase of cell biomass with time is mathematically expressed as

$$dX/dt = \mu X$$

where X = cell mass and μ = the specific growth rate. The rate of increase in cell number may also be expressed as

$$dN/dt = \mu_N$$

where N = cell number.

The amount of time it takes to double the initial biomass may be expressed as

$$t_d = \ln 2/\mu$$

Similarly, the time it takes to double the cell number (mean generation time) may also be expressed as

$$t_g = \ln 2/\mu_N$$

where μ_N = the specific cell number growth rate. The doubling time is highly variable among organisms. Bacteria double every 45 minutes, while plant cells double every 3,600 to 6,600 minutes. The μ and μ_N are equal in the case of bacteria in which each cell division produces two identical cells by binary fission. However, in other organisms (e.g., yeast, molds, plant cells), the two quantities are not always the same. The specific growth rate is unique for each organism. It is influenced by the conditions of the cultural environment (temperature, pH, medium composition, dissolved oxygen levels).

Stationary Phase

The specific growth rate is a function of the concentration of the limiting substrate. When a critical growth substrate (e.g., carbon dioxide or nitrogen source) is exhausted, the increase in cell mass slows down and eventually ceases, marking the onset of the stationary growth phase. This phase may be ushered in by an accumulation of toxic substances representing metabolic by-products. The amount of biomass remains constant during the stationary phase. This phase is of economic importance. Antibiotics are usually produced during the stationary phase of microbial growth. The duration of this phase depends on the organism and the cultural conditions.

Death Phase

For most commercial bioreactors, the cells are harvested before the death phase begins. This phase is characterized by cell lysis or other mechanisms that cause loss of cell viability. The metabolic activities of cells cease.

■ *MICROBIAL CULTURE SYSTEMS IN BIOREACTORS*

There are three basic ways in which microorganisms are grown in bioreactors—batch, fed-batch, and continuous fermentation cultures.

1. **Batch fermentation**

   Batch fermentation is the most widely used system for growing microorganisms in commercial bioreactors. In batch culture, the culture medium is inoculated only once, at the beginning of the process, with the appropriate amount of inoculum. The growth patterns described in Figure 18.1 pertain particularly to batch fermentation.

2. **Fed-batch fermentation**

   In fed-batch fermentation, the volume of substrate is periodically increased during the reaction process. The consequence of this approach is an extension of both the exponential (log) phase and stationary phase, leading to an increase in biomass production. The yield of metabolites (antibiotics) is also increased. However, this phase is often characterized by the synthesis of proteases by the microorganisms that can attack any proteins in the reaction chamber. Consequently, it is not suited for production of bioengineered protein from a recombinant microorganism, unless the process is altered before the fermentation reaction reaches the undesirable stage. Fed-batch fermentation systems require closer monitoring to determine when fresh substrate should be added to the reaction chamber.

3. **Continuous fermentation**

   In continuous fermentation, the reaction is brought to, and maintained at, a steady state in which the rate of increase of cell biomass with time (dX/dt) is equal to zero. The loss of cells due to product harvesting is balanced by a gain in production of new cells from new growth by cell division. At a steady state, the total number of cells and the total volume in the bioreactor are constant.

■ *BATCH VERSUS CONTINUOUS FERMENTATION*

There are similarities and significant differences between batch and continuous fermentation systems:

1. Researchers are more familiar with batch fermentation than continuous fermentation systems.
2. Bioreactors used to produce a specific amount of product are smaller in size for batch fermenters.
3. Batch fermenters produce a large amount of product at the end of the reaction, requiring large capacity equipment for processing. Product harvesting in continuous fermenters is periodic and hence results in small amounts, requiring smaller capacity equipment.
4. Batch fermenters require a period of no production during which the equipment is cleaned and prepared for the next production cycle. Continuous fermenters remain productive for a longer period of time.
5. Product uniformity is more uniform under continuous fermentation because the physiological state of cells is more uniform. Possible variations in the time of harvesting of product can lead to variations in cell physiology and consequently product from batch fermentation.
6. Continuous fermenters are more susceptible to contamination because it is more difficult to keep a large-scale operation sterile over a longer period of time than for a shorter one.
7. Because continuous fermentation occurs over a longer period of time (500 to 1,000 hours), cells are susceptible to a loss of recombinant plasmid constructs and consequently a decrease in yield of desired product with time. Organisms in which the recombinant construct is spliced into the genome rather than plasmids are less susceptible to this instability.

■ *FACTORS AFFECTING THE FERMENTATION PROCESS*

Temperature

The efficiency of conversion from carbon source to cell mass depends on temperature; it declines as temperature declines. Microorganisms have a preference for the optimal temperature for growth. **Psychrophiles** are microbes that prefer an optimum temperature of about 15° C. **Mesophiles** prefer about 37° C, while **thermophiles** prefer about 55° C. Temperatures below optimal levels induce slow growth and reduced productivity. By the same token, higher-than-optimal temperatures may induce adverse effects (such as premature expression of the target gene or induction of undesirable cellular proteins). It should be pointed out that the optimal temperatures for growth of microorganisms and formation of desired products are not necessarily identical. Heating is deliberately provided but can also be generated as a result of the fermentation process itself, stirring of the reaction tank, or from aeration.

Dissolved Oxygen

Oxygen is supplied to a commercial bioreactor in a continuous stream as sterilized air that produces bubbles in the medium. The maximal oxygen demand in fermentation (Q_{max}) depends on the cell mass (X), the maximal specific growth rate (μ_{max}), and the growth yield based on oxygen consumed (YO_2). Mathematically, this is expressed as

$$Q_{max} = X_{\mu max}/YO_2$$

pH

The pH of the reaction environment in a bioreactor is monitored regularly because it is susceptible to change as cellular metabolites are released into the medium. The ideal pH for microbial growth is between 5.5 and 8.5. As needed, pH may be adjusted with an acid or base.

Degree of Mixing

The bioreactor medium must be thoroughly mixed by stirring to ensure a homogeneous growth environment for optimal productivity. When the medium is aerated with oxygen or if pH is adjusted, it is critical that the medium be stirred. Mixing also prevents the accumulation of toxins in certain parts of the reaction chamber. It promotes efficient heat distribution through the medium as well as uniform mixing of other medium additives (e.g., antifoaming agents). Thorough mixing is more difficult to achieve when the fermentation is on a large scale. Further, overmixing may increase shearing and increase temperature, with both events being detrimental to cell viability.

■ *BIOREACTORS*

There are two basic bioreactor designs: stirred tank and pneumatic reactors.

1. **Stirred tank**

    Most of the aerated commercial bioreactors have the stirred tank design. The commercial tank is normally constructed of stainless steel. The standard stirred tank is described in Figure 18–2. The medium level in the tank is equal to the diameter of the tank. This leaves headroom to accommodate the foaming which develops from the stirring of the medium. Antifoaming agents are often added to reduce foaming. Such additives must be removed before downstream processing. Air is introduced under pressure through a **sparger** (which may be in a form of a ring with many holes or a tube with a single orifice). Whereas this causes some agitation, agitation is mainly caused by an impeller (or several impellers), an internal mechanical agitator that is propelled by a motor. Baffles prevent vortexing or swirling of the tank. Impellers vary in design (e.g., 6-bladed turbine, helical ribbon, and propeller). An important consideration in a bioreactor design is gas-liquid mass transfer (transfer of

FIGURE 18–2
A stirred tank bioreactor.
An impeller is used to stir
the culture in the tank
while fresh air is
introduced through a
sparger.

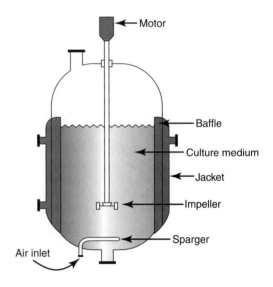

FIGURE 18–3
A bubble column
bioreactor. The culture
medium is stirred by
introducing air through an
inlet in the bottom of the
tank.

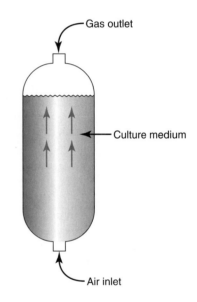

oxygen from the gas phase to microorganisms suspended in gas-liquid dispersion).
This goal is effectively accomplished in stirred tanks. Tanks are designed to allow
thorough sterilization of all parts to eliminate contamination.

2. **Pneumatic reactors**

Pneumatic reactors use air under pressure for stirring the fermentation tank.
They are more energy efficient than stirred tanks because they utilize air stream instead of mechanical devices. The absence of a shaft also minimizes the opportunities
for contamination. These reactors also subject the cells to less shear stress and are
suitable for tissue culture and genetically engineered microbes. They are easy to
clean. However, they are limited in operational flexibility.

a. **Bubble column**

The pressured air (oxygen) is introduced into a bubble column near the bottom of the tank. The consequences of this design include excessive foaming. Furthermore, as the air bubbles rise, they coalese into bigger bubbles, leading to
nonuniform gas distribution.

Unlike stirred tanks, the height-to-diameter of gas-agitated reactors is unequal
(see Figure 18–3). The gas flow has to be set at a velocity that promotes bubble

flow (homogeneous bubbling) rather than the chaotic turbulent flow (churn-turbulent). Height-to-diameter ratios of four or higher are commonly used in designs of gas-sparged reactors for optimal accommodation of gas-holding (volume fraction of gas in dispersion) produced under various conditions.

b. **Airlift bioreactors**

In airlift bioreactors, the liquid is induced to circulate. The liquid pool has two distinct zones, one of which is usually sparged by gas introduced in the bottom of the vertical channel called the riser. Gassing causes the liquid to be less dense. The gassed and ungassed zones have a different gas holdup that produces differences in bulk densities of the liquid in these two zones, thereby inducing a circulation of the fluid in the reactor. As the less bulky gassed liquid flows up the riser, it reaches an open space at the top where it becomes at least partially degassed. The bulkier degassed fluid descends in a separate channel called the **downcomer**.

Airlift bioreactors are produced according to one of two basic designs—internal loop or external loop. Modifications of these basic designs occur. Internal loop designs accommodate both the riser and downcomer in the same reactor container (see Figure 18–4). In the external or outer-loop configuration, the culture liquid circulates through two separate channels that are linked near the top and the bottom (see Figure 18–5). Whereas the internal loop design is simpler, it is less flexible to operate. Its volume and circulation rates are fixed for all fermentation processes. However, the mass transfer is higher in internal loop than external loop designs, while shearing is lower. On the other hand, when the culture liquid is of high viscosity and requires higher turbulence and shear for adequate mixing and mass transfer, external loops may be preferable.

FIGURE 18–4
An internal loop airlift bioreactor. Air is introduced into the tank from the bottom such that an internal cyclic flow is created. The downward flow is the downcomer.

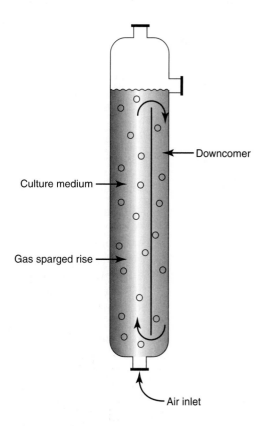

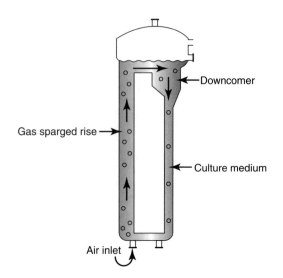

FIGURE 18–5
An external loop airlift bioreactor. The structural design of the tank causes the air that is introduced to create a cyclic flow in an external loop.

Downcomer

Gas sparged rise

Culture medium

Air inlet

■ *DOWNSTREAM PROCESSING*

After the bioreaction phase, the microbial cells are harvested so that the product that was synthesized during the microbial fermentation may be retrieved, purified, and packaged for use or sale. A number of processing steps are involved at this stage: cell disruption, debris removal, fractional precipitation, ion exchange chromatography, gel filtration, affinity chromatography, and crystallization. From the economic standpoint, the fewer the steps, the better. Some applications do not need high purity of product and hence only a few steps are required in purification.

Solid-liquid Separation

The first step in the purification process is to separate the cells (solids) from the liquid. Different technologies are available for solid-liquid separation; the ones commonly used in bioprocessing are centrifugation and filtration.

Centrifugation

High-speed centrifugation is commonly used in biotechnology as the method of choice. Centrifuge designs are variable, including the simplest tubular bowl and more complex disc-stack and scroll discharge decanter types. Centrifuges operate like sedimentation tanks, except that gravitational force is applied to accelerate the rate of sedimentation. The suspension is fed into the inner chamber of the centrifuge in a continuous flow. Cells are retained while the liquid collects in an external chamber. It is necessary to interrupt the operation periodically to remove the packed solids inside the centrifuge. Some designs automatically discharge accumulated solids.

Filtration

Membrane filtration is also used in biotechnology to separate solids from the culture medium (see Figure 18–6). Like centrifuges, there are different designs of membrane filters, the rotary vacuum filter being the most commonly used in the biochemical industry. In the conventional dead-end filter designs, cells accumulate on the membrane surface, decreasing the filtration rate progressively. An alternative design, cross-flow filtration, reduces the accumulation of solids by feeding the cell suspension in a flow direction parallel to the surface of the membrane. Separation membranes are usually constructed out of polymers. However, ceramic types (which have lower porosity) are available.

Cell Disruption

The microbial product of interest that was synthesized in the bioprocess may be excreted into the culture medium (extracellular product) or retained in the cell (intracel-

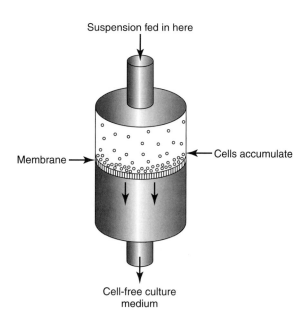

Suspension fed in here

Membrane →

← Cells accumulate

Cell-free culture
medium

FIGURE 18–6
Filtration of culture
medium following a
bioreaction. The
suspension is fed into a
cylinder with a filter
membrane installed in the
midsection. The membrane
causes cells to accumulate
while cell-free culture
medium is drained out of
the tank.

lular product). If extracellular, the cells are discarded while the medium is concentrated and subsequently purified. In the case where the target protein is intracellular, solid-liquid separation is followed by microbial cell disruption or lysis. Cell disruption methods may be categorized into two broad groups—nonmechanical and mechanical. The cell wall characteristics differ according to the species, the cellular growth rate, the growth phase in which cells were harvested, storage of the cells after harvesting, and other factors.

Nonmechanical Methods

Microbial cell walls may be disrupted by using various chemicals, including alkali, organic solvents, and detergents. It is critical that the pH of the environment into which the target product will be released upon lysis of the cell will permit the product to remain stable. Some detergents can denature proteins. Biological disruption of microbial cells by enzymatic lysis has high specificity. Common enzymes used alone or in combination include lysozyme, EDTA, β-1,3-glucanase, mannanase, and chitinase. However, these enzymes and detergents are expensive for large-scale application. Other nonmechanical disruption methods include physical methods like osmotic shock and application of pressure.

Mechanical Methods

Mechanical cell disruption methods are most commonly used in industrial applications. There are two general categories of mechanical methods—solid shear and liquid shear. Solid shear methods involve a grinding action of extrusion of frozen cells through narrow gaps under high pressure. Examples include wet milling in which a concentrated cell suspension is pumped into a high-speed agitator bead mill. The mill contains steel beads or glass beads, both inert abrasive materials, and fitted with a central shaft with attached agitator discs or blades. Liquid shear methods involve procedures such as acceleration of the suspension at high velocity and under high pressure to impact a stationary surface. Sometimes, two parallel streams of cell suspensions are impinged against each other to achieve disruption of cells. The device that allows this interaction between multiple liquid jets is called a microfluidizer™. In another technique, called high-pressure homogenization, the cell suspensions are drawn into a pump cylinder and then forced through an adjustable discharge valve with a restricted orifice. The pressure is rapidly decreased as the suspension leaves the valve seat, causing cavitation bubbles that formed to implode. The shock energy released and the turbulence which develops cause the cell walls to disrupt.

Precipitation

Precipitation converts the soluble protein product to an insoluble form. This is accomplished by altering the solute-solvent interactions by manipulating the pH-dependent charges on protein molecules. At the isoelectric point, the protein molecule carries a net charge of zero and becomes insoluble in a polar solvent and consequently precipitates. Organic solvents tend to denature proteins. The most commonly used protein precipitation technique is salting-out, in which certain salts are added to protein solutions to decrease protein solubility. Both anions and cations are used in salting-out procedures. The lyotrophic series of cations that may be used is $Al^{3+} > Ba^{2+} > Ca^{2+} > Mg^{2+} > NH^{4+} > Na^+$, and for anions is citrate $^{3-} >$ tartarate$^{2-} >$ $PO_4{}^{3-} >$ $SO_4{}^{2-} >$ acetate- $>$ Cl^-. Generally, the higher the charge, the more effective the ion is in protein salting. The commonly used salts include ammonium sulfate, sodium sulfate, potassium phosphate, and sodium phosphate. Ammonium sulfate is cheap and soluble at low temperatures, but it is corrosive and releases toxic ammonia in alkaline conditions. For a particular salt, protein salting depends on the ionic strength of the solution.

■ CHROMATOGRAPHY

Chromatography is a separation technique in which a mixture is separated based on differential physiochemical interaction between the components of the mixture and a stationary phase. The key step in chromatography is the choice of the stationary phase, which is usually a solid packed in a column. The materials used in packing the column for bioseparations are porous and hydrophilic substances like agarose, cross-linked polyacrylamide, and cellulose. Alternatively, it may be liquid supported on a solid. The mixture to be separated is introduced at the top of the column in a small volume. A solvent is then applied to wash it down the column. Because the mixture components associate differently with the stationary phase, those that bind strongly move slowly down the column, thereby effecting mixture separation based on velocities. The mobile phase is always a liquid. Different types of chromatography exist based on the nature of the interactions between the two phases. These include affinity chromatography and ion exchange chromatography. The former is very useful for separating biologically active substances, while the latter is widely used for protein purification.

■ DRYING

Drying of the product is required for safe storage, packaging, and transportation. Dehydration is required in the production of vaccines, enzymes, pharmaceuticals, and many other products for their preservation. Drying may be accomplished by several techniques. Products that are thermolabile or biologically active are spray- or freeze-dried.

KEY CONCEPTS

1. A bioprocess may be broadly defined as an industrial operation in which living systems (microorganisms, animal or plant cell cultures or their constituents, or enzymes) are used to transform raw materials (biological or nonbiological) into products.
2. Converting science into products entails: (1) the presence of an engineering innovation, (2) the development of appropriate enabling technologies, and (3) an economic opportunity.
3. There are three key steps in a bioprocess: pretreatment, bioreaction, and downstream processing.

4. There are two main kinds of biofluids and slurries used in bioreactors: Newtonian fluids and non-Newtonian fluids.
5. Contamination is a major enemy of a successful bioprocess.
6. The growth medium is formulated to maintain microbial growth for economic production of the target product. Its design should be made based on a good understanding of the relationship between microbial growth kinetics and product formation.
7. Microbial growth typically progresses through four basic phases—lag, exponential, stationary, and death phases.
8. There are three basic ways in which microorganisms are grown in bioreactors—batch, fed-batch, and continuous fermentation cultures.
9. There are three basic bioreactor designs: stirred tank, bubble columns, and airlift reactors.
10. Downstream processing occurs after the bioreaction phase. This is the stage in which the microbial cells are harvested so that the product that was synthesized during the microbial fermentation may be retrieved, purified, and packaged for use or sale.

OUTCOMES ASSESSMENT

1. Define a bioprocess. Give an overview of its importance in biotechnology.
2. Describe the bioreaction phase of a bioprocess.
3. Describe the downstream processing phase of a bioprocess.
4. Discuss the importance of sterilization in a bioprocess.
5. Describe microbial growth patterns in a bioprocess, indicating the phase in which the economic product is produced.
6. Distinguish between batch fermentation and continuous fermentation.
7. Discuss the factors that affect the efficiency of a bioprocess.
8. Describe how a stirred tank bioreactor operates.

INTERNET RESOURCES

1. Journal of bioprocess and biosystems engineering: *http://link.springer.de/link/service/journals/00449/*
2. How a bioprocess engineer operates: *http://vega.soi.city.ac.uk/eu790/BioprocessEngineering.htm*

REFERENCES AND SUGGESTED READING

Christi, Y., and M. Moo-Young. 1991. Fermentation technology, bioprocessing, scale-up, and manufacture. In *Biotechnology: The science and business.* V. Moses and R. E. Cape. London, UK: Harwood Academic Publishers.

Moses, V., and S. Moses. 1995. *Exploiting biotechnology.* London, UK: Harwood Academic Publishers.

National Research Council. 1992. *Putting biotechnology to work: Bioprocess engineering.* Washington, DC: National Academy of Sciences.

Shuler, M. L. 2002. *Bioprocess engineering: Basic concepts.* New York: Prentice Hall.

SECTION 2
Microbial-Based Bioprocessing and Pharming

PURPOSE AND EXPECTED OUTCOMES

Industrial applications of biotechnology are centered significantly around the use of microorganisms in whole or their components (e.g., enzymes) to make products. Applications of microbes in food biotechnology were discussed in Chapter 17. This section focuses on selected non-food applications of microbes. Also, the use of organisms as bioreactors to produce pharmaceuticals will be discussed.

In this section, you will learn:

1. The application of whole microbes in producing industrial chemicals.
2. The use of plants and animals to make pharmaceuticals (pharming).

INDUSTRIAL ETHANOL

The energy crisis of the 1970s provided the much-needed impetus for large-scale production of ethanol for use as fuel (biofuel). A blend of 10 percent ethanol and 90 percent gasoline, called gasohol, was heavily promoted in the United States and other parts of the world (e.g., Brazil) as economic and environmentally friendly fuel. Industrial ethanol is produced primarily from biomass, the major ones being corn (the primary substrate in the United States) and sugar cane (mainly used in Brazil). Organic waste products from industrial facilities (e.g., whey, wastes from breweries, wastes from meat processing plants) are also used to a lesser extent. Sugar cane is the most productive biomass source for industrial ethanol, yielding about 4,000 liters per hectare per year, compared to about 1,150 liters/ha/yr from corn.

Large-scale producers use the wet milling process, which entails the fractionation of the milled product to remove oil and gluten before the starch is fermented. Prior to fermentation, starch is cooked to gelatinize, then hydrolyzed by enzymes (or less desirably, acids). The enzyme α-amylase used is derived from *Bacillus lichenformis*. The cooking process, followed by incubation, converts starch to dextrins. Dextrins are converted to fermentable sugars by a reaction catalyzed by amyloglucosidase derived from *Aspergillus niger* at a temperature of 60° C. A producer may use *Rhizopus niveus* for fermentation and thereby omit the conversion stage.

In terms of design, commercial ethanol producers in the United States utilize one of three types of fermentation systems—batch, single-vessel continuous, or multi-vessel continuous vessels in cascade.

BULK CHEMICALS

Two bulk chemicals for which there are established industries are **acetone** and **butanol.** These chemicals are important solvents. Acetone is a starting point for chemical synthesis while butanol is used as a paint solvent, in hydraulic fluids, and for extraction of antibiotics from culture broths. The major microbe involved in the commercial production of these bulk chemicals is *Clostridium acetobutylicum*. This microbe and others are able to ferment glucose to produce acetone and butanol as primary products.

INDUSTRIAL ENZYMES

One of the successful industrial applications of biotechnology is in the area of development and production of industrial enzymes. Not only are enzymes industrially produced in their natural states, but biotechnology also enables scientists to modify and tailor enzyme proteins to meet customer needs. It is estimated that industrial enzymes in 2002 would be a $1.8 billion industry. The dominant application of industrial enzymes worldwide is in the area of food biotechnology where the enzymes are used in food and animal feed. Common areas of applications are in the manufacture of starch-derived syrups, alcoholic beverages, dairy products, and animal feed. Other applications in food biotechnology are in bakery products, fruit and vegetable processing, protein processing, and vegetable oil extraction.

Next to food biotechnology, the most important use of industrial enzymes is in the detergent industry where these enzymes are used in manufacturing more efficient dish and laundry detergents. Industrial enzymes also find significant application in the textile industry where enzymes are used for processing cotton and cellulosic textiles. Enzymes are also used in the leather, as well as the pulp and paper, industry.

WASTE TREATMENT

Raw sewage is laden with microorganisms, some of which are pathogenic. Because domestic sewage is usually discharged into rivers or the sea, resources that are used by humans in a variety of ways including drinking, fishing, and recreation, humans are exposed to pathogens and toxins they may carry. Consequently, it is customary for municipalities to establish sewage treatment facilities to remove organic materials, suspended solids, toxic compounds, and pathogens before discharging the treated sewage into these water bodies. The steps in a typical waste purification process are summarized in Figure 18–7. The sludge is subjected to anaerobic digestion by a mixture of microbes to less-harmful products prior to burying or disposing of it in the seas. The liquid portion is usually aerobically oxidized. A variety of microbes are involved in waste treatment processes, including heterotrophic bacteria (e.g., *Achromobacter, Flavobacterium*, and *Zooglea, Beggiatoa*, and *Leptothrix)*, heterotropic fungi (e.g., *Ascoidea, Fusarium*), autotrophic bacteria (e.g., *Nitrosomonas, nitrobacter*), algae (*Phormidium, Chlorella)*, and protozoa (*Amoeba, Aspidisca, Chilodonella)*. The composition of microbial population depends on the composition of the sewage.

Anaerobic digesters operate in phases. The first phase involves hydrophytic fermenters (e.g., *Bacteriodes, Clostridium*, and *Eubacterium)* that hydrolyze polymers (e.g., polysaccharides) to monomers. These monomers are then fermented to short chain organic acids (e.g., alcohols, esters). This phase is followed by the action of acetogenic bacteria (e.g., *Peptococcus, Propionibacterium)* that convert the products of the first phase to acetic acid, H_2, and CO_2. The third phase involves two different groups of bacteria, the slow growing acetoclastic methanogens (e.g., *Methanosarcina, Methanospirillum)* that produce CH_3 and CO_2 from acetate and hydrogenotropic methanogens (e.g., *Methanobacterium)* that produce methane from H_2 and CO_2.

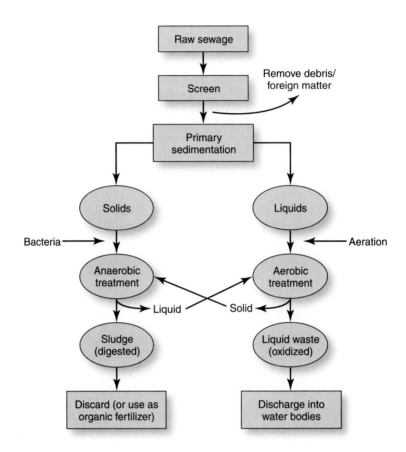

FIGURE 18–7
A summary of a waste management scheme.

PHARMING

In addition to genetically engineering animals and plants to improve their food qualities (e.g., leaner meat, increased milk production, enhanced nutritional content), another fledging biotechnology industry is using animals and plants as chemical factories to produce commercial quantities of certain pharmaceutical products. Dubbed **biopharming, animal pharming,** or **plant pharming** (or simply **pharming**), certain farm animals or plants are transformed with transgenes from humans that encode simply inherited gene products that are used to treat diseases or improve quality of life.

■ *ANIMAL PHARMING*

Animal pharming for drugs is applicable to drugs that are made of protein. The production of human proteins outside of humans is not exactly a recent discovery. In 1982, Genentech, Inc. received an FDA approval to market genetically engineered human **insulin.** This was the first approved genetically engineered drug. It is commercially produced by using *E. coli* bacteria transformed with the gene that encodes insulin in humans. Prior to this, commercial production of insulin supplement for use by diabetic patients was from slaughtered pigs. Similarly, human cadavers were the sources of human growth hormone. In 1987, tPA (tissue plasminogen, used to treat blood clot) was produced in transgenic mice.

Whereas some human protein drugs can be produced in prokaryotes, there are others (e.g., tPA, erythropoietin, **blood clotting factors** VIII and IX) that require certain modifications that can be provided through the expression of the corresponding genes in eukaryotic cells. However, mammalian cell cultures are expensive to maintain, and they yield small quantities of the drugs to make the technology economically viable. Animal pharming is the current state of the science whereby *in vivo* production has replaced *in vitro* production of drugs. Genetically engineered animals can multiply rapidly to produce the desirable population of animals. These animals are fed and raised like conventional types. An attractive feature of using mammals is that most of the protein drugs can be targeted for expression only in the mammary glands. This provides a very convenient way to harvest the product through the milk of the animal. Furthermore, the ability to produce protein drugs is heritable through the transgene.

Commercial animal pharmers have developed protocols around selected animals for the production of specific drugs. Organisms to use in pharming are selected based on technical and economic factors. Organisms vary in their response to/and expression of transgenes, drug yield per unit of milk, success rate of transformation, techniques for transformation, and cost of maintenance, among other factors. For example, the success rate of producing transgenic embryos in mice is about 10 to 30 percent, compared to only 5 percent in larger animals like cows, sheep, and goats. However, a rabbit can produce about 10 L of milk per year whereas a dairy cow can produce about 10,000 L a year. The yield of the transgenic protein varies from 1 to 20g/L depending on the species and the protein. The demand for certain drugs is low (in terms of volume), whereas others are required in large quantities.

The food processing agent called **chymosin** (also called **rennin**) is an enzyme used in the manufacturing of cheese. It was the first genetically engineered food additive to be used in food biotechnology. The traditional method of obtaining chymosin is from rennet (a preparation derived from the fourth stomach of milk-fed calves). It is estimated that more than 60 percent of U.S. hard cheese producers use genetically engineered chymosin.

A sample of organisms being used in commercial animal pharming and the kinds of proteins produced in them is presented in Table 18–1. One of the most valuable protein drugs is **human protein C**, a drug used to treat blood clots. These drugs are produced via the transgene milk technology. A notable exception is human hemoglobin that is expressed in the blood of pigs. Transgenic blood technology is likely to remain an exception, not the norm, because the animals have to be slaughtered to obtain the blood, requiring a replacement herd after each production cycle.

TABLE 18–1

Selected products of animal pharming. Most products are produced in the milk of animals.

Drug	Animal System	Function/Use
AAT	Sheep	Alpha-1-antitrypsin is used to treat the inherited deficiency that leads to emphysema
tPA	Goat	Tissue plasminogen activator is used to treat blood clots
Factors VIII, IX	Sheep	Blood clotting factors for treating hemophilia
Hemoglobin	Pig	Blood substitute for use in blood transfusion in humans
Lactoferrin	Cow	Used as an additive in infant formula
CFTR	Sheep, mouse	Cystic fibrosis transmembrane conductance regulator is used for treating cystic fibrosis
Human protein C	Pig	An anticoagulant for treating blood clots

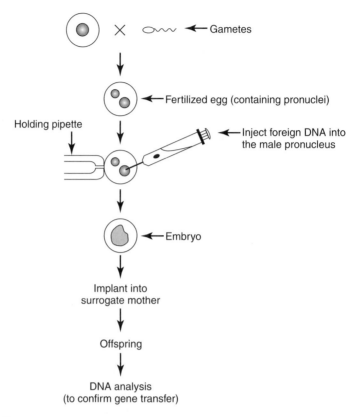

FIGURE 18–8

A summary of the method of microinjection in the production of transgenic animals. An egg is fertilized *in vitro.* The target DNA is prepared and injected into the male pronucleus, using a specialized pipette to hold and stabilize the fertilized egg. The resulting embryo is implanted in a surrogate female for development to full gestation.

In producing the transgenic animals for these products, the gene(s) for the protein drug of interest is isolated and cloned as described elsewhere. The gene is coupled with a mammary directing signal for expression in the mammary glands. The transfer of the transgene into the organism of choice may be accomplished by one or two basic techniques as summarized in Figure 18–8. Method "a" involves microinjection of transgene DNA into a fertilized egg. An early embryo is implanted into a surrogate mother. Method "b" uses the same nuclear fusion technique that was used in the cloning of the first mammal, "Dolly" the sheep (see Figure 18–9).

In terms of safety concerns, all drugs require scrutiny and approval from the FDA. Pharmaceutical production from transgenic animals is no exception. Animals will have to be pathogen-free to avoid transmission of any communicable disease from animals to humans. On a continuing basis, animal pharming is believed to be 5 to 10 times more economical and 2 to 3 times cheaper in starting costs than systems using cell cultures.

■ *PLANT-BASED PHARMING*

Attempts are being made to genetically engineer plants to produce new products including pharmaceuticals, specialty chemicals, and bulk industrial products. Early experimental applications include the production of antibodies, human serum albumin, and *Bacillus lichenformis* α-amylase in plants. Uses of antibodies include therapeutic agents, diagnostic immunoassays, and vaccines as active agents in certain commercial

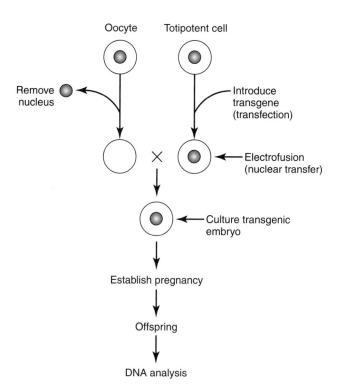

FIGURE 18–9
A summary of the method of nuclear fusion in the production of transgenic animals. This method of nuclear transfer is used for the introduction of the transgene into a totipotent cell. The altered nucleus is transferred into an oocyte that has been emptied of its nucleus. The resulting embryo is implanted in a female to develop to term. This method is the procedure used in the production of "Dolly" the sheep, except that no genetic manipulation was involved.

assays. Alpha-amylases are used in starch processing and in alcohol industries for liquefaction. They are also used in the baking industry to increase bread volume and in the brewery industry for producing low-calorie beers. The α-amylase from *B. lichenformis* is highly conducive for industrial purposes because it is highly thermostable and active over a wide range of pHs. A Canadian company, SemBioSys Genetics Incorporated, has produced the anticoagulant **hirudin** in canola. This compound was originally found in the saliva of leeches.

KEY CONCEPTS

1. Industrial applications of biotechnology are centered significantly around the use of microorganisms in whole or their components (e.g., enzymes) to make products.
2. One of the successful industrial applications of biotechnology is in the area of development and production of industrial enzymes.
3. Other industrial applications include the production of alcohol and bulk chemicals (acetone and butanol).
4. Microbes are used in municipal waste treatment plants.
5. Pharming is the production of pharmaceuticals in living systems (plants and animals).
6. Plant-based pharming is less well developed.

OUTCOMES ASSESSMENT

1. Briefly discuss the production of alcohol by fermentation.
2. Discuss the application of biotechnology in the production of industrial enzymes.

3. Give specific examples of commercial animal pharming.
4. Give specific examples of commercial plant pharming.

INTERNET RESOURCES

1. Industrial enzymes: *http://www.bio.org/er/enzymes.asp*
2. Industrial enzymes and uses: *http://www.dyadic-group.com/wt/dyad/enzymes*
3. Site for novozyme company: *http://www.novozymes.com/cgi-bin/bvisapi.dll/portal.jsp*
4. Production of industrial enzymes: *http://www.pall.com/applicat/biosep/applications/ enzymes.asp*
5. Therapeutic protein production in plants: *http://sbc.ucdavis.edu/Outreach/resource/ pharma_crop.htm*
6. Pharming: *http://www.biotech.iastate.edu/publications/biotech_info_series/bio10.html*
7. Views on pharming: *http://www.biotech-info.net/other-apps.html#pharming*

REFERENCES AND SUGGESTED READING

Biotol Partners Staff. 1992. *Operational modes of bioreactors.* Burlington, MA: Butterworths-Heinemann.

Uhlig, H. 1998. *Industrial enzymes and their applications.* New York: John Wiley and Sons.

SECTION 3
Biosensors

PURPOSE AND EXPECTED OUTCOMES

Diagnostic instruments are depended upon to provide information about biochemical or biological reactions that occur in living organisms or their environment. Such information is critical to determining the conditions of the organism and for prescribing a course of action in case the condition is abnormal.

In this section, you will learn:

1. The design of biosensors.
2. Selected applications of biosensors.

WHAT ARE BIOSENSORS?

Diagnostic instrumentation is used in medical, industrial, agricultural, and environmental applications. Diagnostic laboratories provide services to clients from these fields by processing specimens that are sent to them. Increasingly, there is a demand for diagnostic procedures that yield immediate results that are adaptable to field diagnostics.

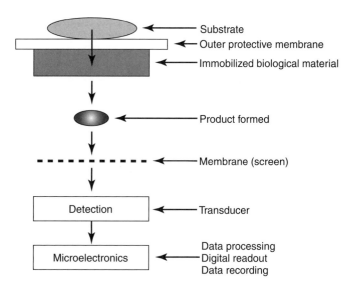

FIGURE 18–10
A schematic presentation of biosensor design and operation. The substrate to be analyzed must make contact with the immobilized biological material through the protective membrane. The substrate and the biological material interact to form a product that is detected by the transducer and relayed to the microelectronic system for display.

Biosensors are diagnostic instrumentations that provide accurate on-the-spot results of biochemical tests. These instruments are designed based on a combination of principles from various areas—biochemistry, membrane technology, and microelectronics. A biosensor is more than a biological probe like a pH meter that measures a biochemical or biological reaction. Integrated into the design and construction of a biosensor is a biochemical component that allows signals from specific biochemical reactions to be detected, measured, and documented. A schematic operation of a biosensor is presented in Figure 18–10.

MEMBRANES USED IN BIOSENSORS

A critical component of this analytical tool is an immobilized biological material (e.g., enzyme, nucleic acid, antibody, tissue, cells). The common techniques of immobilization are covalent attachment and physical entrapment. In covalent attachment, chemical groups on a biologically active material that do not play a role in the activity of the material are attached to chemically activated supports (e.g., synthetic or natural polymers, glass, ceramics). Another effective technique is physical entrapment in which gels or polymers (e.g., polyacrylamide, silica gel, starch) are crosslinked in the presence of the biological material resulting in its entrapment. Membranes play a significant role in biosensors. The substrate to be analyzed flows through an outer protective membrane that may have selective capacity to screen out undesirable materials. In some designs, the product (reaction between the immobilized compound and the biological compound) is selectively screened before reaching the transducer for detection. Membranes of biological origin include cellulose, collagen, and gelatin. Some membranes are made of composites of multilayered structures, or polymers.

A membrane also limits the rate of diffusion of an analyte to the biological element of a biosensor, thereby preventing high concentrations of the analyte from overwhelming the sensor. Furthermore, such restrictions allow the instrument to achieve a linear response to the analyte concentration. Biosensors usually produce a digital electronic signal that is proportional to the concentration of the specific analyte or group of analytes.

THE ROLE OF TRANSDUCERS IN BIOSENSORS

A transducer is basically a device that converts one form of energy to another. A biosensor utilizes a physiochemical transducer that is in intimate contact with the biorecognition layer, essentially forming a biotransducer to convert chemical energy into electrical energy. Transducer combinations used in biosensors include optical, electrochemical, field effect transistors, thermometric, piezoelectric, or magnetic, some of these being more practical than others. They record differences in potential, while thermisistor biosensors record minute variations in temperature. Examples of immobilized enzymes used in electrochemical systems and their substrates are alcohol oxidase (alcohol), glucose oxidase (glucose), urease (urea), and L-aminoacid oxidase (general amino acids).

The glucose meters (electrochemical biosensors) are commercially successful and readily accessible to the public. These meters may be purchased from the local pharmacy, and are essentially enzyme electrodes. The assembly of the unit varies among producers. The Boringher-Roche Glucosan model consists of a freeze-dried layer of the enzyme glucose oxidase, which reacts with the sample of blood to be tested. The reaction produces H_2O_2 that is oxidized directly or via redox mediation at metallic electrodes. In terms of quantification, the magnitude of the current generated is directly proportional to the concentration of H_2O_2 produced. Furthermore, the concentration of H_2O_2 produced is proportional to the rate at which the substrate (blood glucose) is converted to gluconolactone (gluconic acid). Redox-mediated and other direct electron transfer biosensor systems have been widely studied. The design objective in these biosensors is to facilitate the transfer of electrons generated by an oxido-reductase enzyme to the electrode surface. Even though natural mediators exist (e.g., cytochromes), dyes and artificial mediators (e.g., ferrocane) have been most successfully incorporated in biosensor design.

Optical biosensors use optically based transducers that measure changes such as optical absorption reflectance, fluorescence, bioluminescence, and chemiluminescence. The advances in the use of optical fiber sensors have been facilitated largely by the development of inexpensive light-emitting and photosensitive diodes as well as fiber optics systems that are amenable to use in transducer construction.

BIOAFFINITY SENSORS

Bioaffinity sensors utilize biological materials that are capable of molecular recognition. The general principle of these sensors is that a labeled receptor is weakly bonded to a determinant analogue that has been immobilized to a transducer surface. As a sample (free determinant) is introduced, it displaces the receptor in proportion to the concentration of the determinant. The higher the concentration, the more the displacement of the receptor to form a tightly bonded complex, consequently reducing the signal from the labeled receptor accordingly.

WHOLE CELL BIOSENSORS

Biorecognition layers in biosensors may be designed to contain subcellular fragments (e.g., organelles like mitochondria, or receptor-carrying fragments like single whole cells, a small number of cells, or tissue slices). The biosensor configuration is either potentiometric (like a pH meter) or amperometric (like an oxygen electrode). It is usually designed to react to a broad spectrum of substrates, and its interactive response output is slow. A microbial biosensor for conducting the Ames test for mutagenecity of a compound has been developed. The biological materials involved are immobilized whole cells of *Salmonella tymhimurium* revertants or *Bacillus subtilis* Rec. associated with an oxygen electrode. Other sensors include those for alcohols (e.g., ethanol) using immobilized *Trichosporon brassicae* or *Acetobacter xylinium* to detect O_2, gases (methane) using *Methylomonas flagellata* to detect O_2, and for sugars using bacteria from human dental plaque to detect H^+.

GENERAL CONSIDERATIONS

Biosensors may be designed to display continuous digital electronic signals or modified to produce discrete signals. Generally, it is difficult to develop a biosensor to measure an analyte for which there is no preexisting means of measuring. Therefore, many biosensors are based on conventional laboratory protocols for bioassay. Biosensors differ in sensitivity, with some being able to detect minute concentrates of target biological material (substrate). Furthermore, the stability of the biorecognition components of the sensor is variable. The range for stability of glucose in the glucose oxidase system ranges between 20 and 420 days, whereas sucrose is only 14 days. Other issues of concern in the development and application of biosensors is the vulnerability of the biotransducer to materials that interfere with or tend to foul up its sensitive parts. Manufacturers have to develop this equipment to meet the market demands in terms of size, range of analytes, sensitivity, versatility, and price. Sensors need periodic calibration to keep accurate. The biorecognition component has a lifetime during which it has to be replaced.

COMMERCIAL FORMS OF BIOSENSORS

Based on the needs in medical and industrial fields, biosensor designs may be operationally divided into four forms that are adaptable to the practical needs in these fields.

1. **Hand-held models**

     Hand-held biosensors are targeted to home users, doctors, and persons who need to conduct on-the-spot analysis. These devices are designed to be easy-to-operate by unskilled users. They are small in size (e.g., to fit in the palm like a calculator or be held like a pen). Their displays are clear and easy to comprehend. Hand-held biosensors operate at fast speed. The most successful so far are the devices used for monitoring blood glucose levels in diabetics. Victims of the disease can use this device for a quick and accurate evaluation of their blood sugar levels. Another popular biosensor is the pregnancy test kit. It is important that these models be robust in order to accommodate the likely rough handling under domestic use.

2. **Bench-top models**

     Bench-top biosensor models are targeted for use in laboratories. In the clinical setting, the devices enable a variety of measurements to be made on whole, undiluted blood, thus eliminating or minimizing the possibility of contamination and other errors associated with sample preparation (e.g., separation, dilution). They are often designed to measure several parameters of one or many samples simultaneously (multichannel analyzers).

3. **Flow monitoring devices**

     Flow monitoring devices are designed for monitoring selected parameters in continuous processes over a protracted period. One may be located strategically in a production line for online monitoring where a reaction or process is ongoing and requires periodic monitoring for quality control or other purposes. This online monitoring is common in food processing and fermentation reactions. It is also used in environmental monitoring (air and water pollution). For example, a design for monitoring pesticide pollution utilizes immobilized cholinesterase on porous pads. The device is placed where running water flows. If a pollutant (from, for example, an organophosphate pesticide) happens to be present, a reaction is triggered. The substrate is exhausted thereafter, and needs to be replaced.

4. **Implantable monitors**

This is one of the challenging platforms for biosensor design and application. Unlike the other models, the implantable monitor is designed for implanting inside an individual. The immediate challenges include miniaturizing the device to fit unobtrusively in the individual. The device should be able to withstand the harsh environment in which it will reside. It should be sterile and remain so in the recipient. Calibration of an implantable biosensor is also problematic. Early applications have focused primarily on the development of a glucose sensor to be implanted in a diabetic to control the rate of insulin infusion from a miniature insulin pump to be worn outside or carried by the patient.

APPLICATIONS OF BIOSENSORS

Biosensors have potential applications in personal health, monitoring, clinical analysis, environmental screening and monitoring, bioprocess monitoring, agricultural applications, the food and beverage industry, and others.

1. **Medical applications**

Medical applications range from bench-top laboratory analysis (e.g., the Yellow Spring glucose analyzer) to individual-use home testers. Reusable biosensors are superior to those being used in many situations for analysis and that require trained professionals to operate accurately. Health care monitoring will be significantly enhanced if accurate, inexpensive, fast, and easy-to-use metabolic profiles of patients are available. Doctors will also be able to diagnose and prescribe therapy within a short period of time, instead of the often long wait for laboratory results before prescribing a therapy. Other applications of biosensors could be in drug monitoring, cholesterol monitoring for heart patients, fertility monitoring, and invasive monitoring in critical care.

2. **Food and agricultural applications**

Potential applications in food and agriculture include reproductive monitoring (fertility monitoring), as well as monitoring of infectious diseases. In the dairy industry, biosensors could be used to monitor the nutritional profile of milk. Food processing offers opportunities for using biosensors for production and quality control. Biosensors for detecting food poisoning agents (e.g., *Salmonella*) will be immensely beneficial to the food and health industry.

3. **Industrial applications**

Industrial production based on fermentation processes (e.g., alcohol, pharmaceuticals) will benefit from continuous monitoring for high product quality and optimal production at minimum cost.

4. **Environmental applications**

Monitoring of pollution to water and air will help in enforcing environmental control policies and reduce dangers to the public from pollution to the water supply, food from aquatic sources, and air. Biosensors for monitoring the environment have great potential for military applications in detecting bioweapons.

KEY CONCEPTS

1. Biosensors are diagnostic instrumentations that provide accurate on-the-spot results of biochemical tests.
2. A biosensor consists of a membrane (synthetic or natural) to which a biological material (enzyme, nucleic acid, antibody) is immobilized.
3. Also, a biosensor utilizes a physiochemical transducer to convert chemical energy into electrical energy.
4. There are different types of transducers used in biosensors (e.g., optical, electrochemical, thermometric).
5. Glucose meters are electrochemical biosensors.
6. Biosensors differ in sensitivity (some detect minute concentrations of compounds); some biosensors display continuous digital electronic signals, whereas others produce discrete signals.
7. Commercial biosensors may be hand held, designed for the bench (bench top), utilized in flow monitoring, or implantable.
8. Biosensors are applied in the medical field (e.g., health care monitoring), food and agriculture (e.g., detection of food poisoning), industrial (e.g., fermentation process monitoring), and environmental (e.g., pollution monitoring).

OUTCOMES ASSESSMENT

1. Discuss the types and importance of membranes in biosensor design.
2. Discuss the types and roles of transducers in biosensors.
3. Discuss the differences in biosensor design and operation.
4. Give three specific applications of biosensors.

INTERNET RESOURCES

1. Biosensors: how they function: *http://www.devicelink.com/ivdt/archive/97/09/010.html*
2. Biosensors and other medical probes: *http://www.ornl.gov/ORNLReview/rev29_3/text/biosens.htm*
3. Intro to biosensors and types of sensors: *http://www.eng.rpi.edu/dept/chem-eng/Biotech-Environ/BIOSEN2/biosensor.html*
4. Biosensors: past, present, and future: *http://www.cranfield.ac.uk/biotech/chipnap.htm*

REFERENCES AND SUGGESTED READING

Bard, A. J., and L. R. Faulkner. 1980. *Electrochemical methods.* New York: John Wiley and Sons.

Blum, L. J., and P. R. Coulet. (eds.). 1991. *Biosensor principles and applications.* New York: Marcel Dekker.

Gronow, M. 1991. Biosensors—"A marriage of biochemistry and microelectronics." In *Biotechnology: The science and business.* V. Moses and R. E. Cape (eds). London, UK: Harwood Academic Publishers.

Ho, M. Y. K., and G. A. Rechnitz. 1992. An introduction to biosensors. In *Immunochemical assays and biosensor technology for the 1990s.* R. M. Nakamura, Y. Kasahara, and G. A. Rechnitz (eds). Washington, DC: American Society for Microbiology, 275–290.

Rechnits G. A. 1988. Biosensors. *Chem Eng News,* 5:24–36.

SECTION 4
Recovering Metals

PURPOSE AND EXPECTED OUTCOMES

Certain plants and microorganisms have the capacity to extract and retain large quantities of metals from the environment without adverse consequences. This unique capability of certain microbes is capitalized upon to recover precious metals from the soil.

In this section, you will learn:

1. Examples of organisms that have the capacity to extract metals from the environment.
2. The phenomenon of bioaccumulation.
3. The application of bioaccumulation in metal recovery.

WHAT IS BIOACCUMULATION?

The phenomenon of uptake of metals by organisms has been known for a long time. Unfortunately, some discoveries occurred as a result of the accumulation of toxic levels of certain metals that entered the food chain. Specifically, fish is known to accumulate mercury, causing mercury poisoning in humans. When mercuric sulfate from industrial processing is discharged into a water body, the microorganisms transform it into a more toxic form—dimethyl mercury. This compound accumulates in fish and poisons people who consume contaminated fish. The deliberate exploitation of the phenomenon of uptake of metals by organisms to accomplish the specific task of metal recovery in specific situations is recent and expanding. Specific application of this biotechnology includes mining of mineral ores and cleaning of polluted soils and water bodies.

EXAMPLES OF BIOACCUMULATION

Examples of bioaccumulation of metals include aluminum (in *Aspergillus niger*), arsenic (in *Daniella* sp.), cadmium (in *Zooglea ramigera*), lead (in *Spirogyra*), copper (in *Cladosporium resinae*), mercury (in *E. coli*), gold (in *Chlorella vulgaris*), and iron (in *Heptothrix* sp.).

Most of these metals are required as micronutrients in the nutrition of organisms, playing diverse roles, such as enzyme activators and as structural components of biological molecules (e.g., magnesium in chlorophyll). Organisms generally possess mechanisms for detoxifying metals. This includes exclusion from the cell (e.g., ferromanganese-depositing bacteria and fungi which accumulate Fe and Mn oxides), isolating the metal in the cytoplasm by concentrating it in granules, precipitating it in the cell wall (e.g., lead by *Micrococcus luteus*), or transforming it (e.g., by oxidation or reduction) into a harmless form in the organism (e.g., Mo, Hg by *Microccocus* sp.).

Microbes are able to adsorb or complex metals because of the presence of a large number of chemical sites to which these metals may bind. Metallothioneines are proteins that strongly bind metal ions (e.g., metal components of enzymes and pigments like Mg in chlorophyll). Metal accumulation in organisms is a very rapid process, with over 90 percent occurring within 10 minutes of the first contact with the metal ions. Inactivated (dead) biomass has been shown to be more effective than living cells (e.g., uranium uptake by *Thiobacillus ferrooxidans*).

Organisms that are capable of complexing metal with extracellular polymers and cell wall polymers can survive in aqueous environments that are heavily laden with levels of metallic ions that would be toxic if retained in the cytoplasm. Such organisms are suitable for use in water treatments to remove polluting metals.

METHODS OF METAL REMOVAL

Metal removal may be accomplished by using two strategies: growing biomass or non-growing biomass.

1. **Growing biomass**

   Continuously growing cultures of microorganisms are used in waste water treatment to remove metals prior to discharging into a water course. This activated sludge process uses microbes to oxidize the organic matter in the wastewater, and in the process also removes metals by adsorption by the biomass. The metals are removed as a combination of activated sludge bacterial flocculation and settling to the bottom of the treatment tank, leaving metal-free supernatant that can be discharged into a watercourse. The metal removal efficiency of activated sludge systems can vary from 0 to 100 percent, depending on factors like the type of metal, pH, and concentration of the metals in the wastewater.

2. **Non-growing biomass**

   The biomass in this system is precultured and used as an absorbent for the metals. The advantage of this system is that it is not susceptible to metal toxicity. Biosorbent materials may be immobilized into beads or granules and used to recover metals from wastewater. These materials are generally more effective than conventional treatments (caustic and sulfide precipitation) at removing metals. High performance microbes used in biosorbent materials include *Aspergilus niger* and *Rhizopus arrhizus*.

BIOTECHNOLOGY FOR MINERAL EXTRACTION

The use of microbes to recover metals from mineral ores has its roots in history. Unknowingly, the Romans used this biotechnology to recover copper in 1000 B.C. The microbial involvement in the process was formally described in 1947. The importance of microbes in metal recovery is in the mining of low-grade ores or revisiting ore dumps. Commercial mining is often halted after extraction becomes uneconomical. Low-grade ores are more expensive to mine. It is estimated that the grade of copper has declined from about 2 percent to about 0.6 percent over the last century. A desirable aspect of the use of microbes in mineral ore extraction, sometimes called **biomining** in the biotechnology lingo, is that it is environmentally friendly.

MECHANISMS OF BIOMINING

The extraction of mineral ore by bacteria occurs by the chemical process of leaching and hence is called **bacterial leaching** or **bioleaching.** Bioleaching depends on the microbe and the rock mineral type. The minerals most successfully bioleachable are those that contain iron and sulfur. A variety of microbes are involved in bioleaching, especially heterotrophic bacteria and fungi, yeast, and algae. Some of these microbes are ubiquitous. The most commonly known is *Thiobacillus ferrooxidans.* This bacterium has the capacity to oxidize reduced forms of sulfur and iron under highly acid conditions (pH 1 to 2.5). Other leaching bacteria are *T. thiooxidans, T. acidophilus,* and *T. organoparus.* These microbes are aerobic chemoautotrophs that reduce these metals and fix carbon dioxide from the atmosphere, while satisfying their energy needs from the process. The direct mechanism is enzymatic, and may be summarized for iron as follows:

$$4FeS_2 + 15O_2 + 2H_2O \rightarrow 2Fe_2(SO_4)_3 + 2H_2SO_4$$

Indirect leaching uses ferric iron as an oxidizing agent to transform other metals.

The most widespread application of bacterial leaching is the recovery of copper from ore dumps and heaps. Various forms of copper such as chalcopyrite ($CuFeS_2$), chalcocite (Cu_2S), and covellite (CuS) are commercially leached. Chalcopyrite is the predominant form of copper in ore dump, whereas chalcocite is common in ores. The mechanism for chalcopyrite leaching may be summarized as:

$$4CuFeS_2 + 17O_2 + 2H_2SO_4 - \rightarrow 4CuSO_4 + 2Fe(SO_4)_3 + 2H_2O$$

An estimated 25 percent of the world's copper is produced by the bioleaching process.

TYPES OF BIOLEACHING PROCESSES

Bioleaching is a natural process. Industrial applications employ a variety of techniques for different metals. These include dump leaching, heap leaching, *in situ* leaching, concentrate leaching, and vat leaching.

1. **Dump leaching**

    Dump leaching is applied to leach minerals out of old mining dumps. Acid mine water (H_2SO_4 added to lower pH to 1.5− 3) containing bacteria is sprayed on the dump whose top has been leveled. As the leach solution percolates through the dump, S and Fe in the ore are oxidized. The effluent is precipitated over scrap iron (or by using solvent extraction and electro-winning techniques). The leaching solution can be recycled. This application occurs in the United States, Canada, Chile, and Eastern Europe.

2. **Heap leaching**

    Heap leaching is applied to metals like copper and gold, and is similar to dump leaching. However, higher acid concentrations are used in heap leaching than dump leaching. Heap leaching is usually part of existing mining operations and has a metal recovery rate of between 40 and 80 percent. The ore is crushed to finer particles and heaped and then irrigated with leaching solution. The leacheate is collected and passed over iron to precipitate copper powder by cementation, as in dump leaching.

3. *In situ* **leaching**

    This technique is used in the recovery of uranium. The ore (ore body while still in the ground) is treated with a leaching solution in the mine without first bringing it to the surface, thus being much safer regarding miner's lives. The leacheate is pumped to the surface where the metal is recovered by ion-exchange or solvent extraction. The rates of metal extraction and metal recovery are difficult to predict.

4. **Concentrate leaching**

   This technique was developed for processing imported ores. Developing countries often export ores in impure form to foreign markets. These impure ores are in the form of metal concentrates. Rather than smelting to recover the ore, this method can be used to treat copper, gold, and zinc concentrates to recover the pure ores.

5. **Vat leaching**

   Vat leaching for gold has been successfully developed, but is applicable also to silver, uranium, and copper. The ore is crushed into fine particles and then loaded into vats or open tanks through stirring and air injection. It is used in the United States, Canada, and Brazil. The method requires a higher grade of ore than the previous methods.

   Bioleaching technologies are being improved. Bioengineered organisms are being developed to increase the efficiency of bioleaching and to allow new metals (e.g., lead, zinc, cobalt, and molybdenum) to be recovered by the technology.

DESULFURIZATION OF COAL

One of the major problems of combustion of coal as a source of energy is environmental pollution. Sulfur in coal occurs in two basic forms: organic and pyretic. Sulfur is the chief pollutant from burning coal. The sulfur emission is brought down as corrosive acid rain. Some bacteria are able to solubilize the inorganic pyrite fraction of coal. U.S. coal averages 3.0 percent sulfur. Conventional strategies exist for reducing sulfur-related atmospheric pollution. These include fluidized bed combustion and flue gas desulfurization.

Sulfur can also be removed by microbial desulfurization. Organic sulfur is more difficult to remove, being chemically bonded to the coal. Several thermophilic bacteria, *T. ferrooxidans* and *Sulfolobus acidocaldarius,* are known to solubilize inorganic sulfur, the latter being faster at the job. Pyrite is solubilized at pH of 2.0 to 3.5.

KEY CONCEPTS

1. The technology of using microorganisms to recover metals is ancient in origin.
2. Certain organisms have the capacity to accumulate metals at concentrations that would normally be toxic to many other organisms.
3. Commonly bioaccumulated metals include aluminum, lead, arsenic, gold, cadmium, iron, mercury, and copper.
4. Metal removal may be accomplished by using growing biomass (continuously growing culture of microorganisms), or non-growing biomass (as an adsorbent).
5. Bacteria may extract minerals from mineral ores by the chemical process of leaching (called bioleaching).
6. One of the most commonly bioleached elements is copper.
7. The main types of bioleaching are dump leaching, heap leaching, *in situ* leaching, vat leaching, and concentrate leaching.
8. Coal may be desulfurized by the process of microbial desulfurization.

OUTCOMES ASSESSMENT

1. Describe the phenomenon of bioaccumulation.
2. Give four examples of commonly bioaccumulated metals.

3. Discuss the mechanism of bioleaching.
4. Give four examples of microorganisms and the metals they bioaccumulate.
5. Describe the method of dump leaching.
6. Discuss how coal is desulfurized.

INTERNET RESOURCES

1. Biohydrometallurgy: *http://www.environmental-center.com/articles/article425/article425.htm*
2. Discussion of theory and application of bioleaching: *http://www.imm.org.uk/gilbertsonpaper.htm*
3. *Thiobacillus ferrooxidans* in action and mining photos: *http://www.mines.edu/fs_home/jhoran/ch126/thiobaci.htm*
4. Links to bioleaching: *http://www.glue.umd.edu/~rcheong/classes/bsci223h/links.htm*

REFERENCES AND SUGGESTED READING

Brown, M. J. 1991. Metal recovery and processing. In: *Biotechnology: The science and business.* V. Moses and R. E. Cape (eds). London, UK: Harwood Academic Publishers.

19 Environmental Applications

PURPOSE AND EXPECTED OUTCOMES

Modern chemical-intensive crop production has been intensively criticized for being environmentally destructive in its impact. Agricultural production is implicated in groundwater and surface water pollution. The chemical industries that produce these agrochemicals are implicated in a variety of industrial pollution including air pollution. Little wonder, therefore, that biotechnology should receive such vigorous opposition, because of what some believe to be exacerbating the environmental problems by adding a biological source. Genetic crops have come under fire, especially from environmental activists. It is controversial because agricultural biotechnology also has significant perceived benefits to the environment.

In this chapter, you will learn:

1. The environmental concerns about biotechnology applications.
2. The applications of biotechnology in enhancing the quality of the environment.

ENVIRONMENTAL CONCERNS OF BIOTECHNOLOGY APPLICATIONS

Critics of biotechnology have a long list of concerns about the development and use of genetically modified (GM) crops, regarding their adverse impact on the environment. Some of these concerns are as follows:

1. **Herbicide-resistant crops could persist and become weedy species**

   GM crops such as the Roundup Ready™ products are resistant to broad-spectrum herbicides. If, for example, one Roundup Ready™ crop (e.g., corn) is followed by another Roundup Ready™ crop (e.g., soybean), the herbicide will not control any residual Roundup Ready™ corn. The volunteer corn plants could develop into weeds. Furthermore, if these GM crops are more vigorous and persistent than their non-GM counterparts, they could survive adverse weather (e.g., overwinter) and soon dominate the ecosystem.

   While this may be a possibility, it should be pointed out that there are other herbicides that can control Roundup Ready™ crops. At any rate, it is not a good production practice under any circumstance to use one herbicide repeatedly and exclusively for a long period of time. Furthermore, these domesticated cultivars lack the weedy characteristics to make them real contenders as weeds.

2. **Genetic pollution and the development of "superweeds"**

The concern exists that transgenes might escape and become incorporated into wild relatives of crops through cross-breeding. If this gene escape involves a herbicide-resistant gene, this could produce weeds that are resistant to herbicides, which would eventually become "superweeds" resisting the herbicides used in production.

This could happen if a GM crop is grown in an area where its wild relatives occur. The major GM crops grown in the United States (corn, soybean, potatoes) do not have wild relatives in the United States to pose a problem. However, wild relatives of squash and canola occur in North America.

3. **Antibiotic-resistance transfer**

Genetic engineering methods may utilize vectors that incorporate antibiotic-resistance marker selection systems. The concern is that such resistant genes might be picked up from decaying corn crop residue by harmful microbes, thereby posing a threat to veterinary and medical disease control by rendering pathogens antibiotic resistant.

On the other hand, others point out that the antibiotic-resistance markers used in vectors are common in nature, and do not provide resistance to most of the antibiotics used in clinical medicine. Furthermore, active research is ongoing to discover antibiotic-free selection marker systems in transformation research.

4. **Erosion of biodiversity**

Modern agriculture encourages the monoculture of crops, a strategy that does not promote biodiversity. Biotechnology promotes monoculture by providing pest-resistant crop cultivars to farmers, thereby discouraging crop rotation that is used as a strategy for pest control.

5. **Unexpected effects**

An environmental concern is that the transgenic construct inserted into crops might be unstable and break down, or might even stimulate unexpected effects in the host genome that may have an adverse environmental impact.

This event of transferred gene failure is not unique to transgenic plants. Conventional breeding using wide crosses also may be predisposed to genetic breakdown.

6. **Gene exchange between species**

There is some concern about horizontal gene transfer, inserting bacteria genes into plants (e.g., *Bt* products). However, this gene transfer occurs naturally among bacteria. Furthermore, the popular *Agrobacterium* bacterium used in biotechnology is dubbed nature's genetic engineer because the bacterium naturally transfers the tumor-forming genes into susceptible plants to cause the crown gall disease.

7. **Impact on nontarget organism**

Some laboratory studies have shown that pollen from *Bt* crops reduced the survival of monarch butterfly larvae. However, it is also true that GM products are targeted at specific pests (e.g., *Bt* is targeted only at lepidopteran moth and butterfly larvae).

ENVIRONMENTAL APPLICATIONS

The major areas of environmental applications of biotechnology include bioremediation, waste management, diagnostics, and phytoremediation.

■ *BIOREMEDIATION*

Bioremediation is the use of microorganisms to break down toxic and hazardous compounds in the environment. These toxic chemicals originate from mining activities, oil drilling, chemical manufacturing, and industrial accidents (spills). The fungus *Phanerochaete chrysosporium*, the causal agent of white rot in timber, is able to digest organochlorines (e.g., DDT, dieldrin, aldrin), which have chemical structures similar to that of lignin. The organism is used to detoxify PCB-contaminated sites as well as military dumps where TNT (2,4,6-trinitrotoluene) and other chemical contaminants occur in the United States. Certain *Pseudomonas* spp. are known to degrade benzene, toluene, petroleum products, and some pesticide residues.

The EPA utilized oil-eating bacteria to help clean the massive oil spill covering over 70 miles of Alaska's Prince William Sound shoreline in 1989. The mining industry also used chemicals in processing the minerals that are mined, leaving acidic and metallic wastes at the mining sites. Varieties of bacteria and algae are used to neutralize the acids and absorb metallic ions.

Because many pollutants are synthetic and nonbiodegradable, one of the challenges of biotechnology is to genetically engineer microbes with the capacity to metabolize these man-made materials.

■ *WASTE MANAGEMENT*

Use of microbes in waste management is a major application of biotechnology and involves both domestic applications in the home, or large-scale treatment of municipal wastes as well as industrial wastes. The subject was discussed in detail previously.

Tremendous quantities and varieties of waste products are generated by industrial facilities all over the world. Waste disposal is a major concern in many places. Operations that contribute significantly to wastes include animal production enterprises (e.g., hog and poultry), meat processing plants, breweries, and municipal waste treatment plants. Waste treatment technologies are being developed and perfected to use bacteria to anaerobically digest wastes in order to produce CO_2 and methane (CH_4) gases that can be used as alternative fuels. Waste treatment by this process converts about 93 percent of wastes into gases, leaving only about 3 percent as sludge. Commercial bioenergy plants based on this concept have been built in Europe, with the first in Kent, England. Some U.S. firms, including BioTechnica, are developing concepts for the landfill of the future as one giant bioreactor. This concept is attractive, considering the fact that many landfills process over 5,000 tons of refuse daily. Scientists at North Carolina State University developed the thermophilic anaerobic digesters as part of the effort devoted to discovering alternative fuels in the 1970s. Heat-loving bacteria digest animal wastes in enclosed containers at over 115°F and release biogas.

■ *DIAGNOSTIC TOOLS*

One of the goals of biotechnology is to develop diagnostic tools for early detection and field detection of major diseases that plague humans, plants, and animals. The first and most important step in disease management is the early and correct identification of the problem. Some diseases can be diagnosed by visual examination, whereas others require laboratory tests of varying degrees of complexity for identification. In the early stages of an infection, there may not be enough physical evidence for a visual diagnosis to be made.

The general goals of molecular diagnostic tools are:

1. Quick detection—short duration of procedure (minutes, not hours and days).
2. Field diagnosis—portable kits used outside the lab for on-the-spot diagnosis.
3. High sensitivity—use minute samples for detection.
4. High accuracy—eliminate false positives or false negatives.
5. Simplicity—easy to use.
6. Specificity—able to identify only specific targets of interest from among a mixture.
7. Safety—use non- or less-toxic materials in the procedures.
8. Inexpensive.

Sensors may be designed to monitor air or water quality, as well as detect pollutants in the soil.

PHYTOREMEDIATION

Phytoremediation is the use of plants to remove pollutants from the environment or to render them harmless. This roots-based biological remediation of environmental problems using plants entails the extraction of heavy metals from the soil. Various kinds of plants, including trees, are used in phytoremediation projects. Also called phytoextraction, this removal of heavy metals (e.g., lead, uranium, cadmium) from the soil relies on metal-accumulating plants to transport and concentrate polluting metals in their above-ground shoots. Examples of plants with this capacity include *Brassica juncea* and *Thlaspi sp.*

BIOFERTILIZERS

Modern large-scale crop production depends on the use of agrochemicals, primarily fertilizers. Most fertilizers currently being used are derived from inorganic sources and implicated in various environmental impacts including groundwater pollution and **eutrophication** (the nutrient enrichment of water bodies that promotes excessive growth of aquatic species leading to low water oxygen status). Scientists have known for a long time that various microorganisms are able to significantly impact soil fertility and plant nutrition. The natural ecosystem has several nutrient recycling systems to return nutrients immobilized in plant and animal tissue back to the soils by the process of mineralization when organisms die. Biotic decomposition involves various bacteria including *Nitrosomonas* and *Nitrobacter*.

Certain microorganisms also impact plant nutrition through certain associations, the two widely known being **symbiosis** (in which *Rhizobia* live in plant roots and fix atmospheric nitrogen directly into leguminous species) and **mychorrizae** (in which fungi live on plant roots and help in the uptake of certain minerals).

In addition to native microbes in the soil, scientists deliberately introduce new or additional amounts of beneficial ones through practices such as seed inoculation with artificially prepared *Rhizobium* when planting crops like soybeans. Beneficial bacteria and those with high nitrogen-fixing efficiency have been identified, isolated, and commercially cultured and packaged for field use. A Canadian company has developed a biofertilizer called **Provide** from the fungus *Penicillium bilaji*. This fungal inoculum is applied as seed coating prior to planting to facilitate the use of soil phosphorus that is readily fixed and inaccessible to plants. The fungus produces an organic acid that dissolves the soil phosphate for root absorption.

COMPLEMENTING BIOLOGICAL CONTROL

Biocontrol (biological control) is the use of natural enemies of pests to control the pests. The rationale of biological control is that each organism has its own natural enemies. The conventional biotechnology of biological control entails the breeding and releasing of beneficial insects, the common ones being the European seven-spotted lady beetle (*Coccinella septempunctata*) that preys on aphids, the ladybug (*Crystalaemus montrocizieri*) that preys on mealybugs, and the larvae of the Japanese beetle. There have been large-scale releases of beneficial insects under various conditions to control a variety of agricultural pests.

In addition to using insects, modern biotechnology emphasizes the use of microorganisms (bacterial, fungi, and viruses) in the development of pesticides. The most widely used biopesticide is the bioinsecticide based on the widely known bacterium *Bacillus thuringiensis* (commonly called *Bt*). Over 500 strains of the *Bt* are known. *Bt* is highly selective, as each strain affects only specific species of pests. Microbial sprays of *Bt* are available for controlling pests including the larvae of butterflies, moths, corn borers, cutworms, and cabbageworms. In 1983, a strain of *Bt* was used to eradicate blackflies in West Africa. Fungi-based microbial sprays have also been developed. For example, spores of the fungus *Collectotrichum gloesporiodes* have been used to control the northern joint vetch in rice fields. Another fungus used as a bioinsecticide is *Beauveria bassiana*, which is widely used to control forest pests in China. Virus-based microbial sprays are in various stages of development and deployment against insect pests including corn borers, potato beetles, and aphids. The viruses being used are the **baculoviruses** (rod-shaped viruses).

Microbial preparations are used in other creative ways in crop production. For example, a suspension of the bacterium *Bacillus subtilis* is used to treat fruits to delay brown rot caused by the fungus *Monilinia fruticola*. Scientists have determined that ice nucleation (the formation of ice on plants) promotes frost damage. Bacteria (e.g., *Pseudomonas syringae*) are implicated in this event. Through biotechnology, scientists have developed non-ice-nucleated bacteria that can be sprayed on plants to reduce bacteria-mediated frost injury.

These biopesticides are not harmful to humans or nontarget animals and are environmentally friendly. Because they are living things, they are capable of evolving. Consequently, insects are not able to develop resistance to biopesticides as quickly as they do against conventional pesticides.

INDIRECT BENEFITS THROUGH AGRICULTURAL APPLICATIONS

Applications of biotechnology in agriculture, especially crop production, have positive impacts on the environment, especially the soil, in various ways including the following:

1. **Reduced pesticide usage**

    GM crops with pest resistance (e.g., *Bt* products and viral-resistant crops) require reduced amounts of pesticides in crop production. The genetic modification usually protects the plant against the primary pest. However, GM crops may need supplemental pesticide protection to protect against secondary pests. Overall, the reduced pest usage is good for the environment, reducing the adverse environmental impact.

2. **Reduced soil erosion**

    Conventional tillage practices predispose the land to soil erosion. The conservation farming practice of no till or reduced till reduces soil erosion. GM herbicide-resistant crops fit into the no-till cropping practice by allowing the producer to eliminate the need for pre-emergence herbicides, and rather apply post-emergent herbicides to control weeds in one operation. Use of such GM crops reduces soil erosion as well as pesticides in the environment.

3. **Preventing yield loss and reducing the need for clearing more land for cropping**

    In terms of yield *per se*, GM crops do not necessarily outperform non-GM crops. They produce higher overall yield per unit area from reduced yield losses because of the pest resistance they possess. Furthermore, GM crops increase the productivity per unit area, thereby reducing the need to clear new land to increase crop production.

KEY CONCEPTS

1. The public has certain concerns about the development and application of biotechnology. These include the potential of horizontal transfer of transgenes to pose medical problems in the case of antibiotic markers, and creation of weedy species and superweeds.
2. Bioremediation is the use of microorganisms to break down toxic and hazardous compounds in the environment.

3. Because many pollutants are synthetic and nonbiodegradable, one of the challenges of biotechnology is to genetically engineer microbes with the capacity to metabolize these man-made materials.

4. Use of microbes in waste management is a major application of biotechnology and involves both domestic applications in the home, or large-scale treatment of municipal wastes as well as industrial wastes.

5. One of the goals of biotechnology is to develop diagnostic tools for early detection and field detection of major diseases that plague humans, plants, and animals.

6. In addition to native microbes in the soil, scientists deliberately introduce new or additional amounts of beneficial ones through practices such as seed inoculation with artificially prepared *Rhizobium* when planting crops like soybeans.

7. Biocontrol (biological control) is the use of natural enemies of pests to control the pests. The rationale of biological control is that each organism has its own natural enemies.

8. Applications of biotechnology in agriculture, especially crop production, have positive impacts on the environment in various ways including reduced soil erosion, reduced pesticide usage, and reduced land clearing.

OUTCOMES ASSESSMENT

1. What is bioremediation and what are its benefits to the environment?
2. Describe a specific application of bioremediation.
3. Discuss the application of microorganisms in waste management.
4. Discuss two specific environmental concerns of the public about the application of biotechnology.
5. Discuss two specific environmental benefits of biotechnology.

INTERNET RESOURCES

1. Phytoremediation of soil metals; *http://www.soils.wisc.edu/~barak/temp/opin_fin.htm*
2. Applications of phytoremediation; *http://www.mobot.org/jwcross/phytoremediation/*
3. Biotech and the environment; *http://www.jic.bbsrc.ac.uk/exhibitions/bio-future/cleaner.htm*

REFERENCES AND SUGGESTED READING

King, B., G. M. Long, and J. K. Sheldon. 1997. *Practical environmental bioremediation: The field guide.* Kansas City, MO: CRC Press.

Krimsky, S., and R. Wrubel. 1996. *Agricultural biotechnology and the environment: Science, policy, and social issues.* Kansas City, MO: University of Illinois Press.

Raskin, L., and Ensley, D. (eds.). 1999. *Phytoremediation of toxic metals: Using plants to clean up the environment.* New York: John Wiley and Sons.

Rittman, B., P. McCarty, Y. L. Doz, T. M. Devinney, and W. H. Davidson. 2000. *Environmental biotechnology: Principles and applications.* New York: McGraw Hill.

Terry, N., and G. Banuelos (eds.). 1999. *Phytoremediation of contaminated soil and water.* Albany, GA: Lewis Publishers, Inc.

Valdes, J. J. 2000. *Bioremediation.* New York: Kluwer Academic Publishers.

PART

V

Social Issues

The current and potential applications of biotechnology are very far-reaching, with enormous social implications. Biotechnology is big business and subject to regulation like all other enterprises. Because of the versatility of the technology, some of its applications, especially those involving horizontal gene transfer, are controversial. This part examines some of the major issues associated with the development and application of biotechnology.

20 Rights and Privileges

PURPOSE AND EXPECTED OUTCOMES

In the business of biotechnology, intellectual property rights are necessary to preserve the competitive edge over the competition by a company that owns an invention. They also protect a valuable resource that the property owner can later license to a third party for profit.

In this section, you will learn:

1. The concept and importance of intellectual property.
2. What a patent is.
3. The types of patents.
4. What can be patented.
5. What patent infringement is.
6. Patent issues in biotechnology.
7. Other types of intellectual property provisions.

INTELLECTUAL PROPERTY: DEFINITION AND PROTECTION

Intellectual property consists of principles that a society observes to ensure that an inventor is protected from unfair use of his or her invention by others. These principles are backed by laws and are observed in a variety of forms, the most common of which are **copyrights**, **confidential information**, **breeders' rights**, **trademarks**, and **patents**.

■ COPYRIGHTS

Copyrighting is most appropriate for protecting such things as aesthetic creations, music, paintings, works of literature, and computer software. It is rarely useful in the field of biotechnology other than for the customary protection of published data.

■ CONFIDENTIAL INFORMATION

When an invention is not patentable, the inventor might elect to keep the invention confidential in the hope that it will not be leaked to the public domain. The term "confidential information" is used to apply to the variety of strategies used by companies to protect their unpatented inventions. The common strategies include **trade secrets** and **proprietary information**. To ensure that employees who know the privileged information do not divulge it for any purpose, some companies require that their employees sign confidentiality agreements. Keeping trade secrets is cheaper than applying for a patent. However, such secrets sooner or later leak out into the public domain.

In the area of biotechnology, many companies withhold from the public specific information about their fermentation technologies used in manufacturing pharmaceuticals. Other candidates for trade secret protection are cell lines and soil isolates.

■ BREEDERS' RIGHTS

Developers of new plant varieties or strains of animals may seek protection from unlawful use of their creation by other scientists, competitors, or producers. Exclusive rights to the sale and multiplication of the reproductive material may be obtained, provided the variety is new and not previously marketed, different from all other varieties, uniform (all plants in the variety are the same), and stable (remains the same from one generation to another).

■ TRADEMARKS

Trademarks are important to all businesses, and biotechnology companies are not unique in this regard. However, trademarks are of particular significance to pharmaceutical companies.

■ PATENTS

One of the most widely applicable distinct rights provided by intellectual property is the right to patent an invention. While the previously discussed rights involving intellectual property have somewhat limited relevance in biotechnology, patents have a very significant role in this field. Hence, the balance of discussion in this section pertains solely to patents and what they mean to biotechnology.

PATENTS

One of the most widely applicable distinct rights provided by intellectual property is the right to patent an invention.

■ WHAT IS A PATENT?

A patent may be defined as an exclusive right granted for an invention of a product or process that provides a novel way of doing something, or offers a new technical solution to a problem. It provides protection for the invention to the owner of the patent. A patent is described as a "negative right" because it confers upon the patentee the right to exclude others from commercially exploiting the invention without the owner's authorization. Without permission, no one should make, use, sell, offer for sale, or import the invention. Such exclusive rights are effective for a period of 20 years, after which the invention is released into the public domain. A patent is not a "positive right" because it does not empower or obligate the patentee to do something that he or she would otherwise be prohibited from doing. In other words, in making, using, or selling the invention, the patentee must operate within the limits of existing laws of society. Furthermore, it goes to emphasize the fact that an inventor is under no obligation to patent an invention in order to exploit it commercially. However, without a patent for protection of the invention, a third party can exploit the invention commercially without authorization.

■ *THE IMPORTANCE OF A PATENT*

Patents are pervasive in society. Nearly everything created by humans is associated with patents in some shape or form. Patents are in the general interest of the public. As the U.S. Constitution indicates (US. Const., Article 1, sec. 8, cl. 8), the **patent law** serves to "promote the progress of science and the useful arts." In exchange for exclusive rights to the invention, the patentee discloses the invention and provides information that expands the existing technical knowledge base. Whereas others cannot imitate the invention, they can utilize the divulged information toward further innovations, as well as advance science and arts for enhancing the quality of human life. Patents provide effective incentives for creativity and innovation, recognizing the achievement of the inventor and, in cases where the invention can be commercialized, material reward. Companies invest large sums of money in the research and development (R&D) of new products. Patents ensure that they can enjoy a monopoly for a period in which they can recoup their investments. Patents are hence strong incentives for continued R&D.

Patents are crucial to the success of especially small to medium-sized biotechnology companies. It might take over $100 million to bring a pharmaceutical product to the market, and require over 10 years to develop. This obviously is a giant and risky undertaking that needs to be protected for profitability.

■ *WHAT CAN BE PATENTED?*

Patent law specifies what can be patented and the conditions under which a patent can be granted. One of the key functions of a patent is to define the scope of the protection. Ideas and suggestions, no matter how brilliant and creative, cannot be patented. Mixtures of ingredients (e.g., medicines) are also excluded, unless such mixtures produce synergistic effects or some unique and unexpected advantage. Defining the scope of protection is straightforward in certain cases (e.g., a simple device) and very complex in others (e.g., biotech inventions). There are five basic classes of patentable inventions:

1. Compositions of matter (e.g., a new chemical entity produced from the combination of two or more compounds; common in pharmaceutical and agrochemical research).
2. Processes or procedures (a series of steps that are followed to synthesize a new compound or a new method of making a product).
3. Articles of manufacture (nearly every man-made object).
4. Machines (any mechanical or electrical apparatus or device).
5. Improvements on any of the previous four categories.

■ *TYPES OF PATENTS*

There are three basic types of patents

1. Utility.
2. Design.
3. Plant.

Each type must fall into one of the five categories listed previously.

Utility Patent

A utility patent is the most common, and also the most difficult, patent to get. The scope of protection includes the functional characteristics of machines, electronic devices, manufacturing processes, chemical compounds, composition of medical treatment, and manufactured articles. The applicant is required to submit an exhaustive description of how to make and use the invention, including drawings where appropriate, among other requirements. The duration of this patent is 20 years.

Design Patent

A design patent has a 14-year duration and protects the shape, as well as the ornamental or artistic features, of an article (e.g., unique shape of a bottle, the grill of an automobile).

Plant Patent

Awarded for a period of 20 years, the plant patent protects the invention or the discovery of a distinct and new plant variety via asexual reproductive methods.

■ NATIONAL AND INTERNATIONAL PATENTS

Applicants may seek local or worldwide protection for an invention. There are varieties of general types of patents in this regard: national, European Patent Convention (EPC), and international. National patents pertain to the applicant's country of origin or operation. The EPC allows patent rights to be obtained in one or more European countries that are parties to the EPC by making a single European patent application. If successful, the applicant must then use the patents in the countries that the applicant designated. International patents may be obtained by filing a single patent and designating countries in which protection is desired. Within a specified time limit, this single patent will be copied to the designated countries where they will be processed as national applications.

■ SCOPE OF PROTECTION

The scope of protection may be defined narrowly or broadly by the applicant, with each scenario having its own consequences. Sometimes, in an attempt to prevent the competition from copying the invention, an applicant may make claims to encompass the embodiments of the invention that were not existent at the time of the invention. Such a practice favors the inventor to the disadvantage of the competition. Cases in point where the scope is broad are the highly successful herbicide Roundup® and the tranquilizer valium.

■ CRITERIA FOR PATENTABILITY

Criteria for patentability include:

1. **Conception** This is the mental formulation of the invention that is detailed enough to allow a person knowledgeable about the subject to which the invention relates to make and use the invention.
2. **Reduction to practice** The inventor should make or construct the invention and test it to demonstrate its usefulness.
3. **Utility** An invention must be useful (not merely aesthetic) to the user. That is, the invention must take some practical form (applicability).
4. **Novelty** The invention must not be a copy or repetition of an existing invention. Among other things, it should not have been known, published, or used publicly anywhere previously. In the UK, an invention must not have been in the public domain anywhere in the world prior to application of a patent. Some countries (e.g., United States) allow a grace period during which an invention that has been introduced to the public, under certain conditions, can still be patented.
5. **Obviousness** This is a very difficult criterion to satisfy. The result should neither be expected nor obvious. A person knowledgeable in the subject matter (skilled in the art) should not be able to readily figure out how to piece together component parts to make the product (Section 103 of U.S. patent law).

In view of the previous criteria, it is wise to keep an invention a secret until a patent application has been filed. Alternatively, some measures may be taken to impose obligations of confidentiality upon a party receiving advanced disclosure of information. For example, information may be bound by a Confidentiality Disclosure Agreement.

■ APPLYING FOR A PATENT

An appointed office grants patents. In the United States, patent applications may be submitted to the Patent and Trademarks Office, U.S. Department of Commerce, Washington, DC, 20231. Copyright applications may be submitted to the U.S. Copyright Office, Library of Congress, Washington, DC, 20559. In Europe, applications may be submitted to the European Patent Office. The services of patent attorneys may be engaged in the preparation and filing of an application. An applicant must define the problem or objective addressed by the invention very clearly and in detail. There are certain general steps involved in applying for a patent.

1. **Filing fee**
   An applicant for a patent is required to pay a fee for the processing of the application.
2. **Search and examination**
   The patent examiner will conduct a "prior art" search to ascertain the novelty and non-obviousness of the invention. The claims defining the scope and monopoly being sought are rigorously examined. The examiner may accept or reject the application based upon the search and examination results. The applicant has the right to argue against an adverse judgment, or amend the claims to the satisfaction of the examiner.
3. **Publication**
   Successful applications will be published along with the claims of the applicant.
4. **Maintenance fees**
   Most countries require a successful applicant to pay a periodic maintenance fee to prevent the patent from lapsing.

■ EXPLOITING INTELLECTUAL PROPERTY

The owner of an intellectual property may use or work with it, as well as exploit it in other ways.

1. **Assignment**
   A patent, or a patent application of invention, may be sold or assigned to another party just like a piece of property. Such a transaction may be closed with a one-time payment, or periodic royalties may be paid to the owner of the property.
2. **License**
   Rather than outright sale, the property may be licensed to another party according to specified terms and conditions. The license agreement usually guarantees the licensor royalties from the licensee for use of the invention. License agreements have limitations to the extent of exploitation of the invention permitted. For example, an agreement may limit the application of the invention to certain uses.
3. **Freedom of use**
   Freedom of use consideration may hinder a patent from being exploited without infringing upon existing patents. Such restrictions on free exploitation of an inventor's patent may derive from the scope of the patent. Scopes that are too narrow are susceptible to such infringements. An example of such a situation involving freedom of use may arise when an inventor patents a process that requires a patented compound in order to exploit the invention.

PATENTS IN BIOTECHNOLOGY: UNIQUE ISSUES AND CHALLENGES

Patenting living organisms is a challenge. Just like computer software and copyrighted music, organisms are readily reproduced. Consequently, the early application of patent laws to biology tended to be strongly and broadly interpreted in favor of inventors.

Generally, it is easier to satisfy the patent requirement when inventions are the results of empirical discovery (e.g., in pharmaceuticals and agrochemical industries). Scientists systematically search for compounds that can be used as active ingredients in pesticides, antibiotics, and other therapeutics. However, molecular biology and allied fields are not empirical sciences. Biotechnology thrives on principles from a wide variety of fields. Researchers tend to focus on advancing these principles or applying them to accomplish desired practical objectives. Development of truly new and unexpected phenomenon is not common. Progress is made incrementally. It is often a challenge to satisfy the traditional criteria stipulated by the patent office—obvious phenomenon, specific utility, and teaching others how to make and use the invention. Another limitation is that ideas and properties of nature are not patentable.

Patenting genes and other biological resources became relatively easier following the landmark 1980 U.S. Supreme Court decision in *Diamond v. Chankrabarty* that granted a patent for an oil-dissolving microbe. Thanks to the proliferation of genomic projects, notably the Human Genome Project, the patent floodgates have been opened wide for scientists in the biotechnology area. It is estimated that over 3 million expressed sequence tags (ESTs), representing fragments that identify pieces of genes and thousands of other partial and whole genes, have been submitted for patenting. However, not all players are satisfied with the scope of protection provided by the patent laws. A microscopic view will allow nearly anything novel to be patentable, while opening up the doors for competitors to easily circumvent the narrow claims. Some scientists are opposed to the granting of broad patents to what they describe as the early stages of the biotechnology game. Some of the genes submitted for patents have not been characterized; in addition, the applicants have not determined their functions and specific uses. The concern is that large-scale and wholesale patenting of genes by biotechnology companies who have no clue about the functions of these genes is tantamount to staking a claim to all future discoveries associated with those genes (the so-called "**reach-through patents**").

This concern is a genuine one. Not too long ago, genomics companies had a field day staking claims to the genome landmark (the "genome run"). But with the focus now on understanding gene function, the proteomics companies have their chance to do likewise. This is stirring up new controversies in the patenting of biotechnology inventions. Even though the Human Genome Project has revealed that humans have about 30,000 to 40,000 genes, it is believed that they code for about 2 million proteins. Obviously, each gene does not encode a single protein as had been the dogma. Because cells often have the capacity to splice mRNA together in different ways, different versions of a protein are often produced. These versions can be significantly different in their functions. For example, one mRNA variant makes calcitonin (a hormone that increases calcium uptake in bones), while another version creates a calcitonin-generated polypeptide that prompts blood vessels to dilate. To complicate matters, these proteins are susceptible to further modification whereby the cell can attach to them small chemical groups that are not coded for by genes, resulting in significant changes in their functions.

Another implication of these scientific revelations is that a patent on a specific DNA sequence and the protein it produces may not cover some biologically important variant. It is estimated that the top genomics companies (Human Genome Sciences, Incyte Genomics) have collectively filed over 25,000 DNA-based patents. The business rationale to their strategy includes the potential to receive royalties from third parties that use any of them. But this may not be as simple as it sounds, unless one gene makes one mRNA, which in turn makes one protein, something that appears not to be true anymore. To strengthen their position in the race to stake claims to proteins (the "proteome run"), these genomic companies are also building their own proteomics research arsenal to help them discover any important protein variants linked to any of their patented genes.

It is most likely, and perhaps inevitable, that some protein discovery projects will turn out proteins that correlate better with disease than those for which patent claims are already in existence. In such cases, litigation seems the likely recourse. However, it is also likely that potential litigants may opt for the less costly route of **cross-licensing**, whereby each party can cross-license another's patents.

Another issue with biotechnology patents is "**patent stacking**," a situation in which a single gene is patented by different scientists. This situation is not favorable to product development because users are deterred by the possibility that they would have to pay multiple royalties to all owners of the patent. Furthermore, because patent applications are secret, it is possible for an R&D team in a different company to be working on development of a product only to be surprised at a later date by the fact that a patent (called a "**submarine patent**") has already been granted.

Patent laws protect the public by enforcing "the product of nature" requirement in patent applications. The public is free to use things found in nature. That is, if a compound occurs naturally but it is also produced commercially by a company via a biotechnology method, the genetically engineered product is technically identical to the natural product. However, in the case of *Scripp v. Genentech,* a U.S. court ruled that a genetically engineered factor VIIIc infringed a claim to VIIIc obtained by purification of a natural product. This indicates that a previously isolated natural product had first claim to patent rights over a later invention by genetic engineering. If a company seeks to apply for a patent for an invention to produce a rare naturally occurring compound in pure form, the argument will have to be made for the technique used for extraction, purification, or synthesis, not for the material *per se.*

Biotechnology also faces a moral dilemma in patent issues. Specifically, is it moral to patent any form of life? Furthermore, if the discovery has medical value, should it be patented? Then there is the issue of the poor. Is it moral to demand that the poor pay royalties they can ill afford for using patented products for survival purposes? A debated issue is the plant breeders' rights. Should breeders be permitted to incorporate seed sterilizing technology (e.g., the so-called "terminator technology") in their products to prevent farmers from using seed from harvested proprietary material for planting the field the following season?

The foregoing is only an overview of the complex nature of patenting biotechnology inventions. It should be pointed out that courts in Europe and the United States, as well as other parts of the world, differ in their positions on patent issues, not to mention the fact that patent laws vary among nations.

Patent laws and the way they are enforced may differ among nations. For example, the European Directive on the Legal Protection of Biotechnological Inventions passed in 1998 declared that the mere discovery of the sequence or partial sequence of a gene does not constitute a patentable invention. Genes are not patentable while they are in the body (*in situ*). However, genes isolated from the organism or artificial copies of the genes produced by some technical process may be patentable, provided the novelty, inventive step, and utility are clearly demonstrated. The U.S. laws have been tightened to include a clause to the effect that the utility of the invention must be "specific, substantial, and credible" (i.e., readily apparent and well-established utility).

In addressing the issue of morality, the European Directive also specifically excludes certain inventions from patentability. These include processes for the reproductive cloning of human beings, processes for modifying the germline genetic identity of human beings, and uses of human embryos for industrial or commercial purposes. Essentially, if the publication or exploitation of an invention would generally be considered immoral or contrary to public order, it cannot be patented.

KEY CONCEPTS

1. Intellectual property consists of principles that a society observes to ensure that an inventor is protected from unfair use of his or her invention by others.
2. These principles are backed by laws and are observed in a variety of forms, the key ones being patents, copyrights, confidential information, breeders' rights, and trademarks.
3. There are five basic classes of patentable inventions: compositions of matter, processes or procedures, articles of manufacture, machines, and improvements on any of the previous four categories.
4. There are three basic types of patents: utility, design, and plant.
5. There are certain criteria for patentability: conception, reduction to practice, utility, novelty, and obviousness.
6. The scope of protection may be defined narrowly or broadly by the applicant, with each scenario having its own consequences.

OUTCOMES ASSESSMENT

1. What is a patent?
2. What is the importance of intellectual property rights in biotechnology?
3. All inventions are not patentable. Explain.
4. What is a utility patent?
5. Patents are not always the best protection of an invention against illegal use. Explain.
6. A patent is not a "positive right." Explain.
7. Discuss the concept of "freedom of use" in patent law.
8. List and discuss the basic steps in applying for a patent.

INTERNET RESOURCES

1. Some myths about intellectual property: *http://www.ifla.org/documents/infopol/copyright/ipmyths.htm*
2. General information about patents: *http://www.wipo.org/about-ip/en/about_patents.html*
3. European perspectives on IP: *http://www.3bsproject.com/html/ip.html*
4. Variety of IP information and services: *http://www.ipmall.fplc.edu/*
5. Links to various IP topics: *http://www.brint.com/IntellP.htm*
6. Examples of possible patentable biotech inventions: *http://www.usask.ca/ust/IP/biotech.html*

REFERENCES AND SUGGESTED READING

Crespi, R. S. 1988. *A basic guide to patenting in biotechnology.* Cambridge: Cambridge University Press.

Gaythwaite, D. M. 1991. Intellectual property and technical know-how. In *Biotechnology: The science and business.* V. Moses and R. E. Cape. London, UK: Harwood Academic Publishers.

National Research Council. 1997. *Intellectual property rights and plant biotechnology.* Washington, DC: National Academy Press.

SECTION 2
Ethical Implications

PURPOSE AND EXPECTED OUTCOMES

Development and application of biotechnology raises ethical questions, some of which are serious enough to generate significant opposition from the consuming public to certain technologies and their applications. The purpose of this section is merely to bring to light the ongoing debate on ethics and biotechnology without weighing in one way or another. It appears the debate will only intensify as biotechnology advances and new technologies and applications emerge. Hence, it is imperative that the student become aware of the ethical dilemmas in biotechnology and explore the implications of siding with one view or another.

In this section, you will learn:

1. To distinguish between ethics, morals, and value.
2. How ethics impact the development and application of biotechnology.
3. About public acceptance of biotechnology.

THE BIOTECHNOLOGY DEBATE

It was previously indicated in this book that public perceptions about biotechnology products are rooted in the perceived risks that these products pose to social and personal values. It is also suggested that the raging biotechnology debate is rooted in three fundamental disagreements as follows:

1. **Scientific disagreements**

     These disagreements are about the types and degrees of risk to human, animal, and environmental health. These issues involve empirical questions and are usually resolved by scientific methods. However, they are not exclusively resolvable by the scientific method of enquiry. Sometimes, value judgment is critical in their resolution. For example, handling uncertainties in scientific data and definition of the levels of risk deemed acceptable are both value judgments.

2. **Political disagreements**

     Political disagreements are generally about the social and economic impacts of biotechnology based on the various political viewpoints.

3. **Religious, ethical, and philosophical disagreements**

     These disagreements are often faith based and include issues about morality, whether scientists are playing God, or whether the biotechnology products are natural.

INTRODUCTION TO ETHICS

It will be helpful for the uninitiated reader to read a primer on bioethics at this juncture. An Internet resource has been listed at the end of this section for a quick overview of bioethics. Dictionary-based definitions will now be presented to help the reader put the subject into the proper perspective. **Ethics** is the science of morals in human conduct (i.e., a study of moral principles). **Morals** is concerned with the accepted rules and standards of human behavior in a society. It involves the concept of right or wrong, the goodness or badness of human character or behavior. **Value** is basically the worth attached to something. In other words, ethics is evaluative of the decisions people make and the actions they take as they are presented with dilemmas. Morality depends on values in order to determine the goodness or badness of an action. In a pluralistic society, there are differences in the sense of values (i.e., relativism). Consequently, there is a variety of moral theories that do not necessarily constitute truth. Furthermore, law, religion, and customs should be distinguished from morality. In law, lawmakers define what is right or wrong. Those who break the law are subject to punishment prescribed by the legislature. In religion, right or wrong is based on revelation or scriptural authority. Whatever choice is made has eternal consequences. In the case of custom, tradition determines what is acceptable or not, and society expresses approval or disapproval of an action.

BIOTECHNOLOGY AND ETHICS

One of the major sources of discord in society regarding biotechnology is the notion that scientists are playing God when they fail to respect human limitations. God, humanity, and nature are linked, with God being the creator of the other two. Some people see nature as God's creation for the benefit of humans, who therefore can use plants, animals, and the ecosystem for their purposes, as they deem necessary. Others see nature as a sacred creation that must be respected and not tampered with. Does this respect mean that humans cannot manipulate nature? What cannot be denied is that the Creator has endowed humans with considerable creative genius. The obvious question then is whether exercising creativity through biotechnology is within the scope of this endowment or whether it is tantamount to an infringement on divine prerogative. For those who see nature as a gift to humans for their use, recombining genetic materials may be justified as just another way of using natural resources.

In order for us to be correctly evaluative of our choices, decisions, and acts as they pertain to biotechnology, there is the need for certain basic sets of information to be available. One set pertains to the values we attribute to things and acts we perform, the other set being value-free. Scientists, traditionally, generate value-free information. However, both kinds of information (that tested empirically and experientially) and their impact need to be accumulated for use in making choices and decisions about biotechnology.

The ethical issues and the passion with which they are debated in the public arena vary among applications. Manipulation of the food chain seems to attract more attention than clinical applications (e.g., xenografts). For example, heart valves from pigs have been used in humans without fanfare. However, GM grains have encountered considerable public opposition from certain quarters. In general, the ethical issues of concern to the public are the impacts of biotechnology on human health and safety, environmental impacts, intrusions into the natural order, invasion of privacy, issues of rights and justice, economics, and others. It is important that both benefits and risks of biotechnology be considered in making ethical decisions about the discipline. We need to ask the questions: "Is it ethical to cause harm to someone in order to help another?" "Does the end justify the means?" and "Who decides these issues?"

SOCIAL CONCERNS

The specific issues of concern about the development and application of biotechnology include damage to the environment, injury to human health, food safety, socioeconomics, and infringement on religious beliefs. In each case, there is the need to consider both the benefits and risks of biotechnology to be ethical. The problem is that, at the moment, we are limited in our knowledge about the full benefits and risks of biotechnology. Consequently, we are in danger of either underestimating or overestimating the potential of biotechnology for good or evil. Furthermore, public reaction may be rooted in undue fear or hope stemming from misunderstanding, misinformation, or lack of information about various aspects of biotechnology. It should be pointed out that gathering information about the risks and benefits of biotechnology cannot be accomplished overnight. It is both time-consuming and expensive to undertake.

In terms of the environment, the public is concerned about contamination from bioengineered organisms or products that could pose a threat to human health and damage natural resources (forests, water bodies, land, and so on). There is a concern about destroying the aesthetic value of nature and jeopardizing the survival of wild animals and plants by destroying their habitat. The perception is that "natural" is better, while "artificial" is inferior. We have a moral obligation to take care of and preserve nature. It is sometimes difficult to see what value is at stake. For example, restricting human use of natural resources places a constraint on other aspects of life (e.g., higher cost of living). Also, biological processes are instruments that are used for producing products. But, does it matter if variation arises spontaneously (natural mutation) or is induced artificially (mutagenesis) if both events help humans attain a desirable value?

Regarding food safety and health, the concerns include food allergens and toxic compounds introduced as a result of genetic manipulation. For example, consumers were leery about the use of bST (bovine growth hormone) in the dairy industry because they were not certain about how cow hormones would affect humans. Arming plants to protect themselves against pests (e.g., use of *Bt*) is of concern because the toxins they produce might affect humans. However, it should be pointed out that plants have natural defense mechanisms that are based on the chemical toxins they produce.

New technologies often tend to tip the scales in favor of those with resources to acquire them. The limited-resource producers may be marginalized by new technologies. Small producers may be forced out of business by producers with more resources to acquire new technologies. New technologies are most likely to be adopted if they increase profitability to producers while lowering the cost to consumers. This implies that, with the introduction of new technologies, there will be winners and losers. The question then is, "Is it fair to implement a technology that is advantageous to one person and disadvantageous to another?" Private sector entrepreneurs are profit driven. However, the reality of the matter is that a significant number of new technologies are pioneered in the public sector with taxpayer funds. Is it, then, fair to use a taxpayer's resources to develop and implement a technology that will put him or her at a disadvantage? Is some compensation in order when this happens?

There is also the issue of developing countries. Since they are unable to acquire new technologies, many of these nations are unable to compete fairly in the international markets. Also, many of the germplasm resources used in plant and animal improvement are derived from these regions of the world. Is it fair for biotechnology companies in developed economies to obtain these materials free of charge and then claim intellectual property rights to them and the products they bioengineer using these materials? One could also ask the reverse question: Is it fair to expect a biotechnology company that has invested millions of dollars into developing a product to give it away free of charge or sell at a loss?

At the religious front, some concerns are that scientists are overstepping their bounds by radically manipulating organisms. Transferring genes across natural boundaries is seen as usurping the prerogative of God the Creator. The idea of cloning humans is especially widely viewed as unacceptable. Some in society view the use of stem cells from fetuses as encouraging abortion. Others feel that the objective of stem cell research is laudable and outweighs religious prohibitions. Some who protest the use of stem cells from aborted fetuses tend to support the use of adult stem cells. This prompts the question of what the real issue is that is being protested. The basic controversy is about when life begins. Protesters believe that life begins at fertilization and hence, whether the embryo is destroyed at a 2-, 4-, or 8-celled stage, it is tantamount to destroying life. The fact is that the use of adult stem cells from cloning would follow the protocol used to clone the first mammal ("Dolly" the sheep) in which an adult cell was induced to go through the embryonic stage, and then eventually to full adult. Is the embryonic stage then different if the process excludes the direct union of egg and sperm? In other words, is what makes the embryo descended from the direct union of gametes sacred (spirit and soul) from the religious point of view absent in the embryo developed *in vitro*?

These are but a few of the many dilemmas associated with the development and application of biotechnology. These ethical issues are complicated to resolve, especially in a pluralistic society. There is a need for more knowledge to be gathered from research and experience, as well as objective dialog between the scientific community and the general public.

PUBLIC ACCEPTANCE OF BIOTECHNOLOGY

Notwithstanding the pros and cons of its development and application, and all the debates and protestations it is surrounded by, when it comes to the success of biotechnology in society, the consuming public will ultimately be the arbiter. Public understanding, attitude, and acceptance of the technology will be of increasing strategic significance for the progress of biotechnology.

Many polls have been conducted in various parts of the world on the public acceptance of biotechnology. The general indication is that biotechnology is more readily acceptable to the consuming public in North America than Europe. Developing countries, who are generally looking for any promising way out of their predicament, are also generally indicating a positive disposition toward biotechnology (at least their leadership is). As evidence of this latter fact, China is among the top five users of GM crops in the world.

Biotechnology is pervasive in its effects across many economic sectors, building upon recent discoveries. The advancement of science and technological applications are not completely independent of society. Scientists are members of the general society and their activities are subject to a degree of social control. It is fair to say that where science and technology demand significant resources, or their development and applications threaten established interests, their future development and application will depend on social consent. Whether or not society is readily obliging with its coveted consent would depend on the competitive capability of the science or technology. For example, will it give the society some economic or military advantage over its competitors? It will also depend on whether the benefits are obvious to the ordinary person in society, for example, a cure for deadly diseases (e.g., cancer and AIDS). Where the benefits are not very obvious and the impacts are unforeseeable or indirect, the public will be less willing to give their blessing and to provide the resources needed. Furthermore, where an innovation is controversial, arousing significant apprehension or even threatening established interests or the status quo, be it in economic, social, or intellectual realms, support will not only be forthcoming, but deliberate opposition will be encountered. This is partly what seems to be unfolding with biotechnology. Some benefits have materialized, and many are anticipated. However, the advance of the field threatens traditional establishments like environmentalists.

Modern communities tend to congregate, leading sooner or later to population density issues. To accommodate increasing population numbers on limited land, all modern societies depend on the advances in science and technology. In a democracy, the electorate needs to be well educated in the nature, accomplishments, and potential of science in dealing with issues confronting modern societies. This is critical in order for the scientist to receive the public support and mandate to pursue his or her work.

Scientists and biotechnology companies need to aggressively pursue public education to avoid consequences from uninformed judgment. Polls might be favorable at one point, but issues are hardly settled based on statistics alone. The public needs to have confidence in the regulatory systems and also accept the fact that, like most innovations, biotechnology is Janus-faced. There are benefits and risks that have to be embraced as one package. Efforts should, however, be made to improve the processes to reduce or eliminate the risks.

KEY CONCEPTS

1. The development and application of biotechnology has ethical implications.
2. Society is divided in its view of biotechnology along ethical grounds.
3. Public perception significantly impacts the development and application of biotechnology.
4. Social issues related to biotechnology vary among the regions of the world.
5. Religion plays a significant role in shaping the social response to biotechnology.

OUTCOMES ASSESSMENT

1. Distinguish between ethics, morals, and values.
2. Give two specific public concerns about the development and application of biotechnology.
3. In terms of ethics, what is your opinion about the development and application of biotechnology?
4. What role should the public play in the way the biotechnology industry operates?

INTERNET RESOURCES

1. Biotech and morality debate: *http://www.cid.harvard.edu/cidbiotech/comments/comments117.htm*
2. Ethics resources; variety: *http://www.ethics.ubc.ca/resources/*
3. Journal of ethics: *http://www.journals.uchicago.edu/ET/*
4. Bioethics resources center: *http://www.bfn.org/~bioethic/*

REFERENCES AND SUGGESTED READING

Baumgardt, B. R., and M. A. Martin (eds). 1991. *Agricultural biotechnology: Issues and choices.* West Lafayette, IN: Purdue University Agricultural Experiment Station.
Brill, W. 1985. Safety concerns and genetic engineering in agriculture. *Science, 227:*381–384.
Chargoff, E. 1976. On the dangers of genetic meddling. *Science, 192:*938–940.

21 Risks and Regulations

PURPOSE AND EXPECTED OUTCOMES

Biotechnology is a rapidly evolving scientific discipline that has aroused significant apprehension from certain quarters of the general public. There is considerable pressure on regulatory agencies to scrutinize the activities and the products of the biotechnology industry.

In this chapter, you will learn:

1. The U.S. agencies and their specific roles in the regulation of biotech products.
2. The specific criteria for submitting a product for testing.
3. A practical example of a GM product subjected to environmental assessment.
4. International biosafety regulations.
5. Labeling issues.

It is significant to mention that the regulation of biotechnology, in fact, was started by the scientific community in 1975 at the Asilomar Conference on rDNA Molecules. At this meeting, a self-imposed moratorium was placed on the extension of the technologies of biotechnology until the associated risks could be better assessed.

THE COMPLEXITY OF RISK ANALYSIS

Risk analysis of biotechnology is a very complex undertaking because the situation is unique for the crop species, the genetic modification, and the environment in which the concern is being expressed. A much more useful and fair analysis of environmental impact is obtained if risk analysis of a biotech product is done in comparison with competing products or technologies. For example, it would be desirable to compare chemical pesticides with *Bt* products, use of glyphosate herbicide with glyphosate-resistant crops, or planting GM crops with high productivity with clearing new land to plant conventional lower productivity cultivars. In conducting risk assessment, it is important that the process enhances consumer confidence and trust, without which marketing GM products is bound to be problematic.

Public perceptions and attitudes about biotechnology are shaped by concerns about the risks and safety (acceptability of risk) of genetically engineered foods and other products. These biotechnology products are perceived as posing risks to a variety of social and personal values. An expert panel on the future of food biotechnology commissioned by the Canadian Food Inspection Agency and Environment Canada categorized the values that are perceived by the public as being placed at risk by biotechnology into three categories:

1. **Potential risks to the health of human beings, animals, and the natural environment**

     The risks to human health and the environment are at the top of the list of public concerns about the impact of biotechnology on society.

2. **Potential risks to social, political, and economic relationships and values**

     Commonly, the public is concerned about the monopoly of certain industries (e.g., seed) by multinational corporations to the detriment of small producers and the risk of increased dependency of developing economies on these monopolies. It is the opinion of many experts that the level of risk acceptable by the public depends on the overriding benefits to be achieved (risk-cost-benefit).

3. **Potential risks to fundamental philosophical, religious, or metaphysical values held by different individuals and groups**

     This particular category tends to address the issue the public takes with the *process* of biotechnology rather than the product or *impacts*. The concern is the risk of playing God by implementing processes that are unnatural to alter nature.

The extent to which the public is willing to be exposed to unknown or uncertain risks, and how much risk is acceptable, is influenced by social, economic, and philosophical factors. People will be more willing, for example, to accept a higher risk level if they are strongly convinced about the benefits of adoption of biotechnology products, or, on the other hand, the adverse consequences of not adopting biotechnology products.

REGULATION OF BIOTECH PRODUCTS

In order to protect consumers from product risk and promote and retain their confidence in biotechnology products, as well as promote trade, there are local, national, and international entities charged with regulatory oversight of biotechnology products. The agencies with regulatory oversight in biotechnology in the United States are the USDA, Food and Drug Administration (FDA), and the Environmental Protection Agency (EPA). The products they regulate are summarized in Table 21–1. Manufacturers and developers of biotech products are required to meet certain minimum product standards stipulated in state and federal marketing statutes. These include state seed certification laws, the Federal Food and Drug Cosmetic Act (FFDCA), the Federal Insecticide, Fungicide, and Rodenticide Act (FIFRA), the Toxic Substances Control Act (TSCA), and the Federal Plant Pest Act. Depending upon the product, an agency may review it for its safety to grow, safety to eat, or safety to the environment. The Animal and Plant Health Inspection Services (APHIS) conduct the USDA biotech evaluation. The EPA ensures the safety of pes-

TABLE 21–1
United States regulatory oversight in biotechnology responsible agencies.

Agency	Products Regulated
U.S. Department of Agriculture	Plant pests, plants, veterinary biologics
Environmental Protection Agency	Microbial/plant pesticides, new uses of existing pesticides, novel microorganisms
Food and Drug Administration	Food, feed, food additives, veterinary drugs, human drugs, medical devices

Source: USDA

ticides and enforces the FIFRA (to regulate the distribution, sale, use, and testing of plants and microbes producing pesticidal substances) and the FFDCA (to set tolerance limits for substances used in pesticides, food, and feed). The FDA is part of the Department of Health and Human Services and enforces the FDA (to regulate foods and feeds derived from new plant varieties).

The details of the laws and regulations enforced by these agencies can be obtained by visiting their websites. A summary of how they regulate products is presented next.

■ *USDA-APHIS*

APHIS is authorized to regulate interstate movement importation and field-testing of organisms and products altered or produced through biotech processes that are plant pests or are suspected of being so. An individual or an entity seeking to conduct any of the previously mentioned activities must apply and receive one of the three permits from APHIS before proceeding.

1. **Permit for movement and importation**
   This requires the applicant to disclose the nature of the organism, its origin, and its intended use.
2. **Permit for release into the environment**
   APHIS oversees field-testing of biotech products. The applicant is required to provide information on the plant (including new genes and new gene products), its origin, the purpose of the test, an experimental design, and precautions to be taken to prevent the escape of pollen, plants, or plant parts from the experimental site.
3. **Courtesy permit**
   This applies to nonregulated plants that have been bioengineered in some fashion.

■ *FDA*

In 1997, the FDA decided to subject all biotech products to the same standards of regulation as traditional products. In the Federal Register, vol. 57, the FDA directs that companies or researchers whose products meet one of the following criteria should submit them for testing:

1. **Unexpected effects**   The product produces unexpected genetic effects.
2. **Known toxicants**   The product has higher than normal levels of toxicants than other edible varieties of the same species.
3. **Nutrient level**   The product has altered levels of essential nutrients.
4. **New substances**   The chemical composition of the product is significantly different from existing normal products.
5. **Allergenicity**   The product contains proteins that have allergenic properties.
6. **Antibiotic resistance selectable marker**   The product is produced by a biotech process that utilizes genetic markers that could adversely impact current clinically useful antibiotics.
7. **Plants developed to make specialty nonfood substance**   The plants are engineered to produce pharmaceuticals or other polymers.
8. **Issue specific to animal feed**   The product's chemical composition regarding nutrient and toxins is significantly different from levels in similar products used for feed.

In addition to these federal regulatory activities, individual states are at liberty to develop and implement additional regulations. Exempted from premarket approval are products that are classified as GRAS (generally accepted as safe). Such food products may have been engineered to express proteins. However, a GRAS substance is excluded from the definition of a food additive.

TABLE 21–2
List of viral coat proteins that have EPA tolerance exemptions

Papaya ringspot virus coat protein

Potato leaf roll virus (PLRV) replicase protein as produced in potato

Potato virus Y coat protein

Watermelon mosaic virus (WMV2) coat protein in squash

Zuchini yellow mosaic (ZYMV) coat protein

WMV2 and ZYMV in ASGROW ZW0

■ *EPA*

The EPA regulates pesticides. Its definition for pesticides is any substance or mixture of substances intended for preventing, destroying, repelling, or mitigating pests. A new category of pesticides is called plant-pesticides. These are plants that have been genetically engineered to be resistant to produce pesticides in their tissues (e.g., *Bt* crops). Although plants engineered to be herbicide resistant are not classified as plant-pesticides, they nonetheless are subject to EPA regulation simply because they can affect the use of herbicides. The authority for such regulation is provided under FIFRA and FFDCA. If a plant producing a plant-pesticide is intended to be used for food, the EPA must establish a "safe level" of the pesticide residue allowed. The EPA defines a safe level as a reasonable certainty that no harm will result from aggregate exposure to the pesticide chemical residue, including all anticipated dietary exposures and all other exposures for which there is reliable information. The EPA has exempted several genetically engineered products that utilized viral coat proteins in their development from the requirement of a tolerance prior to being allowed in the food chain. Some of the products are listed in Table 21–2.

THE ISSUE OF FOOD ALLERGENS IN GM FOODS

Food allergy or food sensitivity is an adverse immunologic reaction resulting from the ingestion, inhalation, or contact of a food or food additive. It should be distinguished from food intolerance (e.g., lactose intolerances, the "Chinese restaurant syndrome" caused by monosodium glutamate, food poisoning, wine-induced migraine), even though the tendency is to use both terms interchangeably. The most widely studied mechanism of food allergy is that mediated by **immunoglobulin E (IgE).** Allergic reactions are produced immediately when IgE, an antibody, is exposed to an allergen (usually a protein substance). Such an exposure causes allergy cells in the body (mast cells and basophils) to release different kinds of toxic mediators (e.g., histamine and leukotrienes), which then trigger an allergic reaction. The reaction may manifest itself in a variety of ways, ranging from minor itches to anaphylactic shock and death. There are some allergic reactions to foods that are not mediated by IgE (e.g., celiac disease or gluten-sensitive enteropathy).

The only current treatment for a food allergy is avoidance. An estimated 34 percent of emergency room visits in the United States are for treatment of anaphylaxis related to a food allergy. Because of the rise in allergic disorders, the public is concerned about the potential risk posed by bioengineered foods and food products. The most common foods associated with food allergies (and accounting for over 90 percent of reported food allergies worldwide) are peanuts, tree nuts (e.g., almond, brazil nut, cashew, macadamia, hazelnut, pecan, pistachio, walnut), cow's milk, fish, shellfish (crustaceans and mollusks), soy, wheat, and sesame seed. Some of these food allergies disappear as children grow older (e.g., cow's milk, wheat, egg allergies), whereas others (like allergies to peanuts, tree nuts, and seafood) are usually lifelong problems. It has been shown that some proteins are intrinsically more allergenic than others, and hence different cultivars of the same plant may vary in their allergen content (e.g., peanut, wheat, avocado).

It is known that allergenic proteins can be transferred by genetic engineering from one organism to another. Such a transfer was confirmed in the case of the brazil nut 2S albumin storage protein that was transferred by rDNA technology to soybean to increase its methionine contents (legumes like soybean are low or deficient in the essential amino acid methionine). Tests for allergenic potential to humans by the radioallergosorbent test (RAST) and skin prick test showed that the brazil nut allergen had been transferred. It should be mentioned that commercial development of this food product was summarily discontinued. Furthermore, there are no validated reports of allergic reactions to any of the currently marketed GM foods as a result of transgene protein. Nonetheless, there is a concern that the use of a transgene in a staple food, or a specific transgene in several types of commonly ingested food, may increase the concentration of the particular recombinant product protein in the foodstream to extents that could increase the risk of allergy development in the population. There is evidence that allergic reactions from the consumption of conventional (non-GM) products like peanuts, avocadoes, mangoes, and exotic fruits increased in North America as a result of the increase in total dietary consumption of these foods.

Assessment of allergenicity is accurate and reliable only when dealing with proteins from known allergenic sources. Certain factors should be taken into account when assessing the allergenicity of a transgene protein. It is important to know the source of the donor gene, and also to compare the donor protein with known allergens. Donor genes from common food sources are easier to evaluate because they have been studied more extensively. *In vitro* and *in vivo* immunologic analysis is the most sensitive and specific test of allergenicity. Because allergens tend to have certain characteristics in common (e.g., molecular weight of 10 to 70 kiloDaltons, resistance to acid and proteolytic enzyme digestion), it is important to assess the key physiochemical characteristics that are common to allergenic proteins. Genetic engineering of foods may have potential collateral changes (or pleiotropic effects) resulting from the transgene having an effect simultaneously on more than one characteristic of the host. This may include an alteration of the intrinsic allergenicity of the protein by, for example, glycosylation, or alteration of the amount of allergenic protein produced. It is hence important to consider the potential changes in endogenous host allergens subsequent to gene transfer.

THE CONCEPT OF SUBSTANTIAL EQUIVALENCE IN THE REGULATION OF BIOTECHNOLOGY

How different are GM crops and the non-GM plants from which they descended? A major challenge facing regulatory agencies all over the world is in deciding what constitutes a meaningful difference between a conventional crop cultivar and its genetically modified derivative. The concept of substantial equivalence originates from the general position taken by regulators to the effect that conventional cultivars and their GM derivatives are so similar that they can be considered "substantially equivalent." This concept, apparently, has its origins in conventional plant breeding in which the mixing of the genomes of plants through hybridization may create new recombinants that are equivalent. However, critics are quick to point out that this is not exactly what is obtained with genetic engineering. In conventional breeding, the genes being reshuffled have had the benefit of years of evolution during which undesirable traits have been selected out of most of the major crops. The genes involved in GM crop production have been introduced into unfamiliar genetic backgrounds.

The traditional method of evaluating new cultivars from breeding programs is to compare them with the existing cultivars they would be replacing, if successful. To this end, evaluations include gross phenotype, general performance, quality of product, and chemical analysis (especially if the goal of the breeding program was to improve plant chemical content). It is expected (or at least hoped) that the new cultivar will not be identical to the existing cultivar, because of the investment of time and effort in breeding. Nonetheless, it is not expected that the new cultivar would produce any adverse effects in users of the product.

The World Health Organization (WHO) published a report in 1995 in which the concept of substantial equivalence was endorsed and promoted as the basis for safety assessment decisions involving genetically modified organisms and products. Since then, the concept has attracted both supporters and opponents. Some feel that, as a decision threshold, the concept is vague, ambiguous, and lacks specificity, setting the standard of evaluation of the GM product as low as possible. On the other hand, supporters believe that what is intended is for regulators to have a concept, though not a scientific formulation, which does not limit what kinds and amounts of tests regulators may impose on new foods. The concept of substantial equivalence has since been revisited and amended by the FAO and WHO. It appears that opposition to the use of this concept as a regulatory tool would be minimized if a product is declared substantially equivalent after rigorous scientific analysis has been conducted to establish that the GM product contains no more health or environmental risk than its conventional counterpart.

THE ISSUE OF "NOVEL TRAIT"

Sometimes, conventional plant breeding may introduce a "novel trait" into the breeding program through wide crosses or mutagenesis. This notwithstanding, the new cultivar produced is still considered substantially equivalent to other cultivars of the same crop. On the other hand, even though the presence of a transgene in a GM cultivar is considered an incorporation of a "novel trait," the GM product and the conventional product differ in some fundamental ways. The novel traits incorporated in the major commercially produced GM crops are derived from nonplant origins (mainly from microorganisms). Secondly, only a single gene separates a GM product from its derivative. In conventional breeding, the desired genes are transferred along with numerous other unintended genes.

The question then is whether the more precise gene transfer of genetic engineering means that a GM crop and the traditional counterpart would differ only in the transgene and its products. If this were so, a simple linear model would be adequate to predict the phenotype of a GMO. Unfortunately, because of the role of environment in gene expression, and the complex interactions that occur in a biological system, linear models are seldom adequate in predicting complex biological systems. Furthermore, it is known that single mutations often produce pleiotropic effects (collateral changes) in the organism. Similarly, the collateral effects of a transgene have been demonstrated in the transgenic salmon carrying the transgene coding for human growth hormone, in which researchers found a range of phenotypes. It is also important to mention that the altered phenotypes may appear at particular growth times in the growth cycle of the organism, or in response to specific environmental conditions. Furthermore, these phenotypic changes induced by the transgene may only be minor alterations.

In view of the foregoing, it appears that the best way to assess any adverse effect of a transgene is to directly test for harmful outcomes. In food biotechnology, such assessments should include testing for both short- and long-term human toxicity and allergenicity, among others. There should also be an assessment of environmental impact over time and across relevant sites. Then, a final assessment should be made regarding the extent to which the transgenic cultivar deviates from the parental genotype. It is important that the analysis indicate whether such deviations, if any, are biologically significant. Otherwise, the GM cultivar would be substantially equivalent to the existing cultivars and would not need prior approval for introduction into the food chain.

THE PRECAUTIONARY PRINCIPLE

The **precautionary principle** is a rule about handling uncertainty in the assessment and management of risk. This rule recommends that uncertainty, when it exists, be handled in favor of certain values (health and environment) over others. In other words, when our best predictions turn out to be in error, it is better to err on the side of safety. Another way of putting it is that, *ceteris paribus* (all things being equal), it is better to have foregone important benefits of a technology by wrongly predicting its risks to health or the environment than to have experienced the harmful consequences by wrongly failing to predict them. In statistical terms, if an error in scientific prediction should occur, it is better to commit a Type I error of declaring a false positive (that is, erroneously predict an adverse effect where there is none) than a Type II error (erroneously predict no such effect when there actually is one). However, it is the custom of science that it is a more serious flaw in analysis to commit a Type I error (make a premature claim, such as to reject the null hypothesis that the GM crop poses no significantly greater risk than its conventional counterpart) without adequate scientific evidence.

In view of the foregoing, it is not difficult to see that the precautionary principle has both proponents and opponents. Proponents see it as a proactive and anticipatory strategy for protecting the public, environment, and animals from potential harm that is hard to predict by even the best science available. On the other hand, opponents view the precautionary principle as unscientific, a tool that promotes unfounded fears in the public and mitigates against the research and development of new technologies.

This principle emerged in the 1970s and is currently invoked in numerous international laws, treaties, and protocols (e.g., the Cartegena Protocol on Biosafety of 2000). It is more widely accepted in Europe than the United States. There are certain common criticisms of the precautionary principle. Some feel it is ambiguous and lacks uniform interpretation. Also, it marginalizes the role of scientists in that, whenever it is invoked, it usually tends to relax the standards of proof normally required by the scientific community. Others see the precautionary principle as a veiled form of trade protectionism. Specifically, nations may invoke this principle to circumvent the fundamental rules established by trade agreements and enforced by the World Trade Organization. Such rules generally require that a nation provide reliable scientific evidence to support its decisions (e.g., to ban the importation of a product). For example, the decision by the European markets to ban American and Canadian beef treated with rBST (growth hormone) is considered to be colored by protectionism.

REGULATION AND THE ISSUE OF PUBLIC TRUST

The public is deeply concerned about the integrity of regulatory agencies. There is a need to see these agencies as objective, independent, and transparent. The public needs to trust those who develop and implement regulations that govern the development and application of technologies. It is widely accepted that even the most minimal risks may be unacceptable if levels of public trust in those who manage these risks are low or eroding. The general public apprehension about the risks of GM foods is blamed to a large extent on the loss of public trust in scientists and regulatory bodies resulting from the BSE crisis in Great Britain.

It is claimed that the assessment of biotechnology risks is a science-based activity. Consequently, it is important that the process be above reproach. The science should be high quality, and conduct of the assessment be independent and objective. There should not be a conflict of interest in the regulatory process. Any association between promoters and regulators is bound to cast doubt on the integrity of the process.

One factor in boosting public confidence is transparency of the regulatory process. During the application process, an applicant is required to submit certain data to the regulatory agency. The question, then, is how much of the information should be divulged to the public and how much should remain proprietary information. Because the regulatory process claims to be science-based, the custom of scientific enquiry is to be open and completely transparent. Furthermore, because the decision of the regulatory authority is based on the scientific evidence, it appears that any attempt to withhold information involved in the decision-making process may cast doubt on the integrity of the process.

EXAMPLE OF SAFETY ASSESSMENT

Before a commercial product is made available to the public, it must gain regulatory approval for production and import. The criteria for this vary from one country to another. Genetically modified crops are required to be evaluated for safety as food and feed, and for environmental impact. The transformation event MON810 (YieldGuard™) will be used to demonstrate the process of safety assessment conducted by the developer, Monsanto. The process is summarized in Figure 21–1.

■ *THE PRODUCT*

One of the most economically important pests of corn is the European corn borer (*Ostrinia nubilalis*). This pest usually has one to three generations per year. The first generation larvae feed on leaves, while both generations one and two tunnel the stalk. The second- and third-generation larvae feed on the leaf collar and sheath, and damage the ear. Once the larvae have tunneled into the stem, chemical control is ineffective. The Monsanto Company developed an insect-protected corn line by the transformation event MON810 and marketed it as YieldGuard™.

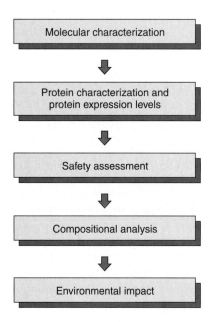

FIGURE 21–1
An example of the steps in a safety assessment process for genetically modified products.

■ THE ENVIRONMENTAL ASSESSMENT OF YIELDGUARD™

Monsanto conducted an environmental assessment of YieldGuard™ that consisted of the following:

1. **Molecular characterization**

    Molecular characterization entailed a disclosure of the source of the transgene, the methods used in transformation, and the goal of transformation. YieldGuard™ was produced by microprojectile bombardment of embryonic tissue using the plasmid PV-ZMBKO7 that contained the *cry1Ab* gene. The gene was isolated from the bacterium *Bacillus thuringiensis kurstaki* HD-1 strain. The gene was modified to increase the level of expression of the *cry1Ab* protein in plants. The expression of the gene was regulated by the enhanced CaMV 35S promoter and hsp70 main intro.

2. **Protein characterization and protein expression levels**

    Plant samples were obtained from four field trials. Expression levels were determined by ELISA. The results were used to define the level of active ingredients as required for the EPA product label and to calculate expected exposure levels. The results also were needed to support the effective-insect-resistance management strategy as well as establish the stability of the insert through breeding.

■ SAFETY ASSESSMENT OF THE CRY1AB PROTEIN

The *cry1Ab* protein produced by the *Bt* gene was subjected to a variety of chemical and comparative analyses as follows:

1. **Mode of action and specificity**

    The protein is insecticidal to only lepidopteran insects. *Cry1*-type toxins selectively bind to specific receptors found on the brush border midgut epithelium. These receptors do not occur on mammalian intestinal cell surfaces.

2. **Digestion of *cry1Ab* protein in simulated gastric and intestinal fluids**

    The trypsin-resistant core (the insecticidally active form of the *cry1Ab* protein) was used in the simulated digestion test. Western blots indicated that more than 90 percent of the initially added *cry1Ab* protein degraded within 30 seconds of incubation in simulated gastric fluids. Other similar tests were also conducted.

3. **Acute mouse gavage study with *cry1Ab* protein**

    This study was conducted to directly assess the potential toxicity associated with the *cry1Ab* protein. The results showed that the protein was not acutely toxic to mammals.

4. **Lack of homology of *cry1Ab* protein with known toxins**

    This test compared the amino acid sequence of *cry1Ab* protein to known toxic proteins to assess potential toxic effects. The *cry1Ab* protein sequence was compared to databases in PIR, EMBL, SwissProt, and GenBank and had no match.

5. **Lack of homology of *cry1Ab* protein to known allergens**

    The allergenic potential of *cry1Ab* protein was assessed. Using the FASTA computer program, the amino acid sequence of 219 allergens in the public domain databases (GenBank, EMBL, PIR, SwissProt) were searched. No homologies were identified.

■ COMPOSITIONAL ANALYSIS OF CORN GRAIN AND FORAGE

This assessment takes into account the uses of the crop and crop products in both animal and human nutrition. Compositional analyses showed corn line MON810 to be substantially equivalent to current corn varieties. Both the grain and forage of corn line MON810 were compared with those of the control line and other values published in various scientific literature. Tests included proximate analyses (protein, fat, ash, crude fiber, moisture), amino acid composition, fatty acid profile, Ca, and P. No statistically significant increases were found between the control and the genetically modified product for proximate analyses. Statistically significant increases were observed for eight amino acids (cystein, tryptophan, histidine, phenylalanine, alanine, proline, serine, and tyrosine) but were not consistent across multiple-year data.

■ ENVIRONMENTAL IMPACT

The leading microbial insecticide in agricultural use is the *Bacillus thuringiensis kurstaki* HD-1. This was the same strain that supplied the gene for YieldGard™. This insecticidal effect is specific and limited to lepidopteran pests. Studies on nontarget insect species (including honey bee, ladybird beetle, lacewing) and earthworms showed no toxic effect at high concentration.

BIOSAFETY REGULATION AT THE INTERNATIONAL LEVEL

Because biotech products are accepted to varying extents in various countries, and because international trade involves crops that are targets for biotech, it is imperative that trading nations develop a consensus for biosafety regulation. An international delegation convened to draft global regulatory guidelines, called the **Biosafety Protocol**. An outgrowth of the Convention on Biological Diversity (at the Earth Summit of 1992), the Biosafety Protocol is designed to provide guidelines on the transfer, handling, and use of what is described as living modified organisms (LMOs) that have the potential to impact the conservation and sustainable use of biodiversity. The Biosafety Protocol is under the auspices of the UN Environmental Program (UNEP). The Cartagena Protocol on Biosafety has been interpreted by some to mean that LMOs intended for food, feed, or processing must be identified as LMOs.

Basically, an exporter of a product will be under the obligation to provide the importer with information about the LMO regarding risk assessment and obtain consent prior to shipment. Critics of the Biosafety Protocol say that its implementation will adversely impact international trade by imposing severe trade barriers on a wide variety of biotech products (bulk grain, processed food, drugs, and so on). Cost of goods will increase as shippers have to segregate products, thereby increasing handling costs. Scientific development progress will also be impacted as scientists are compelled to pay more attention to special interest groups.

Biosafety regulation stringency is variable from one nation to another, being generally more liberal in the United States than the European Union. In Japan, the Ministry of Agriculture, Forestry and Fisheries is responsible for assessing environmental and feed safety, while the Ministry of Health and Welfare is responsible for food safety assessment. Basically, a product is subject to scrutiny if it is developed by rDNA technology. In Canada, the basis of assessment is the safety of the novel traits that have been incorporated, regardless of the technology used to produce the product. Gaining access to the EU market is a complicated task. However, once approved, the product becomes legal in all the member countries of the European Union. The product manufacturer or importer must submit a notification to the competent authority of the member state of the European Union where the product is intended to be marketed. In China, the State Science and Technology Commission has the responsibility of developing a regulatory system for GMOs. Regulations in developing countries are generally lacking. The Biosafety Protocol might be beneficial in this regard to assist the less-industrialized economies in gaining market access to developed economies. There is no denying that a unified regulatory system of GMOs would facilitate international trade involving these products. Unfortunately, a consensus that will be fully acceptable to all nations will be difficult to achieve in the near future.

LABELING OF BIOTECH PRODUCTS

As the battle over the safety of biotech food continues to be waged, the issue of labeling remains in the forefront of the debate. Some propose that consumers should have the right of "informed choice" about the exposure to risks of GM products. This push for labeling is partly because of the perception of lack of transparency from regulatory agencies, and the absence of balanced risk/benefit analyses. Because the first-generation products of biotechnology benefited the food producing industry directly, as previously indicated, consumers tend to view GM crops as geared toward enriching large corporations. Some consumer advocates would like to see all biotech foods labeled as such. The argument against labeling, advanced by the biotechnology industry, is viewed as an attempt to conceal information from the public. Opponents do not see a need for labeling since the FDA has ruled that there is no inherent health risk in the use of biotechnology to develop new food products. The food industry opposes mandatory labeling because of the concern that such labeling could be interpreted to be "warning labels" implying that biotech foods are less safe or nutritious than their conventional counterparts.

The FDA requires a food product (including biotechnology foods) to be labeled if the following apply:

1. It contains a protein known to pose allergenic risk (e.g., milk, eggs, peanuts, tree nuts). Consequently, any genetic engineering involving gene transfer from any of these organisms must be labeled.
2. Its nutrient content as a result of the genetic manipulation is significantly different from what occurs in a normal product. For example, if high protein is engineered into a cereal or root crop, the product must be labeled.

Opponents argue that labeling all biotechnologically produced foods would increase the cost of products as a result of the added cost of product segregation for the purpose of the so-called **identity preservations** of certain products. To avoid contamination, biotech and conventional products must be kept apart at all phases of production, storage, processing, and distribution at additional cost. This would impact bulk or commodity products like grains (corn, wheat, soybean). However, specialty and high-value fruits and vegetables are already identity preserved for premium prices.

Labeling of all products might be helpful to those who practice certain lifestyles or religious beliefs that impose strict dietary observances. A plant with an animal gene may not be acceptable to a strict vegetarian. However, studies have shown that both the kosher (Jewish) and *halal* (Muslim) communities have mechanisms in place to determine which products are acceptable to their adherents. Leadership of both religious groups have ruled that simple gene additions that lead to one or a few components in a species are acceptable for their religious practices. However, the Muslim community has not resolved the issue regarding the acceptability of gene transfer from swine into species, should that happen. Both Jewish and Muslim communities accept the use of bioengineered chymosin (rennin) in cheese production.

Many countries have some form of labeling in force, which may be mandatory or voluntary. The primary forum for the discussion of food labeling at the international level is the Codex Alimentarius Commission. Mandatory labeling has been implemented in the European Union and is currently being implemented in Japan. In Europe, all products containing GMOs must be labeled as such. Even where mixtures of conventional food products and GMOs are concerned, a label must be provided to indicate that GMOs may be present. The United States and Canada require GM food products that pose a health and safety hazard (possible allergens or changes in nutritional content from acceptable levels) to be labeled.

In the Western world, labeling is generally thought to be necessary only when there is some feature of the product itself that needs to be brought to consumers' attention (e.g., health risk or nutritional issue). The process by which the product is produced (e.g., by genetic modification) is considered inconsequential. This is described as product-based regulation. An exception to this approach in the United States and Canada is the requirement that food subjected to the processes of irradiation be labeled. In the United States, the FDA and the courts generally consider reference be made to a "material fact" about the product that concerns pertinent nutritional value or safety. This affirms the concept of substantial equivalence in which a new food product that is substantially equivalent to existing products is exempt from labeling.

ECONOMIC IMPACT OF LABELING AND REGULATIONS

The economic impact of food regulations and labeling on trade depend on the products involved, the cost of labeling, and sometimes how consumers use such information. The cost of labeling will depend on the stringency imposed; that is, whether "zero tolerance" or "minimum tolerance" of GM product is the goal. Implementing the former standard would require expensive safeguards to be implemented to avoid cross-contamination. Harvesting, processing, shipping, and other product handling would require modifications.

Government approval can have severe adverse consequences on trade. For example, sale of U.S. corn in the EU countries was dealt a devastating blow in 1999 because certain GM corn varieties were not approved for sale in the European Union. This action caused U.S. corn exports to the European Union to drop from $190 million in 1997 to $35 million in 1998, and then to a low of $6 million in 1999. Consumer response to labeling has an impact on product demand. Sometimes, products intended for use as feed may not require labeling.

KEY CONCEPTS

1. Biotechnology is subject to regulation like any other industry.
2. The agencies with regulatory oversight in biotechnology in the United States are the USDA, Food and Drug Administration (FDA), and the Environmental Protection Agency (EPA).
3. The criteria that are used by the FDA are: unexpected effects, known toxicants, nutrient levels, new substances, allergenicity, antibiotic-resistance selectable markers, plants developed to make specialty nonfood substances, and issues specific to animal feed.
4. Regulations vary among nations.
5. Food labeling is a controversial issue.
6. The EPA regulates pesticides. Its definition for pesticides is any substance or mixture of substances intended for preventing, destroying, repelling, or mitigating pests.
7. Food allergy or food sensitivity is an adverse immunologic reaction resulting from the ingestion, inhalation, or contact of a food or food additive.
8. The concept of substantial equivalence originates from the general position taken by regulators to the effect that conventional cultivars and their GM derivatives are so similar that they can be considered "substantially equivalent."
9. The precautionary principle is a rule about handling uncertainty in the assessment and management of risk. This rule recommends that uncertainty, when it exists, be handled in favor of certain values (health and environment) over others.
10. The public is deeply concerned about the integrity of regulatory agencies. There is a need to see these agencies as objective, independent, and transparent.

OUTCOMES ASSESSMENT

1. What agencies oversee the regulation of biotechnology in the United States?
2. Describe how the FDA regulates products.
3. How is regulation of biotechnology in the United States different from what is obtained in the European Union?
4. What are the pros and cons of labeling for the biotechnology industry?
5. Discuss the concept of substantial equivalence in risk assessment.
6. Discuss the precautionary rule and its application in risk regulation.
7. Give four common sources of allergens. Discuss the importance of allergenicity in risk assessment.
8. The public must view the regulating agencies of biotechnology as trustworthy. Explain why this is necessary.

INTERNET RESOURCES

1. U.S. biotech regulatory oversight: *http://www.aphis.usda.gov/biotech/OECD/usregs.htm*
2. Links to international regulations on biotech: *http://www.agwest.sk.ca/saras_reg_int.shtml*
3. Codex Alimentarius Commission: *http://www.codexalimentarius.net/*

REFERENCES AND SUGGESTED READING

Brill, W. 1985. Safety concerns and genetic engineering in agriculture. *Science,* *227*:381–384.

The Royal Society of Canada Expert Panel on the Future of Food Biotechnology Report. 2001. Canada, Ottawa: The Royal Society of Canada.

Sanders, P. R., C. L. Thomas, M. E. Groth, J. M. Astwood, and R. L. Fuchs. 1998. Safety assessment of insect-protected corn. In *Biotechnology and safety assessment,* 2nd ed. J. A. Thomas (ed.). Philadelphia: Taylor and Francis USA.

22 Biotechnology as a Business

PURPOSE AND EXPECTED OUTCOMES

As previously discussed in this text, biotechnology is steeped in science and technology. However, it is more than research or an academic exercise; it is about commercializing biology. It is about exploiting biological processes for industrial (profit-making) purposes. In Chapter 1, biotechnology was defined as using living systems to make products for the benefit of humankind. This is largely a business proposition. Products will be made if there is a market demand. Product development and marketing are capital-intensive activities. Investors will support a business proposition that is likely to be profitable to an acceptable degree.

In this section, you will learn:

1. About biotechnology business models.
2. What investors look for in a company.
3. How to start a new company.

THE ORIGIN OF THE BUSINESS CONCEPT

In order for biotech products to become reality, the discoveries will have to be taken out of the laboratory setting and into an industrial setting. Many of the outstanding discoveries originate in publicly funded academic and research environments. These institutes, traditionally, are not business oriented. Consequently, entrepreneurs with business acumen developed the concept of a **biotechnology company** outside these hallowed grounds. Whereas a company structure is not needed to make scientific advances, a commercial structure is needed to make money. Furthermore, companies survive only if they are financially viable.

A feature of the biotechnology industry worth noting is that, just like the computer industry, many markets have evolved in parallel with the technological advances that accompany the growth and development of biotechnology. Researchers in academia are able to tap into the opportunities that emerge, often securing significant private sector funds to acquire sophisticated equipment and other resources to conduct cutting–edge research.

Whereas it might appear that the profit motive of biotechnology companies is being overemphasized, the reality is that biotechnology *is* profit driven, whether the profit accrues to the investor or the general public. Consequently, biotechnology business must be managed like any other business to succeed.

Because it makes business sense to pursue profitable ventures, the private sector tends not to devote resources to the development of products that may be critical to society but are unprofitable. Consequently, governments are compelled to fulfill social obligations by stimulating certain critical activities through avenues like the award of research and development grants to public institutions. This raises the issue of developing countries that could benefit immensely from the commercial exploitation of biotechnology but lack both private and public sector resources to pursue such an endeavor. What obligation, if any, do affluent nations have in assisting economically challenged nations in this regard?

Entrepreneurial activities in life sciences started in the late 1970s. Some of the notable early players were established in 1980. During that year, an estimated $500 million was invested by venture capitalists into biotech start-up companies. In 2000, investment into the biotech industry was estimated at $5 billion. It should be pointed out that the period in between exhibited significant fluctuations in the financing environment for biotech. As previously indicated, there were over 1,200 biotechnology companies in the United States alone in 2002.

It is imperative for biotech companies to realize that they operate in a capital-raising environment, and consequently, they must create companies that can endure the turbulence in the environment.

BIOTECHNOLOGY BUSINESS MODELS

A criticism of biotech companies is that they often tend to be too science oriented, not paying adequate attention to the fundamentals of corporate culture. As one business executive candidly put it, there is too much science masquerading as companies. Many new companies continue to be formed based on only one narrow idea from academia.

It is a fact that the foundation of a business is a **business model**. Needless to say, there is no universal model. Furthermore, considering the diversity and dynamics of the marketplace, the business model may have to be appropriately modified to adapt to the prevailing business climate. There is no universal business plan, but there are universal business principles. In addition, another business executive observes that a good business model or plan should calculate a return on investment as well as a net present value, something many biotech companies fail to do.

There are two biotech business models that characterize biotech companies: (1) product development companies and (2) platform technology development companies.

■ PRODUCT DEVELOPMENT COMPANIES

Product development companies focus on commercializing a product (e.g., insulin). Between 1980 and 1990, a number of life science companies capitalized on new technologies to mass-produce a number of beneficial, small biomolecules that had not been considered by the giant pharmaceutical companies ("big pharma" as they are called). A sample of such companies and their products is presented in Table 22–1.

Advantages

The product-based biotechnology business model has certain advantages.

1. The market for some of the products, especially therapeutic products, is large and sustainable, for the very fact that disease conditions are seldom eradicated and thus sickness persists in the population, generating a perpetual need for medicines.
2. It is often difficult for competitors to encroach because of the favorable conditions for obtaining a patent for the product, thus protecting the investments of the creators of products.
3. It is very profitable, with gross margins often ranging between 85 and 95 percent. Furthermore, there is little pricing pressure while the patent for the product remains valid.

TABLE 22–1
Selected "products" companies and some of their specific products. These companies produce pharmaceuticals developed with biotechnologies.

Company	Selected Products and Some of their Uses
Amgen	Epogen®—for treatment of anemia
	Infergen®—interferon for treatment of hepatitis C
Genentech	Nutropin®—for treatment of growth hormone deficiency
	Pulmozyme®—for treatment of mild to moderate cystic fibrosis
Centocor	Retavase®—for management of acute myocardial infarction
	Reopro®—for reduction in acute blood clot related complications
Genezyme	Ceredase®—for treatment of type I Gaucher's disease
	Carticel™—for reconstruction of knee cartilage damage
Eli Lilly	Humatrope®—for treatment of growth hormone deficiency
	Humulin®—for treatment of diabetes
Centeon	Helixate®—for treatment of hemophilia
	Bioclate®—for treatment of hemophilia A

Disadvantages

Disadvantages of product development companies include the following.

1. The product-based biotechnology business model is high risk. It is estimated that only 1 in 10 companies that are evaluated in clinical trials receive approval for a new drug application. As a case in point, when a monoclonal antibody for sepsis failed in Phase III trials in 1992 to 1993, its manufacturer, Centro, experienced a drop in its stock price from $60 a share to $5 a share. The company subsequently laid off 1,000 employees.
2. Product development is characteristically of long duration (from 5 to more than 10 years) and expensive.

▪ PLATFORM TECHNOLOGY DEVELOPMENT COMPANIES

Platform technology (or tools) development companies focus on making existing technology more efficient (better, faster, and cheaper). Prior to 1990, tools companies engaged in activities such as the enhancement of the delivery of existing therapeutics. Currently, numerous companies have been formed not only in drug delivery but also in "hot" areas like combinatorial chemistry, gene discovery, and proteomics. A sample of such companies and their activities is presented in Table 22–2.

Advantages

The benefits of platform technology development companies include the following:

1. It takes a much shorter time for the product to reach the market. This is partly because FDA approval is often not required, since an existing and previously approved product is being improved.
2. Because the technology is not being developed from the beginning, the risk of product failure is lower.

Disadvantages

The disadvantages of the platform technology development company model for establishing a biotechnology company include the following:

TABLE 22–2
Selected "platform" companies and some of their specific products. These companies produce equipment, techniques, and materials used in biotechnology research.

Company	Selected Products or Tools
Affymetrix	DNA chips
Aurora Biosciences	High-throughput screening
Rosetta	DNA chips
Pharmacopeia	Combinatorial chemistry
Kiva Genetics	SNP genotyping
Oxford Glyosciences	Proteomics, protein chips
Xyomyx	Proteomics, protein chips
Perspective Biosystems	Sequencing instrumentation
Alza	Drug delivery
Élan	Drug delivery
Millennium	Gene discovery
Human Genome Science	Gene discovery
Mycometrix	Microfluidics

1. It is prone to the risk of competition that is almost inevitable, coupled with the risk of commoditization of the company's technology. Both factors diminish the value of the product and consequently reduce the gross margin of returns on investment. The adverse effects are more significant when the technology being used to enhance existing ones is not patentable. Such is the case in combinatorial chemistry, which is essentially a conceptual approach to building molecules, and hence cannot be patented. A similar fate is suffered by the DNA chip area in which the industry leader, Affymetrix, has good company in Corning, Agilent, and others.
2. Since quality is never ending, it is likely that, sooner or later, some better, cheaper, and faster technology will emerge on the market. Because of varying market loyalty, platform technology companies run the risk of being rendered obsolete by advances in technology development. An example is Kiva Genetics in the SNP genotyping field and Gamera Bioscience in microfluidics where competitive alternatives have taken away significant market shares from these pioneering companies.

WHAT IS THE BEST BUSINESS MODEL?

As previously indicated, there is no one best business model. Business executives generally believe that a successful biotech business model is one that works. A business plan should be the plan for a sustainable business. A successful biotech business should develop products or services that create increasing market share, a sustainable market franchise, and sustainable and increasing profits, notes one business executive.

Biotech companies constantly need to be aware that they operate in an arena in which they have to compete for investment capital. Biotech companies need to present themselves more attractively and effectively to the investment community. There is currently a significant diversity in business models in biotechnology. There are biotech business models for inventors only, inventors and producers, and even inventors, producers, and distributors. This diversity can be very confusing to investors.

Another proposition from some business executives is the development of a new biotech business model based on category dominance. Biotech companies should focus on a specific area and create significant value by seeing themselves as part of a structured industry where their skills and products are identified and sold into the larger biotech marketplace. Bottom line, the biotech industry should be structured to create value.

Some business executives note that a critical issue confronting emerging biotech companies is how to bridge the gap between their potential and the goals of the investor. What biotech companies need to do is to go back to the basics—the fundamentals of corporate formation and development. Emerging biotech companies need to spend less on development, and operate "virtually" until sufficient human data accrues to tell them they could begin to build infrastructure. It is also important to calculate a return on investment each step of the way.

Whereas tools-based business models may be the initial thrust of a start-up company and a significant determinant of its success, it is important that plans are made as soon as feasible for the company to evolve and expand into other business segments. However, companies often face a dilemma of how to best extend their original business proposition in order to access large and sustainable revenue growth. Consequently, a tools company in such a situation where there is stagnation of revenue growth and market capitalization often switches to a product business model. To this end, they leverage their core technology into, for example, an internal drug discovery organization. For example, the Millennium Company, originally established as a technology development and target discovery company, used its huge market capitalization to access a number of late-stage development compounds like leukocyte in 1999.

STARTING A BIOTECHNOLOGY BUSINESS

As previously indicated elsewhere in this chapter, it is not necessary to organize a company to embark on a scientific pursuit. However, it is important to have a company structure to make money. In order to effectively commercialize an intellectual idea, it needs to be nurtured in a business culture. How do biotechnology businesses get started? They can be started through new potentially profitable ideas. Traditionally, businesses are created to fulfill needs in a society and in the process financial rewards accrue to the owner(s) of the business. To fulfill a need, a business must either add value to something by making a product or add value to someone by providing a service. This suggests that, for many traditional businesses, the needs existed before the creation of companies to fulfill them. In the case of biotechnology, however, the technologies (e.g., for horizontal gene transfer) are sometimes discovered before their commercial value, targeted to a particular market, is thought of. Furthermore, biotechnology is rapidly evolving, posing a significant challenge to businesses based on it. Let's briefly review the key aspects of how one of the most successful biotech companies was created.

■ BUSINESS PLANNING

A business plan is critical to the success of any business undertaking. It articulates the vision of the owner and serves as a blueprint or map to guide the development of the company. A plan is crucial for the proper allocation of resources and monitoring and evaluating the progress of the company. Unless one is financing the business alone, it is critical to have a viable business plan in order to attract investors.

As previously indicated, there is no single best business plan. The contents of a plan will vary according to the nature of the proposition. However, there are certain topics that are frequently presented in a business portfolio.

1. **Summary**

    The purpose of a summary is to capture the essence of the proposal in a nutshell. This should be well done such that, after reading it, an investor would be adequately informed about and interested in the proposal to form an opinion. It should be brief and concise, touching adequately on all aspects of the project. Usually, a summary should not exceed two pages.

2. **The opportunity**

    A business is about exploiting an opportunity for profit. This portion of the plan should describe clearly the market opportunity that has prompted your decision to start a company. The next logical thing after describing the opportunity is to discuss how you intend to exploit it (i.e., what products or goods and services you intend to launch to capitalize on the market opportunity). For a technology-based venture like biotechnology, it is critical to describe the technology you plan to use. This should include, if applicable, the uniqueness and innovativeness of your approach, and how it is comparable or, better still, superior to existing efforts. It might be advantageous to use graphics to illustrate trends that support your vision. It is also impressive to identify any foreseeable challenges or barriers to the proposed enterprise and then discuss how you would address them.

3. **Business background**

    Just what kind of company are you interested in and what is its structure? Is it incorporated, unincorporated, a partnership, or a limited liability company? What stage of development is it at, and what pertinent assets does it own (e.g., patent)? Does it already have a customer base?

4. **Market**

    Why produce something you cannot sell? This is a fundamental question to consider in business creation. In this section of the plan, it is critical to demonstrate a good understanding of the immediate and future market for the products or services to be provided. Is it a niche market or broad based? Have you conducted a market survey or do you have access to existing data that describe the size and nature of the potential market? It is important to discuss the competition that exists and how competitive your product or service would be. To do this, you should know the strengths and weaknesses of the existing competition and in what ways your product is superior.

5. **Technology**

    What technology is going to drive this business? How sound is this technology? Is it proven or experimental? A detailed enough discussion of the technology should be provided, without giving away any proprietary information. Cite expert opinions to support your scientific protocols, and where applicable, point out the difference between yours and those of the competition, indicating the superiorities and advantages of your methods.

6. **Intellectual property issues**

    It is not only important for you to indicate any intellectual property you may own, but it is more critical to know what intellectual property infringements may occur in the development of your product, and how you propose to address them.

7. **Development plan**

    A development plan should be used to describe specific goals and milestones to be achieved as the company grows. Sometimes, the development may be divided into phases. There should be plans for evaluation of the progress made.

8. **Marketing plan**

    How do you intend to distribute your product? How do you plan to handle the market already captured by your competitors? Do you intend to start with local markets and then expand to national and international markets? Your market plan should be practical.

9. **Management**

Provide an organizational chart showing the levels of management and components of your management team. In the narrative, provide job descriptions for all key personnel and attach their curriculum vitae to show their qualifications for the offices. It is not necessary to fill all positions, but it is necessary to provide a timetable or some explanation as to how the empty slots will be filled.

10. **Financial information**

A business plan should include the sources of resources and how they are allocated. There should be a balance sheet to show profits and losses. A reasonable forecast of company finances should be included.

11. **Appendices**

Appendices may include resumes of key personnel, patent information, market research information, customer interviews, and pertinent information about competitors. These and other materials should be provided as supporting evidence for the presentation in the body of the plan.

■ *THE ROLE OF PEOPLE IN A BUSINESS*

It is needless to say that companies do not materialize out of thin air; people build companies. Consequently, business problems are essentially people problems.

1. **Corporate structure**

The people who work in a company define its character. It is important that a company develop a culture that would promote its success in realizing its purpose. A company should project a certain image, uniquely its own. The culture of the company will, among other things, determine how the company conducts its internal affairs regarding its values, how it treats employees, the attitude of employees, how it rewards excellence or achievement, how it manages and motivates workers for success, and how it promotes employee personal growth and development. The corporate structure is hence the lifeline of a company.

2. **Skills base of employees**

A business goes through phases of development. Each phase requires a different set of expertise. Consequently, a business may start with a certain composition of personnel and modify it as it grows. The personnel should be qualified in the first place, but it is critical that they work together as a team. It is critical that a new business start with the right team, without which the vision of a company can be lost and consequently doomed in its infancy. It is important that the corporate culture be able to keep its key personnel for a reasonably long time for continuity. Any sign of instability would send a negative signal to investors. The founding team should manage the company in a manner that attracts new, and retains competent, workers.

3. **Building the right team**

The way to build the right team is to first define the company needs and the kinds of people needed to fill positions that would help move the company toward realization of its vision. As already indicated, the needs of the company vary from one phase to another. At certain stages, the critical needs might be administrative, whereas technical skills may be paramount at other stages. Whatever the need, it is critical to recruit the right talent by organizing a search. For top-level personnel, the process of "head hunting" whereby outstanding candidates are identified and vigorously courted, may be useful. For other positions, candidates are usually identified through responses to advertisements followed by an interview.

4. **Importance of a human resources system**

Along with building a good team, building good human resources capabilities is necessary to manage the personnel for success and retention. A company should develop a human resources manual that details how the company culture would operate. Such an employee handbook would describe benefits, safety measures, terms and conditions of employment, grievances and disciplinary procedures, a reward policy, and other factors.

A good retention strategy should be included in a human resources system of a company. There should be procedures for performance appraisals and employee feedback. Also, there should be opportunities for on-the-job training or personnel development and mentoring programs to assist new employees.

■ THE ISSUE OF TECHNOLOGY

Whether the company is tools based or product based, the technology involved should be proven to be technically feasible and workable. Investors are not interested in funding scientific ideas or concepts; they will fund what is realistic and can be commercialized for profitability. The technology may be completely developed or may be in various stages of development. It is important to disclose the technical status to potential investors. Another aspect of interest to investors is the range of application of the technology. This will determine whether the market base for the company would be narrow or broad. Furthermore, it is advantageous if the technology has room for further development and expansion to include new product opportunities. A caution with technology in a fast-developing industry like biotechnology is the issue of patents. Whereas the core technology of the company may not be patented, it is important to conduct a thorough search (via the freedom to operate a search as described previously) to determine what proprietary materials may be infringed upon in the conduct of the business, and how they would be addressed upfront to avoid unnecessary and costly litigations at a later date.

■ FINANCING A BIOTECH BUSINESS

Biotechnology companies can access capital from the same pool that other businesses do. In 1997, the U.S. biotechnology industry raised a total of $5,529,500,000 to fund research and product development. In 1998, the amount totaled 9.9 billion dollars (see Table 22–3).

Types of Finance

A business may be financed in one of two ways—**debt financing** and **equity financing**. In debt financing, the company obtains capital by securing it against an asset or group of assets. The company incurs debt through loans, bank overdrafts, and other such transactions. The lender charges interest on any outstanding balance regardless of whether the company makes or loses money. The main sources of capital in debt financing are banks and other commercial lenders. Equity financing entails the element

TABLE 22–3
Selected biotechnology industry statistics, 1998.

Number of biotechnology companies	1,283
Number of employees in biotech (high-wage, high-value jobs)	153,000
Amount spent on research and development	$ 9.9 billion
Amount of money invested in biotech (market capitalization)	$ 97 billion
Total product sales	$ 13.4 billion
Total revenues	$ 18.6 billion

Source: Biotechnology Industry Organization

of risk, with the investor standing to lose the invested capital should the company be unprofitable or unsuccessful. Assets that may be placed at risk include plant, machinery, and stocks. The primary sources of equity capital are venture capitalists and the general public. Equity funders receive dividends only when the company has accrued distributable profits. Such dividends can change over the life of the investment. It should be pointed out that, in addition to these funding sources, a company may incur additional costs associated with the capital through a variety of fees paid to agents such as attorneys, accountants, and financial advisors.

What Investors Look For in a Biotech Company

The following topics represent insights from the institutional perspective.

1. **Business plan**

   This exhibit is of utmost importance in a business portfolio. Unfortunately, many companies lack this most critical requirement. As one portfolio manager observed, the science may be great, but our job is not to fund science; it is to fund companies to get to profitability to satisfy our investors. Investors will critically examine the management team (in terms of qualification, team spirit), past performance where applicable (as well as financial history), future performance (financial projections), and exit strategies.

2. **Quality of the science**

   It is important that the company provide evidence of peer review of the science behind their activities. As one executive puts it candidly, investors do not want to take a leap of faith.

3. **Multiple products**

   Biotech companies are notorious for long product development cycles. Investors have a risk tolerance of only a few years. They become jittery when they have to wait too long to see a viable product on the market.

In addition to the previous factors, funding sources are also interested in the stocks of the biotech company for the following.

a. Amount of cash left.

b. Is share-basis low?

c. Frequent news flow. (Mutual fund performances are based on quarterly numbers. Consequently, companies need to make significant and frequent announcements.)

d. Are expectations modest? Companies with stocks that are extremely high are not attractive.

■ *MARKETING BIOTECHNOLOGY PRODUCTS*

It is often said that the three factors to consider in establishing a new business are market, market, and market. Do not produce and then hope to find a market for the products. It is important to conduct market research to determine market outlets for the product(s). In this exercise, it is important to determine factors such as the size of the market, the distribution, the presence of competitive products, stability, profitability, and opportunities for growth. Because markets are not identical in characteristics, it is often necessary to segment the ones identified into categories such that each one is serviced with the appropriate versions of the product, if necessary. Upon market segmentation, it may not be necessary to pursue all the categories. The standard analytical method of SWOT (strengths, weaknesses, opportunities, and threats) may be employed to help identify which markets are ideal for the company to pursue. In general, it is recommended that before a company selects a market to serve that the following be considered.

1. Is the market measurable? It is important to be able to describe a market in terms of size and value.

2. Is the market accessible? The company should be able to reach the selected markets and serve them readily and in a sustainable fashion.
3. Is the market substantial? The market may be big and valuable, but is it big enough to make the venture profitable enough for the company?
4. Is the market actionable? Does the company have the resources to serve the market well enough to establish a dependable market base?

Once the markets have been firmly decided upon, the company needs to plan on how to actually deliver the goods and services to the customers. This has to do with the mechanics of moving the product. The alternatives are numerous, but the company needs to identify the most effective and profitable ones. Will the company deal directly with customers or through third parties? If competitors are operating in the market, what strategies would be employed for competitive advantage? Would you place ads? If so, where and how? A good product is unprofitable unless it is marketed effectively.

■ THE IMPORTANCE OF BUSINESS IDENTITY

As part of the business, a company should endeavor to define and develop a unique identity by which it wishes to be known in the business community. What does the business stand for? This should be in consonance with its vision and objectives. The identity should be unambiguous and easy for customers to associate with. For example, will the business be known for a certain line of products with high quality and affordability? Business identity builds a certain loyal customer base. It is important to be consistent in keeping the standards set in order for the customer base to grow.

STRUCTURING THE BIOTECH BUSINESS FOR SUSTAINABLE PROFITABILITY

There are certain factors that tend to reduce the chances of building a viable business. Unfortunately, these factors tend to characterize biotech companies.

a. Large unpredictable capital requirements
b. Extremely long product development cycles
c. Uncertainty about the product regulatory environment
d. Rapidly changing market forces
e. High late-stage probability of product failure
f. Rare instances of sustainable profits

In spite of these characteristics, a biotech company that has a good business plan is likely to be successful. Such a plan should incorporate factors for increasing the probability of success of a business, including the following:

a. Low, well-defined capital requirements
b. Well-defined, predictable business milestones
c. Clear, market-oriented business plan
d. The critical mass to compete
e. Management expertise relevant to the strategy being pursued

THE QUESTION OF INTEGRATION

It appears many business executives seem to prefer the approach of doing business whereby companies focus on their strengths, doing what they do best and relying on others for services that they really cannot add much value to. Instead of being bogged down with tasks like order entry, distribution, and customer service, such tasks can be outsourced. This allows companies to handle their product technology in a focused way. Some companies outsource product manufacturing and some upstream activities such as screening, genomics, and combinatorial chemistry. There are many service providers with capabilities for high-throughput services. Some predict that the near future is likely to witness a shift of drug discovery and drug development companies from a wet environment to a dry environment. Investors are similarly going to finance companies that get involved in business collaboration ventures. Incyte is one company that has benefited from such a shift. This highly flexible system of doing business whereby companies are highly segmented operates very successfully in the information technology and computer business. For example, Intel focuses on chips and not disk drives, while Seagate focuses on disk drives and not software.

THE START OF A BIOTECH COMPANY: THE AMGEN STORY

Amgen is the acronym for Applied Molecular Genetics. It was formed in April of 1980 by four venture capitalists, 10 university scientists, and a vice president of a major oil shale company. Others preceded this company like Collaborative Research (1961), Cetus (1971), Genentech (1976), and Biogen (1977). Of these, Genentech was the only company to focus right from the onset on recombinant DNA and human pharmaceuticals.

The first task of the founding partners was to organize a Scientific Advisory Board (SAB) and find a competent Chief Executive Officer (CEO). Dr. Winston Salser, Professor of Molecular Biology at the University of California at Los Angeles, was recruited to organize the SAB. Dr. George Rachman was hired in October as the President and CEO of Amgen, leaving his position at Abbott Laboratories. The law firm of Cooley, Godward, Castro, Huddleson, and Tatum was retained to help with the incorporation of the company in April of 1980.

With the first employee hired (CEO), the next critical business was fundraising. The seed money of $100,000 was raised quickly from a combination of stock sales, venture capitalists, future board members, SAB members, and three candidates for management positions. On October 23, 1980, Amgen's offering memorandum was sent to potential investors. In addition, individual meetings were scheduled. The initial fundraising efforts got off to a rocky, disappointing start. They discovered that they lacked a business plan and did not have an effective and formal presentation for would-be investors.

After heading back to the drawing board, the group put together a business plan, which among other things showed the company would generate a 10-fold return to investors within five to seven years. Unfortunately, their second round of attempts to woo investors was equally disappointing, with both E.I. du Pont de Nemours and Smith Klein turning them down for various reasons. This left them with a $3 million assurance from Tosco and another $200,000 from Assess Management. Later rounds of presentations finally convinced investors, including Abbott Laboratories which purchased $5 million in stocks. On February 2, 1982, Amgen closed their public offering for $18.9 million.

Dr. Salser soon resigned his part-time position as V.P for research and development to devote his full attention to his academic pursuits. By this time the first laboratory space had been leased in Thousand Oaks, California. Dr. Caruther was employed to establish a second lab in Boulder, Colorado, to pursue research in synthetic oligonucleotide chemistry. Their first assignment was to synthesize short genes for β-endorphin and calcitonin. Amgen also decided to develop a subsidiary in San Francisco, headed by Dr. William Rutler, a member of the SAB. Other key appointments made in 1981 included Dr. Daniel

Vapneck, Professor of Genetics at the University of Georgia, as Director of Research and Development; and Dr. Nowell Stebbing, Director of Biology at Genentech, as President for Scientific Affairs. A variety of successful staff was hired to bring Amgen's staff total to 42 by the end of 1981 and 100 by 1982.

In September of 1983, a preliminary analysis showed that Amgen could be out of money by that time. However, armed with a sound five-year plan, three separate negotiating teams were put together and dispatched to companies in the United States, Europe, and Japan to woo prospective investors. About 100 companies were contacted, but no corporate investors were found. Amgen was not prepared to lose its autonomy by offering rights to all products, including those yet to be discovered, as investors were demanding. As a last resort, Amgen management decided to go public in December 1982. In March 1983, Amgen, with the approval of its Board of Directors, retained three underwriters to complete an initial public stock offering. Amgen became a public company in June 1983, with a sale of 2.3 million shares of stock at $18 per share, and a market value of $200 million. In 1988, Amgen's first product, erythropoietin, received approval for marketing in several countries in Europe. By January 1990, nearly 100 patients were using the drug, with Japan granting approval for use in that country.

KEY CONCEPTS

1. The foundation of a business is a business model.
2. There are two biotech business models that characterize biotech companies: (1) product development companies and (2) platform technology development companies.
3. Each of the two basic models has pros and cons.
4. There is no one "best" business model that fits all companies.
5. Every business must have a business plan to be successful.
6. It is also suggested that the raging biotechnology debate is rooted in certain fundamental disagreements—scientific, political, religious, ethical, and philosophical disagreements.
7. A biotechnology business portfolio has certain general contents: summary, the opportunity, business background, market, technology, intellectual property, development plan, marketing plan, management, financial information, and appendix.
8. Companies do not materialize out of thin air; people build companies. Consequently, business problems are essentially people problems.
9. The people who work in a company define its character. It is important that a company develop a culture that will promote its success in realizing its purpose.
10. A business goes through phases of development. Each phase requires a different set of expertise. Consequently, a business may start with a certain composition of personnel and modify it as it grows.
11. The way to build the right team is to first define the company needs and the kinds of people needed to fill positions that would help move the company toward realization of its vision.
12. Whether the company is tools based or product based, the technology involved should be proven to be technically feasible and workable. Investors are not interested in funding scientific ideas or concepts; they will fund what is realistic and can be commercialized for profitability.
13. A business may be financed in one of two ways—debt financing and equity financing.
14. It is often said that the three factors to consider in establishing a new business are market, market, and market. Do not produce and then hope to find a market for the products.

OUTCOMES ASSESSMENT

1. Give the two basic business models in biotechnology. Give their pros and cons.
2. Discuss the importance of a business plan to a biotechnology company.
3. What factors should be considered in developing sustainability in a biotechnology company?
4. Discuss the weaknesses that appear to characterize many biotechnology companies.
5. Discuss the importance of teamwork in the success of a business.
6. Distinguish between debt financing and equity financing.
7. Discuss the concept of business culture.
8. Discuss the role of market in a business plan.
9. Describe the contents of a good biotech business portfolio.

INTERNET RESOURCES

1. Trends in biotech business models: *http://www.thebiotechclub.org/focus/columns/VC1-01.html*
2. Biotech at the Millennium National Biotech and Infotech Summit 2000: *http://www.connectionscorp.com/cgi-bin/biotech/content.pl?Subject= reportSection= opening*
3. Selected facts about biotechnology: *http://www.bio.org/aboutbio/guide1.html*
4. Business fundamentals: *http://www.romwell.com/books/business*
5. Forty concepts for a small business: *http://www.cbsc.org/osbw/concepts.html*

REFERENCES AND SUGGESTED READING

Rachmann, G. B. 1991. Biotechnology startups. In *Biotechnology: The science and business.* V. Moses and R. E. Cape (eds.). London, UK: Harwood Academic Publishers.

SECTION 2
The Role of Management

PURPOSE AND EXPECTED OUTCOMES

A biotechnology business, like other businesses, is comprised of resources—human and nonhuman—that must be managed within a business environment for profitability. The purpose of this section is to discuss how a biotechnology company should be managed for success.

In this section, you will learn:

1. The functions of a manager.
2. The organizational structure of a company.
3. The types of decisions made by a manager of a biotech company.

WHAT IS MANAGEMENT?

Management is a dynamic process. The factors that affect a business are constantly changing. The dynamics of the business environment are magnified several times in the case of entrepreneurial and cutting-edge businesses like biotechnology. Some of these factors are within the total control of a manager. Others can be manipulated, while yet others are outside the control of management.

In a conventional production enterprise, management is simply a decision-making and problem-solving process by which a producer allocates limited resources to a number of production alternatives to organize and operate the production enterprise for profitability or to attain desired objectives. Biotechnology is not much different from a conventional operation except that research and development often precede production of a product. This may entail new product development or enhancing existing techniques and methodologies. Some biotechnology companies are fully integrated to perform all the necessary functions at that stage. Others may form partnerships with public or private research institutes to conduct new research and development activities. Depending upon how this is done, some form of management may be necessary to guide the process.

GENERAL FUNCTIONS OF A MANAGER

A manager performs three basic functions: planning, implementation, and control.

1. **Planning**

    Planning is the key to success of any business undertaking and thus is the most urgent role of a manager. A plan for the enterprise entails the goals and objectives of the enterprise, resources needed, and strategies for identifying and solving problems. The operation of an enterprise is guided by an effective plan.

2. **Implementation**

    Once developed, a plan should be implemented correctly. The resources should be acquired in a timely fashion, and activities commenced according to schedule.

3. **Control**

    The producer should monitor the implementation of the plan by keeping accurate records. These records should be analyzed and used to improve the implementation of the plan.

ORGANIZATIONAL STRUCTURE OF A COMPANY

In the discussion of starting a biotechnology company, we got an overview of how Amgen was put together. This section focuses primarily on the Managing Director or Chief Executive, the person who makes the final decisions below the level of the Board of Directors. It is the role of the Managing Director to formulate a clear strategy for where the biotechnology company is going and how to get there, and to mobilize the other managers and their staff to accomplish the goal(s) set. These are the decisions that "make or break" a company. The Board of Directors is answerable to investors and, consequently, their main objective is to ensure that the company is well managed for profitability.

Below the Managing Directors are a number of Executive Directors of the various divisions or departments in the company (e.g., marketing, personnel, and production). Because biotechnology thrives on great science, a key department in a biotechnology company will be research and development. These departments may be organized into smaller groups or teams with team leaders for higher efficiency.

In addition to this permanent staff, a company may maintain a variety of temporary or associated personnel with unique expertise as consultants.

THE MANAGER AS A DECISION MAKER

The manager of a biotechnology company will be called upon to make decisions related to all aspects of the company that may be categorized into two broad groups.

1. Those affecting the internal environment (internal decisions concern issues like technology, resources, and marketing).
2. Those concerning the external business environment.

Obviously, these are not clear-cut categories since internal decisions are made mindful of external factors.

THE IMPORTANCE OF A STRATEGIC DECISION

A clear and sound strategy is critical to the survival of a biotechnology company. A company needs a clearly defined mission statement that defines its existence. What does the company want to be known for (e.g., a pharmaceutical company, seed company, diagnostics company)? How can the company mobilize its assets (technical, human, physical, and fiscal) to exploit the technical tools that embody biotechnology, to produce a product or render service for acceptable profit? Given the very nature of biotechnology and its rapid rate of evolution, it is conceivable and indeed probable that a company would be compelled at some point in its operation to revisit its strategic plans. In a highly competitive area like biotechnology, new products can readily diminish the market share of a company or render it completely obsolete. One of the qualities of a good and effective manager is to foresee such adverse events and make the necessary adjustments as is feasible to minimize their impact on the company.

TYPES OF DECISIONS

As indicated previously, one of the categories of decision making is based on the external business environment.

1. **Technology-based decisions**

    Biotechnology is a technology-driven business that entails the use of organism-based technologies. Technology-based decision making is a major aspect of the decisions a manager of a biotechnology company faces. We have already indicated elsewhere that the technologies in biotechnology are rapidly evolving. Not only do the ways in which the core science is conducted change, but also the production of a product may require modernization of the production process. As high-throughput technologies become available, a manager must decide whether it is more cost-effective to update the company's equipment or depend on outsourcing to have some routine jobs (e.g., sequencing) done. How does a new technology fit in with existing capabilities?
2. **Resource-based decisions**

    A biotechnology company thrives on quality human and nonhuman resources. Biotechnology depends on principles and concepts from several scientific disciplines. A biotechnology company should have a compliment of qualified scientists, especially biochemists, chemists, geneticists, and microbiologists. To attract high-caliber scientists, the manager should be prepared to pay competitive salaries and provide other incentives. For example, being from academia, such scientists would like to publish their research in peer-reviewed journals. On the other hand, the company may want to keep certain discoveries secret for at least a period of time. Academics will love to have a good library resource. Some companies offer stock options to scientists and other key staff.

Apart from scientists, a biotechnology company needs accountants, marketing personnel, public relations personnel, laboratory technicians, administrative staff, and other personnel for housekeeping. Some companies retrain or contract with other professionals, like lawyers for legal counsel on a variety of issues and doctors to service the employees as part of the benefits package.

Top-notch personnel can be hired only if the company has adequate financial resources. Biotechnology is a very capital-intensive business that depends on investors to start and sustain the operation. Biotechnology business accesses capital in a variety of ways, a major one being raising capital by selling equity. By this method, the company is publicly traded so that investors can own shares in the company. It is not uncommon for biotechnology companies, especially the entrepreneurial ones, to operate in a deficit mode during the early years of their existence. This is so because, during those years, the company is engaged in the development of their first products. Attracting investors for entrepreneurial business activity is difficult because the risk is high. The potential of the product to be profitable may not materialize.

A biotechnology manager should be able to market the company in a very positive light. A shake in investor's confidence can cause volatility in company stock.

MARKETING-BASED DECISIONS

As previously stated, before a product is developed, a company must be certain of three things—market, market, and market! In the real world, a consumer often has choices or alternatives. The producer must compete for customers by producing a superior product at a competitive price. The first critical marketing decision to make is about what to produce. In arriving at such a decision, a company goes through a kind of corporate brainstorming session. The key questions that must be considered in such an exercise including the following.

1. **Is the market real?**

   What this question seeks to answer is whether there is a genuine need for the product to warrant its production. In addition, if a real need does exist, will the customer buy or be able to afford it?

2. **Is the product real?**

   As already indicated, entrepreneurial biotechnology companies spend their early years in product development. There must be a product idea with the potential to satisfy a real market. But the next question is whether this great idea can be marketed. In other words, can the product be made? Is the company equipped for the task? If made, will it have the quality to satisfy the anticipated market?

3. **Can the product be competitive?**

   There are numerous players in the biotechnology industry. The competition is fierce. There is no room for second-rate products. The company should be able to produce a quality product at a competitive price. There should be an effective and efficient marketing and distribution system in place to capture a sizable market share.

4. **Will the product be profitable?**

   When all is said and done, will the product be a rewarding venture for the company and its investors? In addition to merely earning a profit, investors demand a certain profit margin to find it worthwhile for their investments. A company must also make certain returns on their investments to keep the business viable.

5. **Are company needs satisfied?**

As previously indicated, a company should have a clearly defined mission. All products should enhance the mission of the company and further its objectives and goals. Investors are more comfortable with a company that has a focus.

DECISIONS CONCERNING THE EXTERNAL BUSINESS ENVIRONMENT

The external business environment entails factors that are largely outside the control of the manager. This includes decisions that are made outside the business that impact the business. Examples include policies made at the national, regional, or local levels. Another policy impact may be interest rate decisions made by the Federal Reserve that may stimulate business or slow its growth.

External influences on a business may come from investors in the company who may express opinions about company performance or goals. There are also the outsiders who represent special interest groups (e.g., environmentalists) who may campaign against a company's product directly, or against an issue that affects the biotechnology industry in general.

External impact may come in by way of the competition. A competitor may make a business decision that threatens the company. The manager of a biotechnology company will have to take all these extraneous factors into account in making business decisions since they affect the business's ability to accomplish its goals.

KEY CONCEPTS

1. Biotechnology is big business and should be managed as such.
2. A manager is responsible for planning, implementing, and controlling a business.
3. Categories of decisions a manager will make are: technology based, resource based, market based, and decisions concerning the external business environment.
4. Management is a dynamic process. The factors that affect a business are constantly changing.
5. In a conventional production enterprise, management is simply a decision-making and problem-solving process by which a producer allocates limited resources to a number of production alternatives to organize and operate the production enterprise for profitability or to attain desired objectives.
6. A company needs a clearly defined mission statement that defines its existence.
7. Biotechnology is a technology-driven business that entails the use of organism-based technologies. Technology-based decision making is a major aspect of the decisions a manager of a biotechnology company faces.
8. A biotechnology company thrives on quality human and nonhuman resources. Biotechnology depends on principles and concepts from several scientific disciplines. A biotechnology company should have a compliment of qualified scientists, especially biochemists, chemists, geneticists, and microbiologists.

OUTCOMES ASSESSMENT

1. Discuss the role of a manager in a biotechnology company.
2. Discuss specific internal factors that a manager will have to be concerned about.

3. A biotechnology company manager should be concerned about the external business environment. Explain.
4. Management is a dynamic process. Discuss.
5. Discuss the market-based decisions a manager of a biotech company would make.

INTERNET RESOURCES

1. Basic management: *http://www.ee.ed.ac.uk/~gerard/Management/*
2. General management issues: *http://management.about.com/cs/generalmanagement/*
3. Guide to business ethics: *http://www.mapnp.org/library/ethics/ethxgde.htm*
4. Business resources center: *http://www.morebusiness.com/*
5. Principles of scientific management: *http://www.fordham.edu/halsall/mod/1911taylor.html*

REFERENCES AND SUGGESTED READING

Russell, K. A. 1991. Managing a biotechnology business. In *Biotechnology: The science and business.* V. Moses and R. E. Cape (eds.). London, UK: Harwood Academic Publishers.

SECTION 3
Recent Business Dynamics

PURPOSE AND EXPECTED OUTCOMES

Biotechnology is just like any other business. It is subject to the dynamics of the business climate. It has its share of successes and failures, mergers and acquisitions.

In this section, you will learn:

1. The forces behind the dynamics of the biotechnology business environment.
2. Farm-level production issues.
3. The role of the private sector in biotechnology.

BIOTECHNOLOGY AND THE PRIVATE SECTOR

Development and application of technology has economic ramifications. Advances in biotechnology have been possible because of large public and private sector investments into research and development. Whereas private sector agricultural research traditionally tends to focus on mechanical and food processing technologies, public sector research (federal research programs and university research) focus on development of improved animal breeds and plant cultivars. Sustained growth in U.S. agricultural productivity is attributable to the development of improved production inputs (seed, feed, and agrochemicals) and advanced technologies that promote production efficiency, coupled with good management of agricultural production.

Trends in investment into research and development show that the private sector began to outspend the public sector in the early 1980s and has since continued to do so. The difference in spending in 1995 was about $2.8 billion, and the 1995 total of $3.4 billion was about three times the spending by the public sector in 1960 (which was $1.2 billion). Furthermore, private sector investment into biological and chemical research and development rose from 19 percent of total research spending in 1960 to 58 percent in 1995.

The agricultural input industry has also experienced significant restructuring characterized by mergers, acquisitions, and formations of strategic alliances. Some 801 such business transactions were recorded between 1991 and 1996, compared to only 167 between 1981 and 1985. Larger and more established companies, seeking to consolidate their positions in this competitive arena, entered into various alliances with up-and-coming companies. Examples of such major consolidations include the acquisition of Calgene by the chemical giant, Monsanto. In addition, Monsanto acquired major seed companies such as Asgrow, Corn States Hybrids, Dekalb Genetics, Holden's Foundation Seed, the Plant Breeding Institute of Cambridge, Sementes Agroceres, and the foreign business components of Cargill. These consolidations occurred between 1996 and 1998, and transformed Monsanto from a company associated with plastics and petrochemicals to a principal player in the arena of life sciences. In 1999, Monsanto merged with the pharmaceutical giant Pharmacia and Upjohn, sparking a suspicion that perhaps Monsanto was reverting to its familiar roots.

Other agbiotech-oriented consolidations also occurred in the 1990s. DuPont purchased Pioneer Hi-Bred International, while two Swedish pharmaceutical and agrochemical companies, Ciba and Sandoz, merged to form Norvatis in 1996. Similarly, Hoechst and Rhone-Poulenc merged to form Aventis. Also, in 1999, an international merger between the Swedish pharmaceutical company Astra and the British bioscience company Zeneca produced AstraZeneca.

THE FORCES BEHIND THE CHANGES

Crop improvement or plant breeding has been the single most important factor behind all the reshuffling that has gone on in the private sector engaged in agricultural inputs production. The private sector investments into agricultural research have increased because of four specific factors.

1. **Advances in science**

    Many of the companies involved in the agricultural input industry have pioneered and developed some of the modern major technologies being used to facilitate modern plant breeding. These advanced technologies accelerate plant breeding and make it more profitable to develop new crop cultivars. Furthermore, some of the biotechnologies provide additional advantages to these companies because they were the ones that developed them. An example is the herbicide and insect-resistance genes that have been genetically engineered into crops by companies that produce the pesticides as well. This way, the seed and chemicals are marketed as an inseparable combo.

2. **Intellectual property rights**

    The private sector is profit oriented. Consequently, legislation that enforces intellectual property rights encourages the private sector to pursue basic research and invest more in research and development. Examples of such protection are provided by the Plant Variety Protection Act of 1970 and authorization of utility patents for microbes (in 1980) and plant and animals (1985 and 1987), which provide impetus for private sector involvement in research and development. Currently, private sector companies hold utility patents for multicellular living organisms and most of the plant variety protection certificates.

3. **Public-private sector research collaboration**

Various key pieces of legislation promote collaboration between federal research institutions and public sector companies. The Bayh-Dole Act of 1980 not only allows federal institutions "certainty of title" for their inventions, but also allows these institutions to issue exclusive licenses (as opposed to open licenses) for patents of their inventions. Furthermore, legislations like the 1980 Stevenson-Wydler Technology Innovation Act and the 1986 Technology Transfer Act were designed to promote dissemination of innovations originating in the public sector.

4. **Globalization of agricultural input markets**

Various global trade agreements, namely the General Agreements on Tariffs and Trade (GATT), the World Trade Organization (WTO), and the North American Free Trade Agreement (NAFTA), increase foreign markets and trade opportunities for companies. Consequently, companies are encouraged to invest in research and development in other countries.

FARM-LEVEL PRODUCTION ISSUES

Farmers, like biotech companies, are profit oriented. They will embrace a new technology if it will translate into acceptable returns on investment. To this end, what are the main benefits that have been promised to farmers for growing genetically modified (GM) crops, and have farmers' expectations been met so far? In terms of specific promises, farmers are encouraged to embrace GM crops because of several key advantages. The most important reasons cited by producers for adoption of GM crops with herbicide tolerance in cotton and corn production is increased yield through pest control. These advantages were apparently so attractive to U.S. farmers that, between 1996 and 1998, adoption of GM crops increased dramatically. Acreage of GM cotton doubled, while planting of GM corn (*Bt* and herbicide tolerant) increased to 40 percent of the total corn acreage. Similarly, acreage devoted to herbicide-tolerant soybeans rose from 7 percent to 44 percent of total acreage. On the world scene, an estimated 99 million acres of GM crops were produced in 1999, representing a 43 percent increase over the previous year. The United States is currently the world leader in GM crop production, accounting for 72 percent of the 1999 total.

■ APPROVED GM AGRICULTURAL PRODUCTS

The GM products approved for growing or use in the United States by May of 1999 are presented in Table 22–4. The most widely used GM products are those with herbicide-tolerant and insecticide-tolerant genes. The advantage to farmers is that they are now able to utilize a wider variety of herbicides in their weed management programs.

■ REALIZED PROFITABILITY OF GM CROPS TO THE PRODUCER

Preliminary results from research by the Environmental Research Service (ERS) indicate that the economic impact of GM crop production was dependent upon the specific crop and the specific technology. For example, the use of herbicide-tolerant cotton resulted in significantly higher yields and net returns, but did not change the use of herbicides. On the other hand, herbicide-tolerant soybeans increased yield only slightly, but were associated with decreased herbicide use. The adoption of *Bt* cotton in the Southeast, however, produced significant yield and net return increases, as well as a decrease in the use of insecticides.

TABLE 22–4

Selected transgenic products approved for marketing in the U.S.

Trait Engineered	Crops	Product/Company
Bt crops (with gene from *Bacillus thuringiensis*) for protection against lepidopteran insects	Corn Cotton Potato	YieldGard/Monsanto; NatureGard/ Mycogen; DeKalBt Bollgard/Mosanto NewLeaf/Monsanto

Herbicide tolerance (with gene resistant to specific herbicides)	Corn	Roundup Ready/Monsanto;
		    IMI-corn/Mycogen LibertyLink/
		    AgrEvo; CLEARFIELD/American
		    Cyanamid

	Cotton	Bollgard with BXN/Calgene
	Soybean	Roundup Ready/Monsanto/Norvartis
	Canola	Roundup Ready/Monsanto; SMART/American Cyanamid
	Sugar beet	Roundup Ready/Monsanto
Disease resistance	Corn	Gray leaf spot resistance/Garst Seed
Disease/herbicide resistance	Corn	G-Stac/Garst
High-performance oil	Sunflower	High oleic acid/Mycogen/Optimum Quality Grains
	Peanut	High oleic acid/Mycogen
	Soybean	Low linolenic/Optimum Quality Grains
Delayed ripening	Tomato	Fresh World Farms/DNAP Holding
Increased solids	Tomato	Increased pectin/Zeneca
Hormone	Dairy cows	Prosilac/Monsanto
Food enzymes	Cheese	Chymogen/Genecor International

■ CONSUMER-LEVEL ECONOMIC ISSUES

In the final analysis, producing biotech products is as economical as the level of consumer acceptance and patronage. Whereas agricultural products from conventional genetic manipulation are accepted wholesale and without any apprehension, many people view biotech food products as risky and unsafe. Such attributes vary from one place to another. Generally, U.S. consumers are more tolerant of GM food, while European Union consumers generally oppose GM products, sometimes very vehemently. Generally, consumers accept the use of biotechnology in medical applications more than in food applications. Also, when researchers surveying public opinion suggested in their questionnaire that biotech foods would have desirable traits, the respondents tended to be more approving of biotechnology. The lower acceptance of biotech products by the European Union is attributable to the recent incidence of potentially deadly disease outbreaks associated with the food supply system (e.g., bovine spongiform encephalopathy (BSE) in the UK, and dioxin in Belgium) in recent times.

Another factor that impacts consumer acceptance is biotechnology awareness. A significant number of survey respondents (37 percent) in the United States did not think GM foods were available on the market; 23 percent did not know whether or not biotech products were available on the store shelves. Similar results were found in the EU surveys.

KEY CONCEPTS

1. The biotechnology business environment is dynamic.
2. There are both public and private sector interests in the biotech industry.
3. Factors responsible for the dynamics in the business include technological and market forces (local and global).
4. The profitability of GM crops to producers is variable and depends on the specific technology and crop.
5. The agricultural input industry has also experienced significant restructuring characterized by mergers, acquisitions, and formations of strategic alliances.
6. Whereas private sector agricultural research traditionally tends to focus on mechanical and food processing technologies, public sector research (federal research programs and university research) focus on development of improved animal breeds and plant cultivars.
7. The most important reasons cited by producers for adoption of GM crops with herbicide tolerance in cotton and corn production is increased yield through pest control.
8. Producing biotech products is as economical as the level of consumer acceptance and patronage. Whereas agricultural products from conventional genetic manipulation are accepted wholesale and without any apprehension, many people view biotech food products as risky and unsafe. Such attributes vary from one place to another.

OUTCOMES ASSESSMENT

1. Discuss the adoption of GM crops by U.S. producers.
2. How profitable are GM crops to producers?
3. Describe the associations among biotechnology companies in the private sector.
4. Discuss the impact of public perception of biotechnology in the European Union on agricultural trade.
5. Contrast the private sector and public sector traditions regarding investment into agriculture.
6. Producing biotech products is as economical as the level of consumer acceptance and patronage. Explain.

INTERNET RESOURCES

1. Biotech business activities: *http://faculty.washington.edu/~krumme/readings/biotech.html*
2. Biotech industry news: *http://biz.yahoo.com/news/biotechnology.html*
3. Biotech industry news: *http://biotech.about.com/*
4. The Bulletin, biotech business: *http://www.alumni.hbs.edu/bulletin/2000/june/biotech.html*
5. List of best biotech business sites: *http://www.biotactics.com/bestbiotech.htm*
6. Biotech business: *http://www.biofind.com/business/*

REFERENCES AND SUGGESTED READING

USDA. 2001. *Economic issues in agricultural biotechnology.* Washington DC: Economic Research Service, AIB-762.

Kalaitzandonakes, N., and B. Bjornson. 1997. Vertical and horizontal coordination in the agro-biotechnology industry: Evidence and implications. *J. Agric. Appl. Econ., 29(1):*129–139.

Pray, C. E., and K. O. Fuglie. 2000. The private sector and international technology transfer. In *Public-private collaborations in agricultural research: New institutional arrangements and economic implications.* K. O. Fuglie and D. E. Schimmelpfennig. (eds.). Ames IA: Iowa State University Press.

SECTION 4
Jobs in Biotechnology

PURPOSE AND EXPECTED OUTCOMES

The biotechnology industry offers a broad spectrum of job opportunities, some of which are specialized whereas others are not. There are opportunities to work for established businesses or other kinds of employers, or to start one's own business. In this section, we shall explore the variety of job opportunities available in the biotechnology field in a general way.

In this section, you will learn:

1. The academic training required for jobs in biotechnology.
2. The types of jobs available in biotechnology.

DIRECT VERSUS INDIRECT JOBS

Jobs in the biotechnology industry may be described as either **direct** or **indirect**.

■ DIRECT JOBS

Direct jobs are those involved in conducting the actual scientific research and inventing products. The personnel in direct jobs range from highly trained senior research scientists to entry-level research technicians or assistants. Some of the specific jobs and their job descriptions are provided in Table 22–5. People who seek these jobs, especially at the top level, usually prepare by taking the appropriate college courses and acquiring higher academic degrees and other career-enhancing experiences.

■ INDIRECT JOBS

Indirect jobs exist in the support or service industries where companies produce products and provide services to direct jobs personnel. Biotechnology depends on the chemical industry for a wide variety of products used in research. People with training in chemistry

TABLE 22–5
Selected jobs in biotechnology.

Science Experience Required

Professor—teach biotechnology and related courses and conduct research in biotechnology at a university.

Research Scientist—full-time research in biotechnology.

Post-doctoral Fellow—conduct research in a specific program.

Research Associate—conduct research in a specific program.

Research Assistant—assist with research in a specific program.

Research Technician—provide laboratory and field assistance to researchers.

Science Experience Not Required

Personnel Manager—manage employees in a biotech company.

Accountant—manage company finances.

Administrative Assistants—responsible for various office tasks.

Marketing Specialists—market products.

Public Relations Officer—spokesperson for company.

and related basic sciences may obtain employment in such chemical companies. There are indirect jobs in the area of equipment manufacturing, product distribution, advertising, and many other areas.

JOBS IN A BIOTECHNOLOGY COMPANY

As previously discussed, the personnel in a biotechnology company are drawn from a wide variety of backgrounds that can be placed into the following general categories.

1. **Scientists**

    The personnel in this category are engaged in research and product development. The life of the company depends on the discoveries made and products developed by this critical mass of highly trained and technical workers, who produce what the company commercializes for profit.

2. **Administrators**

    There are different levels of administration in a biotech company, depending on the size and complexity of operations. Administration in a company may involve senior scientists. However, persons trained in business administration and other business disciplines such as accounting, marketing, and personnel management have special skills that are needed to effectively manage a biotech company.

3. **Support staff**

    This is a broad category of workers that may include clerical, maintenance, legal, engineering, statistical (analytical), and other support systems, according to the nature of the company. Of course, depending on the focus of the company, engineers or other groups of employees could play a more central role in the operation.

SCIENCE-BASED ACADEMIC TRAINING FOR BIOTECH JOBS

Academic training for biotech jobs may be obtained through a formal college education or acquired via informal training avenues.

■ *FORMAL TRAINING FOR A DEGREE*

If a student is training with the hope of finding a direct job in the biotechnology industry, the academic preparation should of necessity be targeted or specialized. Biotechnology is a science-based field. It draws on the principles, concepts, and techniques from
basic science disciplines, especially genetics, biology, chemistry, biochemistry, microbiology, physiology, and physics. The more recent academic discipline called molecular biology or molecular genetics is essential for anyone seeking a research position in
biotechnology. Microbiology is important because microbes play a significant role in
biotechnology as living tools or sources of biomolecules for a variety of applications (e.g.,
enzymes for research, vectors, or sources of genes for improving other organisms).

These academic courses are offered at various levels, depending on the academic degree
being sought. A graduate degree (preferably a doctoral degree) is desired for employment as
a researcher. Graduate training equips students with the skills of enquiry (research design,
analysis, interpretation) for the discovery and application of new ideas and products that a
company may commercialize. Post-doctoral training provides additional focused and extended practical preparation for success as a researcher. Such an experience often boosts the
intellectual capital of the recipient and is advantageous in a competitive job market. A bach-
elor's level degree is usually good for an entry-level job as a research assistant or technician.

■ *INFORMAL TRAINING*

There are numerous opportunities for on-the-job training to acquire new skills for research. Biotechnology is a very dynamic field in which new and improved methodologies and equipment are being introduced at a feverish pace. Short-term courses are useful
for even the experienced scientists to acquire additional skills as a continuing education
activity. These training activities are offered by a variety of providers. Federal research institutes, universities, and private companies organize workshops all year round. Some of
these may be free, whereas others charge a fee.

■ *SPECIALIZED TRAINING*

Formal training in the foundational principles and practices of biotechnology, and a general familiarity with recombinant DNA technology, may suffice for some jobs. Some employers will provide additional on-the-job training specifically suited to their purposes.
However, some employers may be at a particular stage in their operation where they need
certain specialized expertise. This may be in the form of someone familiar with a specific
methodology or equipment. Some of this specialized training may be acquired during
post-doctoral training, or through participation in short-duration targeted workshops.

An area in biotechnology that requires specialized training beyond the traditional basic sciences is bioinformatics. This fledgling area requires advanced knowledge in computer science in addition to biology. Computer majors may qualify for such jobs by
including an adequate number of courses in biological sciences and mathematics in their
training. Alternatively, biology majors may acquire experiences in computer science to
equip them for such jobs.

■ *DISCIPLINE-BASED ACADEMIC TRAINING IN SCIENCE*

Biotechnology is applied in the fields of agriculture, health, environment, and industry.
Applications in these areas require a general understanding of the principles and concepts
of biotechnology. Just like the DNA is universal, the principles of rDNA are also universal. However, a person trained in animal science, in addition to training in molecular biology or genetics, would be more suited for a job in animal biotechnology than a person
with training in plant science or general biology. The general principles of breeding are the

same, but the methodologies differ significantly between animal and plant breeding. It should be pointed out, though, that for a biotechnology job, the emphasis would normally be on the training and experience in rDNA technology. It is known that applicants with superior training in molecular techniques from a plant science background have been selected from a highly competitive applicant pool to work on the Human Genome Project.

■ NON-SCIENCE-BASED ACADEMIC TRAINING FOR BIOTECH JOBS

As already indicated, there are many job opportunities for persons without specialized training in biotechnology. The biotechnology industry employs business majors to manage the business aspect of a biotech company. As previously stressed, a biotechnology company is more than just a scientific idea. It must be run like the business that it is to succeed. If the company has a commercial product, it needs marketing and advertisement specialists to promote and market the products. In this very highly competitive field, a biotech company should protect its assets and conduct its business in a manner in which it would not fall victim to a competitor. There is the need for the involvement of legal experts to ensure that the company is operating within the law and is fairly treated during business negotiations. Whereas such employers do not need formal training in science, a lawyer who is familiar with the nature of the biotechnology industry would be more effective at negotiations and litigations (should there be the need) than one who has no such familiarity with the field. For example, a lawyer with a good understanding of rDNA technology will be more effective at utilizing DNA evidence in court.

■ TRAINING IN RELATED SCIENCES FOR BIOTECH JOBS

Biotech companies employ a variety of personnel with training in related sciences. For example, a company producing food products may need to have a person trained in food science on staff. Social scientists and economists are needed to conduct market research to ascertain consumer acceptance of a biotech product. Biotech companies that produce seed for sale need plant breeders to incorporate the new genes into appropriate ecotypes where the product is intended for sale, and agronomists to produce or supervise the production of the seed on a large scale. Companies involved in producing pharmaceuticals may have veterinary doctors (in the case of animals) and medical doctors (in the case of humans) to evaluate their products.

■ TYPES OF BIOTECH EMPLOYERS

Biotech employment is available in both the public and private sectors, and in small and large settings. Some operations are localized, whereas others are regional, national, or even international in scope.

Biotech in Academia

In the public arena, a major part of the contribution to the field of biotechnology occurs in academia. To start with, the personnel who are recruited to staff private sector biotech companies are trained in the universities and colleges in the public sector. Some of the major discoveries and innovations in biotechnology occur on academic campuses. Most of the Nobel Prize winners and other winners of outstanding recognitions come from academia. It is from this pool of intellectuals that many private companies lure expertise to staff their programs. Some biotech companies are built upon ideas that were conceived and born in academic research laboratories. Because of the infusion of federal funds into academic research at public institutions, discoveries at these places of higher learning are usually accessible by the general public.

Jobs available at academic institutions include teaching positions with or without research duties (professors, assistant professors, and instructors), research scientists, research associates, research assistants, and technicians. Research at public institutions is nonprofit oriented. Researchers may patent their inventions and publish their findings in public journals.

Researchers in academia often have partnerships with industry. The nature and the benefits of such partnerships are variable from one institution to another. Industry may fund some research of interest that they lack the expertise or manpower to conduct. Some collaborations may involve joint research involving scientists from the university and a private company working on a specific problem. Sometimes, private biotech companies provide sabbatical work experiences for professors.

There are numerous private universities that are engaged in biotech research. These institutions provide training to students in addition to conducting research for profit. Because of the nonprofit orientation of public researchers, scientists in these institutions are able to tackle research problems that are not of economic value to the institutions.

Public Research Institutes

There are numerous public research institutes or stations engaged in biotech research. Some of these are affiliated with universities. For example, the USDA has various research units affiliated with land grant institutions. These facilities provide employment opportunities for college graduates trained in biotechnology, especially in agriculture and related fields. There are also national research institutes that may be specialized in terms of their focus (e.g., National Cancer Research Institute), whereas others have a broader mission (National Institute of Health). Employment at these institutes is often competitive.

Private Sector Biotech Companies

As previously indicated, biotechnology is big business. Private companies are established primarily for profit. Consequently, all their undertakings are selected first for their profit value. There were over 1,200 biotech companies in the United States alone in 2002, providing more than 150,000 high-wage, high-value jobs. These companies may be small or big in size. An estimated one-third of biotech companies employ fewer than 50 people, while more than two-thirds employ fewer than 135 people. Biotechnology is research intensive. In 1998, the U.S. biotech industry spent $9.9 billion on research and development, with the top five companies spending an average of $121,400 per employee.

KEY CONCEPTS

1. Jobs in the biotechnology industry may be described as either direct or indirect.
2. Direct jobs are those involved in conducting the actual scientific research and inventing products.
3. The indirect jobs exist in the support or service industries where companies produce products and provide services to direct jobs personnel.
4. Academic training for biotech jobs may be obtained through a formal college education or acquired via informal training avenues.
5. Formal training in the foundational principles and practices of biotechnology, and a general familiarity with recombinant DNA technology, may suffice for some jobs.
6. There are many job opportunities for persons without specialized training in biotechnology.

7. Biotech companies employ a variety of personnel with training in related sciences.
8. Biotech employment is available in both the public and private sectors, and in small and large settings.

OUTCOMES ASSESSMENT

1. Give two specific direct jobs in biotechnology.
2. Discuss, giving specific examples, jobs that may be obtained in the field of biotechnology without special training in biotechnology.
3. Discuss the subjects that students would need in acquiring formal training in biotechnology.
4. Discuss the job opportunities in both public and private sectors in biotechnology.

INTERNET RESOURCES

1. Jobs in agbiotechnology: *http://www.nationjob.com/ag/*
2. Jobs in biotechnology: *http://www.medzilla.com/biotech-jobs/*
3. Jobs with biotech companies by name: *http://www.bio.com/*
4. Jobs in biology and biotech: *http://www.science-jobs.org/*
5. Science jobs: *http://www.scijobs.org/*
6. Recruiters for biotech jobs: *http://www.mrbiotech.com/*
7. Agbiotech jobs: *http://www.agbiotechnet.com/jobs/*
8. Biotech employment: *http://www.biotechemployment.com/*

REFERENCES AND SUGGESTED READING

Brown, S. S., and M. Rowh. 2000. *Biotechnology careers.* New York: McGraw Hill.

23 Biotechnology and Developing Economies

PURPOSE AND EXPECTED OUTCOMES

Whenever the subject comes up, the role of biotechnology in the humid tropics is often identified to be that of alleviating hunger. There is also an ongoing debate about whether or not developing economies and donor nations and agencies should exploit biotechnology in addressing the food security of developing nations.

In this chapter, you will learn:

1. The reasons why biotechnology is being promoted as a viable alternative to addressing world food security.
2. The barriers to commercializing biotechnology in developing countries.
3. The status of biotechnology in developing countries.
4. International initiatives in exploiting biotechnology in developing countries.

OVERVIEW OF WORLD FOOD ISSUES

Because of the expected population expansion and increasing land erosion, food security in developing countries is a major concern to the international community. Whereas population growth is leveling off in developed countries, most of the estimated 5 billion additional people on Earth by 2030 will inhabit the poor regions of the humid tropics. It should be pointed out immediately that it is an oversimplification to equate hunger alleviation with food security. Associated issues such as effective and efficient distribution networks, effective management of production resources (land, water), and government pricing policies critically impact the success of any food security undertaking. Furthermore, food security in these economically disadvantaged areas is intertwined inextricably with disease and environmental degradation. Poor soils and poor production management result in low crop yield, malnutrition, and a variety of health issues. Some observe that the medical problems of Africa are inseparable from that of lack of food. Needless to say that tackling Third World food security is a challenging proposition that requires careful planning and an integrated approach.

Promotion of agricultural biotechnology in developing economies should be accompanied by a promotion of improved agricultural practices. This way the ecological limits of population growth can be expanded by bringing more land into production and also improving crop harvest. It has been pointed out by some experts that the current agricultural biotechnologies do not increase the productivity *per se* of plants. Instead, they lower pre- and post-harvest losses by up to 25 percent. In terms of strategy, it is suggested that in view of the problems with food distribution, agriculture in the humid tropics must be indigenous and very productive. Consequently, the infusion of foreign technologies must proceed cautiously. Furthermore, the technology of gene transfer must be developed *in situ,* at least in some of the tropical developing countries, to ensure that it responds to local conditions. This strategy will also ensure that the technology is more readily acceptable to the local government, the local scientific community, plant breeders, and the local population.

Biotechnology is very capital and knowledge intensive. Such commitments are woefully inadequate in many developing economies. However, because the private sector plays a dominant role in setting biotechnology research and developing targets, and because economic return on investment is critical to investors, crops that benefit developing countries receive little attention because they are of little commercial interest (Figure 23–1). Consequently, the talk of improving tropical crops falls to other agencies (e.g., the United Nations) with little pressure to make a profit. The need to infuse biotechnology into the agriculture of developing countries is further underscored by the fact that most of these countries have agricultural-based economies. Improving agriculture is hence a major avenue for improving such economies. As illustrated in Figure 23–2, the productivity potential of major food crops of the developing world is far from being attained. Doubling the current level of productivity of these crops will make a significant impact on the food security of these nations.

BARRIERS TO THE COMMERCIALIZING OF BIOTECH IN DEVELOPING COUNTRIES

Developing countries are as diverse as they are numerous. The differences common among them are based on political, socioeconomic, and geographic factors. The existing technology capacity for biotechnology in these nations ranges from none to fairly adequate. The levels of resource (human, capital) endowment also vary widely, and so do domestic conditions regarding politics, government policies, scientific knowledge base, and macroeconomics. It is inaccurate, therefore, to lump all developing nations into one category. Rather, two general groups may be identified: (1) countries in economic transition (moving toward market economies) and (2) those at the bottom (with nothing or very limited existing infrastructural capacities to exploit biotechnology). Most of the countries in the bottom tier are located in Africa. Even so, some countries in Africa have begun to put in place mechanisms for commercially exploiting biotechnology. These include Kenya, Zimbabwe, Nigeria, and South Africa. In the Caribbean region, Cuba has

FIGURE 23–1
Two major developing country food crops are shown in the figure, plantains in the background, and piles of cassava in the forefront. These and other crops are often described as "orphaned crops" for the lack of interest they are shown by the major research institutions in developed countries.

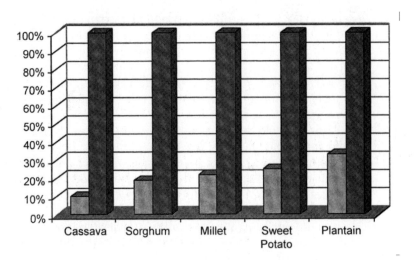

FIGURE 23–2

Productivity potential of selected major food crops. None of the five important food crops, cassava, sorghum, millet, sweet potato, and plantain, has been developed to even 50 percent of its potential productivity.

Source: Courtesy of Claude Forquet, ILTAB, Danforth Plant Research Center, St. Louis, Missouri. Used with permission.

implemented significant biotechnology programs. There are also the newly industrialized nations in Asia (e.g., China, India) and Latin America (Brazil, Mexico).

There are several major barriers to commercializing biotechnology in developing countries.

1. Lack of appropriate technology.
2. Limited infrastructure for exploiting biotechnology.
3. Intellectual property rights.
4. Biosafety issues.
5. Lack of market mechanisms.
6. The biotech debate being waged in the potential donor countries.
7. Local and regional politics.
8. Poverty and disparities.

It might appear that appropriate technology would be a major barrier to attempts at applying biotechnology to benefit the needy in the developing world. This is so because most of the research and product development occurs in developed countries and are targeted to solving problems in their regions. Existing technologies may be adapted for use in developing countries, while some new and unique technologies may have to be developed *in situ* in these nations to be effective. The issue of intellectual property rights is also a potential key barrier. Companies in developed countries own most of the patents for the technologies that would be deployed in poor regions. The commercial companies would have to be adequately compensated, in most cases, to have access to their inventions. These two factors notwithstanding, some experts believe that the primary barrier to successful exploitation of biotechnology in developing countries is the lack of market mechanisms that normally constitute the driving force behind the research and development process. In terms of agriculture, one of the clearly accessible markets is the seed market, especially that for cash crops. Major seed companies in the United States (e.g., Monsanto) and Europe (e.g., Sandoz) have interest in accessing this market. If profitable markets exist for biotechnology, companies in industrialized nations with resources will be enticed to invest in the Third World oriented projects. However, if the objective of biotechnology exploitation in developing countries is to benefit the poor and needy, then other avenues beside business ventures need to be sought. Developing countries also need to implement biosafety guidelines in order to conform to international regulations for conducting biotechnology research. The ongoing debate in developed countries about biotechnology serves as a distraction to the involvement of external partners, both private and public, in Third World biotechnology efforts. Some opponents of biotechnology tend to think that multinational corporations are only profit oriented, and look for opportunities to exploit developing countries.

Apart from barriers that may originate outside the developing world, local and regional politics in developing countries pose a significant barrier to the adoption of biotechnology. Local governments are responsible for developing or implementing biosafety regulations, honoring intellectual property rights, supporting local research and development efforts, accepting biotechnology as a viable tool for helping local agriculture, and putting in place the environment for overseas partnerships to be successful. The issue of poverty is important in the adoption of any technology. Most of the agricultural production in developing countries is undertaken by the rural poor. The concern always is how they can afford new technologies. The other critical concern is about the distribution of benefits or the impacts of technology. A criticism of the Green Revolution is that it marginalized the poorer producers, while bringing most of the economic benefits to the already richer producers.

THE ROLE OF INTERNATIONAL INITIATIVES IN BIOTECHNOLOGY

With proper caution and good planning, biotechnology can be successfully implemented in developing countries to improve agricultural production. It is important that any effort be approached from the angle of partnerships and collaboration. Overseas partnerships should include the public and private sectors, as well as the international entities. More importantly, every partnership should involve the developing countries directly and be implemented in the social context. Including the developing countries would make the technology more readily acceptable and facilitate its adoption. It would also make the developing countries feel they are not being taken advantage of, or being forced to accept what they do not want.

Because of the prohibitive costs of participation in the exploitation of biotechnology by many developing countries, a variety of international initiatives exist for supporting countries to plan or implement research and extension programs in biotechnology. Currently, most of these efforts are directed toward food-based biotechnology, and involve bilateral initiatives by governments from the developed world, private foundations, tripartite arrangements, and efforts by the UN system.

■ EFFORTS BY INDEPENDENT NATIONS (BILATERAL DONORS)

Major international involvement in biotechnology in developing countries exists from countries including the United States, United Kingdom, Switzerland, and the Netherlands. These efforts involve research (e.g., cassava biotechnology) in donor countries that are geared specifically toward solving targeted problems in developing countries, support of research conducted in selected countries, training of graduate students, provision of technical assistance, and the awarding of grants, among others.

United States
Efforts by the United States are channeled through the U.S. Agency for International Development (USAID). A team from academia, private sectors, and federal agencies empanelled by the National Research Council recommended the strategy of product-oriented

programs. These programs are to be executed collaboratively with both the public and private sectors and integrate a variety of issues, namely research, human research development, biosafety, and intellectual property rights. In 1992, the United States initiated the Agricultural Biotechnology for Sustainable Productivity project to address these recommendations. The entities involved were from the United States, Egypt, Indonesia, and Kenya. The specific research projects focused on the transformation of various crops: potato (for resistance to tuber moth), cucurbit (for resistance to potyvirus), tomato (for resistance to geminivirus), and corn (for resistance to stem borers). Activities in micropropagation focused on banana, pineapple, coffee, and ornamentals.

The University of Florida is home to the International Program on Vectors and Vector-borne Diseases. The focus of this program is to generate genetically engineered vaccines and disease diagnostics for the improvement of livestock production. The primary focus is on tick-borne diseases of livestock (cowdriosis, anaplasmosis, and babesiosis) in sub-Saharan Africa. The developing nations' partner in this effort is Zimbabwe. The International Laboratory of Molecular Biology for Tropical Disease Agents at the University of California is another USDA-supported program, focusing on developing genetic vaccines for renderpest and hoof-and-mouth disease. Funding of these efforts by the USAID has exceeded $20 million to date.

United Kingdom

International biotechnology efforts by the United Kingdom are channeled through two programs. In 1990, the Plant Sciences Research Program was established and located at the University of Wales to work collaboratively with public and private British institutions and International Centers to address advanced plant breeding and crop physiological issues impacting agriculture in developing countries. Researchers focus on molecular marker development such as RFLP mapping of pearl millet. Other objectives are transformations of the peanut for virus resistance, and the incorporation of the *CpTi* (cowpea trypsin inhibitor) gene into sweet potato for insect resistance.

The Netherlands

In 1992, the Dutch government initiated a biotechnology program for developing countries, targeting Kenya, Zimbabwe, Colombia, and India. The primary focus of such efforts is the small-scale producers and women. The strategy adopted by the donor is the active engagement of grass roots participation in determining program objectives. Furthermore, country committees administer the program. Specific projects being undertaken include marker-assisted selections of drought-tolerant and insect-resistant corn in Kenya and Zimbabwe (in collaboration with CIMMYT) and production of biofertilizers and biopesticides involving Kenya and India. This Dutch program supports other international efforts like the Cassava Biotechnology Network at CIAT. Funds committed to the program between 1992 and 1997 totaled $27 million.

■ *TRIPARTITE*

Tripartite agreements in which the technology owner, international institute agency, and donor work together to implement biotechnology programs in developing countries is one of the significant ways in which international initiatives in biotechnology are realized. For example, the U.S. government, through the USAID, has transferred proprietary transformation technologies from Monsanto, a private company, to Kenya for the development of a genetically engineered sweet potato resistant to feathery mottle virus. This agreement also gave birth to the International Laboratory for Tropical Agricultural Biotechnology (ILTAB) at the Scripps Research Institute in California. The ILTAB focuses on transferring resistance to African, South American, and Asian viral diseases of cassava, rice, tomato, and sweet potato.

Another example of tripartite agreement is provided by the activities of the Agricultural Biotechnology Support Program based at Michigan State University. This program brings together private corporations and public institutions to work collaboratively with developing countries to implement biotechnology programs. An example was the development of a virus-resistant potato that was undertaken jointly by Monsanto, the USAID, and the Kenyan Agricultural Research Institute (KARI).

The golden rice project is currently the most widely known international collaboration to use biotechnology to address the food needs of developing countries. The details of this project were discussed previously in this book. However, one of the crops that is widely grown and used in developing countries but has not as yet received the level of attention that it deserves is cassava. Cassava is highly heterozygotic in nature and is propagated vegetatively. It is subject to severe biological constraints, with an estimated 35 to 50 million tons lost to diseases and pests annually (see Figure 23–3). It is not readily amenable to conventional methods of breeding. Efforts to change the world status of cassava research are being coordinated under a plan called the Global Cassava Improvement Plan (GCIP). Founding institutions of this effort were CIAT in Colombia, IITA in Nigeria, EMBRAPA in Brazil, and the Danforth Plant Science Center in St. Louis, Missouri. Participating members include India, Nigeria, Ghana, Ivory Coast, Uganda, Kenya, and Congo. The GCIP proposes to use four biotechnological approaches to improve cassava (see Figure 23–4). Micropropagation is a relatively low-tech tool that can be used to rapidly produce disease-free plant materials for farmers. The strategy of double haploids will facilitate breeding by reducing heterozygosity in the species. High-end biotechnology may be used to generate high-density PCR-based markers and physical maps using the

Biological constraints for cassava

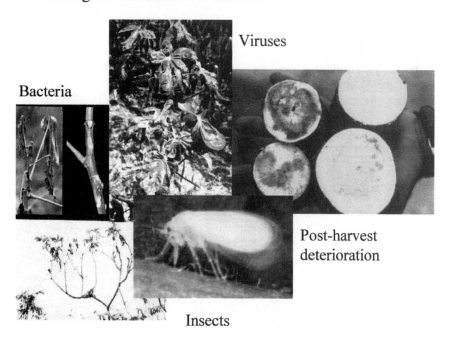

FIGURE 23–3

Biological constraints for cassava. Numerous pests, including those of bacterial, viral, and insect origins, plague cassava. Large amounts of this root crop are annually lost to post-harvest deterioration.

Source: Courtesy of Claude Forquet, ILTAB, Danforth Plant Research Center, St. Louis, Missouri.

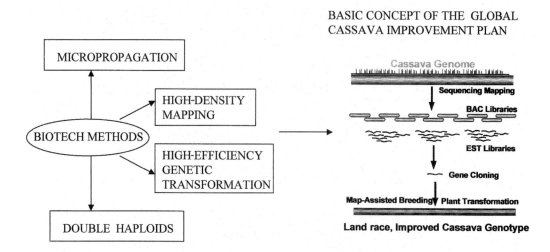

FIGURE 23–4

The basic concept of the Global Cassava Improvement Plan. The efforts proposed under the plan entail the use of four biotechnological tools, namely, micropropagation, genome mapping, genetic transformation, and double haploids, to develop improved and adapted farmer-preferred genotypes for use in the various cassava-producing regions of the world.

Source: Adapted from material produced by Claude Forquet, ILTAB, Danforth Plant Research Center, St. Louis, Missouri. Used with permission.

5'-end BAC sequences. The fourth strategy is to use genetic transformation techniques to transfer appropriate genes into adapted, farmer-preferred genotypes.

■ *PRIVATE FOUNDATIONS*

Private foundations also participate in tripartite agreements. For example, the Rockefeller Foundation of New York supports the efforts of ILTAB as well as the International Service for Acquisition of Agri-biotech Applications (ISAAA) in Mexico, where Monsanto proprietary technology is being used to develop virus resistance in potatoes. The McKnight Foundation and the Resources Development Foundation of Washington, D.C., both support ISAAA with sizable grants. It should be noted that the Rockefeller Foundation is the principal private foundation donor of international biotechnology, with most of the investment going into the International Rice Biotechnology Program. This program is partly responsible for the "golden rice" product. The major recipients of the funding from the Rockefeller Foundation are rice research institutes in developing countries, especially in India, Thailand, and China. These research stations are engaged in transformation for disease resistance (tungro virus, stripe virus, yellow stem borers), marker-assisted selection for resistance to blight and blast, and hybrid rice production.

■ *THE UNITED NATIONS SYSTEM*

The United Nations Development Program (UNDP) supports international biotechnology through the International Agricultural Centers (IRACs) operated under the auspices of the Consultative Group on International Agricultural Research (CGIAR). These centers include both plant and animal research programs. One of the newest additions to the United Nations Systems is the International Center for Genetic Engineering and Biotechnology (ICGEB) located in Italy and India, which is pursuing both basic and applied research. The UNDP supported the Regional Program of Biotechnology for Latin America and the Caribbean that ended in 1996. The FAO is involved in this region through the Technical Cooperation Network on Plant Biotechnology that was initiated in 1990.

STATUS OF BIOTECHNOLOGY IN SELECTED DEVELOPING COUNTRIES

This brief discussion is designed to indicate the variability in levels of development and commitment of national governments of developing countries to biotechnology.

BRAZIL

The Brazilian government has strongly embraced biotechnology and significantly supports biotechnology programs through the Ministry of Science and Technology (PADCT). The ministry supports basic research and encourages collaboration between private companies and universities. In the first phase, which ended in 1991, the ministry disbursed $172 million. Another $120 million was spent in the second phase, which ended in 1996. The third phase, which started in 1997, has a more modest budget of $20 million. Brazil has had national biosafety laws in place since 1995 to regulate the development and use of GMOs. The Brazilian Agricultural Research Enterprise (EMBRAPA) has established its own internal biosafety committee. There are laws in place to regulate intellectual property rights and biodiversity.

Brazilian biotechnology efforts are concentrated in health and agbiotech. The Oswaldo Cruz Foundation is a key player in health research associated with tropical medicine, identifying parasites and microbial pathogens using molecular methodologies. EBRAPA and the National Research Center for Genetic Resources and Biotechnology (CENARGEN) are involved in agricultural research. Private sector research involves entities like the Brazilian Association of Biotechnological Research Institute (ABRABI), which focuses on vaccines, food products, and pharmaceutical substances, among others. Their human insulin from rDNA is commercially marketed.

CHINA

China has one of the well-established and longstanding biotechnology programs in the developing world. The State Science and Technology Commission of China and the China National Center for Biotechnology Development jointly published a National Blueprint Book for Biotechnology Development in China in 1987. The first high-tech initiative, called the 863 Program, was inaugurated in 1986. China's focus is on producing high-yielding cultivars and animal breeds that are resistant to abiotic stresses; developing new medicines, vaccines, and gene therapy; and advancing protein engineering. The central government contributes about 25 million annually to biotechnology research in addition to financial infusions from the private sector. There is a biosafety regulation in place (China Biosafety Regulation). The key research entities include the Life Science College of Beijing University, the Basic Medical Research Institute, the Beijing Institute of Virology, and the Shanghai Institute of Biochemistry. Private sector efforts have yielded commercialized pharmaceuticals (e.g., the recombinant interferon aIB, interleukin IL-2, the recombinant hepatitis B virus vaccine). The Monsanto company has entered into agreement with a seed company in Hebei to develop GM cotton. China ranks among the top countries in the world in acreage devoted to GM cotton. Recently, China has emerged as a powerhouse in biotechnology with the astonishing achievement in the sequencing of the rice genome, as previously discussed.

◼ *INDIA*

On the whole, the biotechnology infrastructure of India is quite well established. India set up its Department of Biotechnology in 1986 and established its rDNA safety guidelines in 1990 (with subsequent revisions). The intellectual property rights registration in India does not permit the patenting of GMOs. Both biomedical and agricultural biotechnology are being vigorously pursued. Diagnostic tools (both DNA based and antigen/antibody based) have been developed for most major infectious diseases (TB, malaria, cholera, and livestock and poultry diseases). Other projects are also being pursued. In agricultural biotechnology, several centers of plant molecular biology have been established in universities, as well as nationally. In addition, wheat and cotton transgenic projects have been initiated. Major entities include the National Institute of Immunology, the National Center for Cell Science, and the Center for DNA Fingerprinting and Diagnostics. Tissue culture prospects have been successful in providing commercial products.

◼ *KENYA*

The National Advisory Committee on Biotechnology and Its Application was established to guide the nation's biotechnology efforts. Several international initiatives involving governments such as the Netherlands, the WHO, and the World Bank are ongoing. An intellectual property rights protocol is in place. The major research players include KARI, ICIPE, ILRI, and KEMRI. Private sector investment is very limited. The situation in Kenya is characteristic of the current status of biotechnology in most economically challenged developing countries. Much of the effort in biotechnology depends on foreign support. Government support is limited or non-existent; private sector investment is absent or very limited.

KEY CONCEPTS

1. Whereas population growth is leveling off in developed countries, most of the estimated 5 billion additional people on Earth by 2030 will inhabit the poor regions of the humid tropics.
2. It should be pointed out immediately that it is an oversimplification to equate hunger alleviation with food security.
3. Associated issues such as effective and efficient distribution networks, effective management of production resources (land, water), and government pricing policies critically impact the success of any food security undertaking.
4. Food security in these economically disadvantaged areas is intertwined inextricably with disease and environmental degradation. Poor soils and poor production management result in low crop yield, malnutrition, and a variety of health issues.
5. It is needless to say that tackling Third World food security is a challenging proposition that requires careful planning and an integrated approach.
6. Promotion of agricultural biotechnology in developing economies should be accompanied by a promotion of improved agricultural practices.
7. The infusion of foreign technologies must proceed cautiously. Furthermore, the technology of gene transfer must be developed *in situ*, at least in some of the tropical developing countries, to ensure that it responds to local conditions. This strategy will also ensure that the technology is more readily acceptable to the local government, the local scientific community, plant breeders, and also the local population.
8. Biotechnology is very capital and knowledge intensive. Such commitments are woefully inadequate in many developing economies.

9. The need to infuse biotechnology into the agriculture of developing countries is further underscored by the fact that most of these countries have agricultural-based economies. Improving agriculture is hence a major avenue for improving such economies.

10. Developing countries are as diverse as they are numerous. The differences common among them are based on political, socioeconomic, and geographic factors.

11. Two general groups of developing countries may be identified: (1) countries in economic transition (moving toward market economies) and (2) those at the bottom (with nothing or very limited existing infrastructural capacities to exploit biotechnology).

12. There are several major barriers to commercializing biotechnology in developing countries: (1) lack of appropriate technology, (2) intellectual property rights, and (3) lack of market mechanisms.

13. If profitable markets exist for biotechnology, companies in industrialized nations, with resources, will be enticed to invest in the Third World oriented projects.

14. Because of the prohibitive costs of participation in the exploitation of biotechnology by many developing countries, a variety of international initiatives exist for supporting countries to plan or implement research and extension programs in biotechnology.

15. Currently, most of these efforts are directed toward food-based biotechnology, and involve bilateral initiatives by governments from the developed world, private foundations, tripartite arrangements, and efforts by the UN system.

OUTCOMES ASSESSMENT

1. Developing countries cannot be lumped together; there are significant differences. Explain.
2. It is an oversimplification to equate hunger alleviation with food security. Explain.
3. Explain why it is important to develop some of the new technologies *in situ* in developing countries.
4. Give the three key barriers to commercializing biotechnology in developing countries.
5. Give an overview of the role of international initiatives in the exploitation of biotechnology by developing countries.
6. Discuss the role of private foundations in the exploitation of biotechnology in developing countries.

INTERNET RESOURCES

1. Agbiotech and developing countries: *http://www.agbiotechnet.com/topics/Database/Developing/developing.asp*
2. Help or harm: *http://www.twnside.org.sg/title/harm-cn.htm*
3. See July 9 issue for position on biotech by Monsanto: *http://www.monsanto.com/monsanto/layout/media/default.asp*
4. Viewpoint from Director General of IFPRI: *http://www.ifpri.org/media/innews/102799.htm*
5. Labeling of food in developing countries: *http://www.cid.harvard.edu/cidbiotech/comments/comments47.htm*
6. UNDP report: *http://usinfo.state.gov/topical/global/biotech/01071001.htm*
7. GMO, world trade, and the Third World: *http://www.twnside.org.sg/title/ciel-cn.htm*
8. CGIAR on biotech partnerships: *http://www.cgiar.org/biotechc/lewis.htm*

REFERENCES AND SUGGESTED READING

Holn, T., and K. M. Leisinger (eds.). 1999. *Biotechnology of food crops in developing countries.* New York: Springer.

Komen, J. 2000. International Initiatives in Agri-Food Biotechnology. In *Biotechnology in the developing world and countries in economic transition.* New York: CAB International.

Toenniessen, G. H. 1995. The Rockefeller Foundation's International Programme on Rice Biotechnology. In *Plant biotechnology transfer to developing countries.* D. W. Altman and K. N. Watanabe (eds.). Austin, Texas: R.G. Landes Company, 193–212.

Tzotzos, G. T., and K. G. Skryabin (eds.). 2000. *Biotechnology in the developing world and countries in economic transition.* New York: CAB International.

24 Perceptions and Fears About Biotechnology

SECTION 1
Contrasting Public and Scientific Perceptions

PURPOSE AND EXPECTED OUTCOMES

We live in a world in which technology plays a significant role in all aspects of living. Older technologies are being constantly improved or replaced with modern and more sophisticated ones. While some embrace technological advancements, others see them as an unnecessary intrusion in nature. Technology may be adapted for unintended purposes. Furthermore, the proper application of some technologies may have unintended adverse consequences. Consequently, it is not uncommon that the introduction of a new technology is embroiled in public debate.

In this section, you will learn:

1. About some of the common perceptions of biotechnology by the public.
2. About scientific responses to the public perception of biotechnology.

THE DIFFICULTY OF CONTRASTING PERCEPTIONS OF TECHNOLOGY

The public has become increasingly vocal in the biotechnology debate with time as new applications of the technology have been proposed. The truth of the matter is that all stakeholders acknowledge that no technology is without potential for adverse impacts. Another truth is that the perception of technologies varies between the scientists who develop them and the public who use them. It is tempting to contrast public perception, which is largely subjective, with scientific knowledge, which is largely objective. Furthermore, public perception is viewed by some as basically illogical, irrational, and based on emotions rather than scientific facts. There is the temptation to condescend and disregard public opinion as uninformed. However, scientists also have perceptions of technologies they develop that are not necessarily rooted in facts, because the knowledge available may be scanty or inconclusive. There is the tendency, therefore, to either overestimate or underestimate the benefits and risks of technology.

Because of the imperfections that plague both camps, there are some misconceptions about biotechnology in the general community that hinder the development and application of the technology. A few of these will be discussed below.

THE TECHNIQUES OF BIOTECHNOLOGY
ARE ALIEN AND TOO RADICAL

It was made clear in Chapter 1 that the definition of biotechnology can be broad or narrow. In the broad sense, organisms have been used to make products for thousands of years (yeast in bakery products and bacteria in fermented products). In the narrow sense, biotechnology allows genes to be transferred unrestricted among living things, in effect disregarding natural genetic barriers. Whereas such widespread gene exchange is not the norm in nature, there are examples in nature of various degrees of such gene transfers. Cross-pollinated species propagate through gene mixing, normally within the species. A more dramatic natural mixing occurred in wheat. Common wheat (*Triticum aestivum*) is an allopolyploid (hexaploid) consisting of three genomes of three different species. Certain microbes have the capacity to transfer some of their genetic material into the hosts they infect, even though the outcome is undesirable.

In biotechnology, the gene transfer system of choice in plants is *Agrobacterium*-mediated transfer. This bacterium naturally transfers a portion of its genome into the plant it infects, creating an animal-to-plant gene transfer. Scientists capitalize on this natural process in biotechnology to transfer genes of choice, only this time the result is not a disease because the bacterium is disarmed prior to use.

The normal direction of genetic information transfer is from DNA to RNA to protein (the so-called central dogma of molecular biology). However, there are certain viruses that have RNA as genetic material. In retroviral infections, the single-stranded RNA is reverse transcribed to single-stranded DNA and then doubled to double-stranded DNA. This is incorporated into the host genome. Scientists can go a step further, being able to synthesize the corresponding gene from a protein product.

Tissue culture is used widely in biotechnology. In somatic embryogenesis, plants can be raised from callus. Callus forms naturally as part of the healing process of wounded plants. Scientists are able to induce callus under laboratory conditions.

Mutations or heritable genetic changes occur naturally as spontaneous events. Such natural gene alterations produce variability for evolutionary processes to occur. Instead of haphazard and random events, scientists are able to induce genetic changes that are desired.

It is obvious from these selected examples that science merely imitates nature after studying to understand it. Rather than random events, scientists attempt to nudge nature purposefully to the advantage of humans. One may argue that just because nature does something does not mean that humans should do the same. However, another may argue, "Why not?"

GENETIC ENGINEERING IS AN EXACT SCIENCE

In an attempt to draw a distinction between conventional biotechnology (e.g., conventional breeding) and genetic engineering, some scientists inadvertently convey the notion that genes are transferred with surgical precision. It is true that specific genes can be identified, isolated, and characterized. However, the current gene transfer systems leave much to be desired. Once the DNA is delivered into the cell, scientists are not able to direct or predict where it will be inserted in the genome. Consequently, scientists cannot predict precisely the outcome of a transformation event. Where the gene inserts itself in the genome has a bearing on its expression. Even though this appears to be a shot-in-the-dark process that can literally create a monster, scientists screen the products of transformation to identify the individual(s) in which the transgene apparently has been properly inserted and is functioning as desired. It should be pointed out that, compared to traditional breeding in which transfer of a desirable gene is usually accompanied by the transfer of numerous others, genetic engineering is relatively very precise. Furthermore, scientists continue to refine the transfer technique. Newer techniques (e.g., site-directed mutagenesis) are precision tools in the arsenal of the biotechnologist.

NONTARGETED PESTICIDE RESISTANCE IN THE ECOSYSTEM AS A RESULT OF THE USE OF BIOTECH CROPS IS UNAVOIDABLE

There are major crops in production with engineered targeted resistance to pests and herbicides. These include corn, cotton, soybean, and potato, all of which have *Bt*-cultivars in production. There are herbicide-resistant cultivars of major crops (e.g., Roundup Ready® soybean). The use of nontargeted pesticide resistance (or "collateral resistance") centers around four main aspects: (1) creating weeds out of cultivated cultivars, (2) creating "superweeds" from existing weeds, (3) creating resistant pests, and (4) creating antibiotic resistance in harmful microbes.

The problems are real to varying extents. The fact is those pests always manage to eventually adapt to any pest management strategy that is implemented repeatedly over a long period. This is especially true when the organism has a short life cycle (e.g., bacteria and many insects). The fear of creating a weed out of cultivated crops stems from the fact that most modern herbicides are broad spectrum in action (i.e., kill many plant species). Bioengineered herbicide-resistant crops are consequently resistant to broad-spectrum herbicides. If, for example, Roundup Ready® soybean follows Roundup Ready® corn, volunteer corn plants will resist Roundup Ready® and be a weed problem. Whereas crop rotations are desirable, it should not involve crops engineered with identical herbicide resistance. Furthermore, using the same herbicide repeatedly for a long time is not a recommended agronomic practice. Should one decide not to heed this advice, there are herbicides that can control Roundup Ready® corn or soybean "weeds."

On the issue of biotechnology contributing to the development of "**superweeds**," the potential exists for gene escape from cultivated species engineered for herbicide tolerance to interbreed with wild relatives, thereby creating more competitive and difficult-to-control weeds (the so-called superweeds). While the movement of transgenes into wild relatives of transgenic crops is possible, this would occur only if the cultivated species of crops are grown where their weedy relatives with which they can interbreed also occur. This is not the case for the major crops that are transgenic for herbicide resistance in the United States (e.g., corn, soybean, and potato). However, it is the case for squash and canola. It should be pointed out that the development of resistance to herbicides by weeds and other plants do occur, following prolonged exposure to certain herbicides. Furthermore, irrespective of the herbicide or weed management tactic used, resistance to the chemical over the long haul is inevitable. This is why new herbicides will continue to be needed. There is no evidence to suggest that development of resistance is more problematic with the use of transgenic crops than direct use of herbicide. The issue of pest developing resistance to transgenic crops is similar to the weeds. Pests routinely overcome management tactics used against them.

To reduce the incidence of genetic drift of transgenes into the wild, producers of transgenic crops require customers to grow a refuge of nonbiotech and untreated crops. This allows the refuge to attract the natural pests of the crop and provides an opportunity to interbreed with any pests on the biotech product that may develop resistance.

Certain biotechnology techniques utilize antibiotic resistance markers in developing transgenic crops. Consequently, the products that contain these genes can possibly be transferred to microbes in the environment. The fact is that the antibiotic resistance used in crop development does not provide resistance to most of the antibiotics used in the clinical setting.

BIOTECHNOLOGY CROP AND ANIMAL PRODUCTION AND USE HAS POTENTIALLY ADVERSE EFFECTS ON BIODIVERSITY

Erosion of biodiversity occurs via diverse avenues. One such avenue is crop/animal breeding. Breeders invest time and other resources into developing breeding lines or stocks from which parents are drawn for hybridization to create new cultivars or breeds. Consequently, it is not uncommon to have a situation whereby only a handful of elite germplasm dominates the genetics of all commercially utilized crops and animals. For example, most cattle in major production areas can be traced to only one or a few stock animals. As a tool for genetically manipulating organisms, biotechnology could contribute to a loss of biodiversity if used to develop breeding lines or stocks. On the other hand, biotechnology can also be used to expand existing biodiversity through creating new genotypes.

BIOTECHNOLOGY PRODUCTS ARE UNNATURAL AND UNSAFE

Nature, and therefore anything natural, is perceived by some as superior to anything artificial. These people perceive that modern foods that have been processed and modified in sundry ways are making us sick. The result is the booming organic food and health food markets. Herbal medicine is being actively promoted in Western societies. The public is concerned about biotechnology inadvertently introducing undesirable and unnatural chemicals into the food chain. The fact is that the public has embraced artificial components in food and medicine for a long period of time. Western therapy is almost exclusively dependent upon synthetic pharmaceuticals. Food additives and coloring are used routinely in both home and industrial food preparations. Instead of adding these materials to food during preparation, biotechnology seeks to make plants and animals produce nutrition-augmenting materials via natural processes to be included in the plant and animal tissues. For example, instead of vitamin enriching of the product (e.g., rice) for value added, the plant is engineered to produce vitamin-enriched grains (golden rice).

POSTSCRIPT

Genetic engineering of animals and plants is a new technology. Like all new technologies, it attracts periodic or sustained debate on its pros and cons. It also has its fair share of proponents and opponents, optimism or skepticism about its potential, and all the customary apprehension that often greets the emergence of a new technology. On certain occasions, it is easy to see clearly the risks and disadvantages of a technology. The risks may even outweigh the benefits. Other times, the benefits clearly outweigh the risks. Still, on other occasions, we have a mixed bag of potential benefits and risks. Objective debate is always a welcome and fruitful exercise in a democratic society. It can help set the records straight. It also provides checks and balances on the development and application of new technologies. Unfortunately, when an issue becomes politicized, it quickly degenerates into what I call "a Summit at the Grand Canyon." People on opposing sides of an issue position themselves on opposite sides of the canyon, separated by a deep chasm representing mistrust, misinformation, selfishness, and even ignorance and arrogance. They attempt to discuss the issue by shouting at each other, instead of reasoning together. Each one is right in his or her own eyes and is determined to win the debate. Naturally, they speak to themselves and against the wind, ending the summit without any resolution of the issues. Meanwhile, the people on whose behalf they purport to deliberate continue to drift to their doom in the river, deep in the canyon and out of sight of their self-appointed representatives. Biotechnology is certainly not a cure all or fail-safe technology. It is important that its risks and benefits be discussed in an objective forum.

KEY CONCEPTS

1. No technology is without potential for adverse impacts.
2. The perception of technologies varies between the scientists who develop them and the public who use them.
3. There is the tendency to either overestimate or underestimate the benefits and risks of technology.
4. There are some misconceptions about biotechnology in the general community that hinder the development and application of the technology.
   a. The techniques of biotechnology are alien, unnatural, and too radical.
   b. Genetic engineering is an exact science.
   c. Nontargeted pesticide resistance in the ecosystem as a result of the use of biotech crops is unavoidable.
   d. Biotechnology crop and animal production and use has potentially adverse effects on biodiversity.
   e. Biotechnology products are unnatural and unsafe.

OUTCOMES ASSESSMENT

1. List and discuss two common misconceptions about biotechnology.
2. No technology is without the potential for adverse impacts. Explain.

INTERNET RESOURCES

1. Monarch butterfly debate: *http://news.bbc.co.uk/hi/english/sci/tech/newsid_1298000/1298397.stm*
2. Biotech debate in Europe: *http://www.csmonitor.com/2001/0911/p6s1-woeu.html*
3. Raging debate over biotech food: *http://www.enn.com/enn-features-archive/2000/03/03052000/gefood_5991.asp*
4. Debate over morals in biotech: *http://www.cid.harvard.edu/cidbiotech/comments/comments117.htm*
5. GM foods debate: *http://uk.fc.yahoo.com/g/genetic.html*

SECTION 2
Bioterrorism

PURPOSE AND EXPECTED OUTCOMES

Over the ages, humans have always sought new ways of gaining advantage over an enemy during a war. The art of waging war is shifting closer to a science in which sophisticated navigational systems are used to guide the so-called "smart bombs" to deliver their deadly load to a target hundreds of miles away, reducing face-to-face combat. Unfortunately, the power of deadly diseases to inflict casualty on people has not gone unnoticed. One of the tactics in waging war is to keep the enemy on edge, unsettled, distracted, and terrified. In peacetime, such tactics can be exploited for political purposes.

In this section, you will learn:

1. What bioterrorism is and how it differs from other unconventional methods of war.
2. The history of biowarfare.
3. The weapons of bioterrorism.
4. International efforts to eliminate biological weapons.
5. How biotechnology may be used in both positive and negative ways in bioterrorism.

WHAT IS BIOTERRORISM?

Biowarfare (biological warfare) may be defined as the harnessing of the disease-causing capability of disease agents as weapons to cause disease epidemics against an enemy in a war or conflict. When an individual or a group with criminal intent conducts such an activity in peacetime, it constitutes **bioterrorism**. The use of disease-causing organisms in a war is also called **germ warfare**. While not a conventional means of prosecuting a war, there is evidence that the strategy of hostile use of biological agents and their potential threat to health has its roots in history. The art of germ warfare, over the years, has developed into a science whereby potent species and strains of disease-causing agents (microorganisms like bacteria and viruses) are deliberately cultured, processed, and packaged for effective spread through the environment.

Biological warfare should be distinguished from **chemical warfare** that relies on chemicals, not biological agents. Chemical weapons may be classified into four basic types—nerve, blister, blood, and incapacitating agents. Their effect is dose dependent and may kill or incapacitate the victim. Chemical weapons are quick acting; victims sometimes display symptoms within seconds of exposure. Examples of potent known chemical weapons are **sarin** and **mustard gases**. Pound-for-pound, chemical weapons are less deadly than biological weapons. It is estimated that it would take 10 g of anthrax spores to kill as many people as a ton of sarin nerve gas.

NATURE'S VERSION OF BIOWARFARE

While they cannot be described as warfare in the true sense, major natural events of catastrophic proportions involving devastation of life, humans, animals, and plants by disease epidemics punctuate our history. The **bubonic plague** occurred between 1347 and

1351 and claimed an estimated 25 million lives in medieval Europe. **Smallpox**, attributed to European explorers, decimated the Native American population in the New World. The **Spanish flu** of 1918–1919 decimated an estimated 50 million people worldwide. The modern era epidemic of **acquired immunodeficiency disease syndrome (AIDS)**, caused by the human immunodeficiency virus (HIV), has claimed over 40 million lives as of 2000, with no end in sight. The **ebola virus** periodically erupts with such fury, killing thousands of people in Africa. In the case of animals, the modern-day devastation of the European livestock industry by mad cow disease and **hoof-and-mouth** (foot-and-mouth) **disease** serve to remind us of how devastating pathogenic epidemics can be to economies. In plants, notable disease epidemics include the infamous **potato blight**, responsible for the Irish Famine of 1846–1850 that claimed the lives of millions of Irish nationals and spurred new waves of Irish immigration to the United States. There are also outbreaks like **Dutch elm disease**, a fungal disease that plagues North American elms. Taking a cue from nature, humans are harnessing the destructive power of disease agents for initiating artificial epidemics for warfare or creating terror.

BIOLOGICAL WEAPONS

Biological weapons are classified under weapons of mass destruction because of their capacity to kill or injure many people when deployed. They injure or kill in a variety of ways—inhalation, contact, and ingestion, depending on the type of weapon. They are usually designed to be dispersed into the air by being delivered through special artillery shells that explode in midair. Biological weapons can also be dispersed by aerial spraying using airplanes. However, terrorists can use more creative ways to deliver germs, as was learned in 2001 when anthrax spores were delivered via the mail in the United States. Domestic water supply sources may also be susceptible to bioterrorist attacks.

Biological weapons have certain unique characteristics that make them particularly dangerous. Because living systems are involved, they are uncontrollable and unpredictable once released. Furthermore, because natural forms of the pathogens used occur, it is possible for artificial preparation of these organisms to be released into the environment for diabolical purposes while mimicking natural events. This offers perpetrators of such activities room for denial when suspected. Biological agents can persist in the environment and may multiply over time to become even more deadly. For example, the anthrax bacterium can persist as spores in the environment for a long period of time.

Whereas toxins from living organisms may behave like a chemical warfare agent, most microbial pathogens require an incubation period to establish and replicate in the host before disease symptoms appear. This may take a few days to weeks and even longer. In terms of potency, it might take only a few thousand microbes (as in tularemia) or tens of thousands of microbes (as in anthrax) to kill an infected person. *Botulinum* toxin is the most potent toxic substance known to man, requiring as little as 10^{-9} mg inhalation to cause death.

Biological warfare agents do not always kill their victims. Potential warfare agents like tularemia, Q fever, and yellow fever offer chances of good recovery, whereas the chance of recovery from bubonic plague and smallpox without therapy is slim.

POTENTIAL BIOWARFARE AGENTS

Biological agents with high potential for use in the development of biological warfare include anthrax, smallpox, Q fever, botulinum toxin (from *Clostridium botulinum*), brucellosis, plague, and tularemia.

■ *ANTHRAX*

Anthrax bacterium (*Bacillus anthraxis*) are soil borne and primarily infect grazing hoofed mammals. They infect humans when they are released into the air and inhaled (inhalation anthrax), ingested through contaminated food (intestinal anthrax), or when they enter the body through the skin (cutaneous anthrax). The symptoms of the disease depend on the mode of infection. Inhalation anthrax is often fatal and resembles the common cold initially, until it causes severe breathing problems in the afflicted person, leading to shock. Intestinal anthrax causes acute inflammation of the intestinal tract, causing nausea, loss of appetite, vomiting of blood, and severe diarrhea. Person-to-person spread of anthrax is extremely unlikely, as it is not known to occur. Early administration of anitibiotics (e.g., penicillin, deoxycycline, fluoroquinolones) is a successful treatment of the infection. Anthrax vaccines can be used to prevent the disease.

■ *SMALLPOX*

Smallpox is a viral infection (*Variola* virus). Unlike anthrax, smallpox is an infectious disease that spreads readily from person to person, especially through the saliva. The spread is most serious during the first week of illness. The incubation period of the virus is between 7 and 17 days. Afflicted people develop characteristic rashes on the body, along with fever and aches. Most people recover from the infection but a death rate of about 30 percent is possible. Routine vaccination against smallpox was terminated in 1972. The disease was eliminated from the world in 1977.

■ *PLAGUE*

Commonly called the bubonic plague, this infectious disease of animals and humans is caused by a bacterium called *Yersinia pestis,* found in rodents and their fleas. Pneumonic plague occurs when the bacterium infects the lungs. The patient has a fever, headache, general weakness, and may cough out watery or bloody sputum. Death may follow without early therapeutic intervention with antibiotics (e.g., streptomycin, tetracycline, chloramphenicol). The disease spreads through respiratory droplets during face-to-face contact with patients.

■ *Q FEVER*

Q fever is transmitted to humans from infected domestic animals. It is an acute infection caused by *Coxiella burnetti.* It is contracted through the inhalation of infected dusts.

■ *BOTULISM*

Botulism is a disease caused by the release of potent neurotoxins by *Clostridium botulinum.* The disease may be contracted from eating contaminated food, or the organism may gain access through wounds, among other modes of infection. The toxin interferes with the presynaptic release of acetylcholine at the neuromuscular junction, leading to acute paralysis, vomiting, abdominal pain, and other symptoms.

■ *BRUCELLOSIS*

Brucellosis is commonly associated with animals (e.g., sheep, cattle, deer, pigs, and dogs). It is caused by the bacterium *Brucella suis.* The disease spreads to humans through contact with contaminated animals. Infected humans develop fever, headaches, and physical weakness, among other symptoms.

▪ *TULAREMIA*

The causative agent of tularemia, *Francisella tularemis,* is one of the most infectious bacteria known. The bacterium occurs widely in diverse animal hosts and habitats. Humans may be infected through inhalation, bites from infected arthropods, handling of infected animal tissue, and ingestion. Symptoms of tularemia include fever, fatigue, chills, and headache.

HISTORICAL PERSPECTIVES OF BIOWARFARE

Accounts of using biological weapons in warfare are historically few and far between. The stench from decomposing bodies of war victims that were deliberately hurled into enemy territory has reportedly been the reason for the surrender in a 1340 war in northern France. A more authentic warfare use of cadavers as offensive weapons was reported to have occurred in 1346 when the Tartar launched a siege of the Russian Crimea. They suffered a plague during their mission. However, before they retreated, they catapulted plague-infested cadavers over the walls into enemy territory. The fleeing residents carried the pathogens into Italy, thus beginning a second wave of epidemic that became known as the "Black Death" in Europe. During the era of British rule in America, they distributed smallpox-infested blankets from the infirmaries to Native Americans (Delaware Indians) in 1763, under the pretext of giving a peace offering. This act caused numerous deaths in the Indian population. In the modern era, historical records indicate the use of biological weapons by Japan against China during World War II in 1936 and 1940. Other accounts indicate that Japanese scientists used Chinese citizens as human subjects to test the lethality of several pathogens (e.g., anthrax, cholera, typhoid, and plague), causing the deaths of an estimated 10,000 people. World War I accounts indicate that German scientists adopted an unconventional war tactic by distributing infected livestock (sheep, cattle, horses, and mules) to enemy countries, but with minimal impact.

During the Cold War, both the United States and the former Soviet Union embarked on the research and development of bioweapons. However, such programs were discontinued in 1972 when the United States and 100 other nations became signatories to the Biological and Toxin Weapons Convention, banning this class of weapons. On the terrorist front, members of the Rajneesh cult in Oregon in 1984 placed salmonella bacteria in salad bars of several restaurants, causing 750 people to get sick. The most recent bioterrorism occurred on September 13, 2001, soon after the infamous terrorist attack on the World Trade Center in New York on September 11, 2001. Anthrax spores were placed in several envelopes and addressed to prominent news media executives and Congressmen.

REQUIREMENTS FOR MAKING A BIOLOGICAL WEAPON

In order to adapt microbes for use as weapons in conventional warfare, three elements are required—a disease agent, appropriate munitions, and a delivery system. A list of potential disease agents was presented previously. Various weapon programs in various parts of the world have researched numerous other agents. Microorganisms are delicate and susceptible to harsh environments.

In terms of munitions, the most efficient way of spreading the agent on the battlefield is through the air. This requires the particles to be airborne for a long period of time. The particle size range that can be airborne in this manner is between one and five microns. Because organisms are delicate, they cannot be delivered through conventional munitions (e.g., artillery rounds, cluster bombs, or missile warheads). Technical issues like excessive heat, ultraviolet light effect, humidity, and other environmental factors impact the potency and persistence of these weapons, a critical factor to the success of a biological weapon. Because various successes have been reported and demonstrated by various political regimes, it is reasonable to assume that such problems have been successfully addressed to varying degrees.

Commercial agricultural sprayers can be readily adapted for use in delivering biological weapons in an unsuspecting manner. The events of September 11, 2001, in New York brought this fact to light when suspects were discovered to have made enquiries about these pieces of equipment and sought training in their use. On the war field, remotely piloted vehicles may be used to deliver such weapons into enemy territory.

HOW REAL IS A THREAT FROM BIOLOGICAL WARFARE?

It is relatively easier and cheaper to develop biological weapons than, for example, nuclear-based weaponry. Individuals or groups of individuals can accomplish their production in small facilities. The protocols are relatively unsophisticated. As previously indicated, existing equipment being used for legitimate purposes can be adapted for such purposes. The anthrax threat of September 13, 2001, at least indicates that bioterrorism is a real threat.

In terms of large-scale state-sponsored biological weapons programs, nations that are known or suspected to have programs for offensive purposes include Iran, Iraq, Israel, North Korea, China, Libya, Syria, and Taiwan. Certain countries, including the United States and Russia, have dismantled offensive bioweapon development programs. In 1969, the United States unilaterally eliminated its biological arsenal.

INTERNATIONAL AGREEMENTS

Three international agreements currently address the issue of global development and use of biological weapons. The first is the 1925 Geneva Protocol that prohibits the use of poison gas and bacteriological weapons. The 1972 Biological Weapons Convention has a broader reach, banning the development, production, and stockpiling of biological and toxin weapons. The third, the 1993 Chemical Weapons Convention, is the chemical equivalent of the Biological Weapons Convention. Many countries are signatories to these treaties. Diplomatic measures and international pressure are tactics that are also used to discourage the development of biological weapons. For example, offending nations could be denied access to equipment and technologies that can be used in biological weapons programs. This strategy, however, is difficult to implement, given the fact that technologies often have a dual-purpose use, for legitimate and illegitimate purposes. Following the Gulf War of 1990–1991, the UN inspectors failed to find any definitive proof of weapons violations by Iraq.

PROTECTING THE PUBLIC AGAINST BIOLOGICAL WEAPONS

A government may implement the following steps to reduce the effects of a biological weapons attack on its citizenry.

1. **Detection system**

    The first step is to detect and correctly identity the offending biological agent, a task that is made more difficult by the fact that these agents are colorless, odorless, and tasteless. Air samples can be tested using technologies like the Biological Integrated

Detection System of the U.S. military (effective for anthrax and plague bacteria). Once detected, the contaminated building should be evacuated. Workers may wear protective masks. More effective and versatile detectors are also being developed.

2. **Decontamination**

   Biological agents may be eradicated from the building by using chemicals or radiation.

3. **Treatment**

   Antibiotics may be used to treat infected persons (i.e. post-exposure) but may also be used for prevention of infection (i.e., pre-exposure). The treatment depends on the pathogen and the nature of the infection. For example, anthrax has two forms of attack—inhalated (affects the lungs) and cutaneous (affects the skin).

4. **Prevention**

   People at risk (e.g., military personnel) may be immunized against high-probability biological agents (e.g., anthrax). However, it is a sound strategy to stockpile vaccines that are ready to use in case of an emergency.

THE ROLE OF BIOTECHNOLOGY IN BIOTERRORISM

Biotechnology may be used to engineer organisms to be more effective for use in a bioweapon development. More importantly, biotechnology may be used more constructively to develop new diagnostic tools to detect a wider variety of biological weapon agents. The next generation equipment should be very sensitive, provide quick and reliable diagnosis, and be usable in the field for on-the-spot diagnosis. The anthrax episode of 2001 is likely to speed up such efforts.

KEY CONCEPTS

1. Biowarfare is a nonconventional method of waging war.
2. Biowarfare (biological warfare) may be defined as the harnessing of the disease-causing the capability of disease agents as weapons to cause disease epidemics against an enemy in a war or conflict.
3. Some countries have biowarfare programs.
4. When an individual or a group with criminal intent conducts such an activity in peacetime, it constitutes bioterrorism.
5. There are certain natural agents that can be used in developing bioweapons.
6. Biological weapons are classified under weapons of mass destruction because of their capacity to kill or injure many people when deployed. They injure or kill in a variety of ways—inhalation, contact, and ingestion, depending on the type.
7. Biological agents with high potential for use in the development of biological warfare include anthrax, smallpox, plague brucellosis, Q fever, botulinum toxin (from *Clostridium botulinum*), and tularemia.
8. In order to adapt microbes for use as weapons in conventional warfare, three elements are required—a disease agent, appropriate munitions, and a delivery system.
9. The technology for developing and using bioweapons is often dual use.
10. Biotechnology can be used to develop more potent agents or diagnostic equipment to detect agents.
11. Three international agreements currently address the issue of global development and the use of biological weapons—the 1925 Geneva Protocol (which prohibits the use of poison gas and bacteriological weapons), the 1972 Biological Weapons Convention

(which bans the development, production, and stockpiling of biological and toxin weapons), and the 1993 Chemical Weapons Convention (the chemical equivalent of the Biological Weapons Convention).

OUTCOMES ASSESSMENT

1. How real is the threat of bioterrorism?
2. In what ways can biotechnology be used to foster or prevent bioterrorism?
3. Discuss the role of international efforts in preventing bioterrorism.
4. Give a brief history of biowarfare.
5. Give four of the natural agents that may be used in bioterrorism.
6. Discuss how a nation may protect itself against bioterrorism.

INTERNET RESOURCES

1. Pathogens that can be used in biological warfare by the National Institute of Health: *http://www.sis.nlm.nih.gov/Tox/biologicalwarfare.htm*
2. Countering bioterrorism by the Food and Drug Administration: *http://www.fda.gov/cber/cntrbio/cntrbio.htm*
3. Centers for Disease Control on anthrax: *http://www.cdc.gov/ncidod/dbmd/diseaseinfo/anthrax_t.htm*
4. Information from CDC on bioterrorism and related issues: *http://www.bt.cdc.gov*

REFERENCES AND SUGGESTED READING

Anon. 1999. How would you handle a terrorist act involving weapons of mass destruction? *ED Management, 11(11)*:121–124.

Arnon, S. S. et al. 2001. Botulinum toxin as a biological weapon: Medical and public health management. *Journal of the American Medical Association, 285(8)*:1059–1070.

Committee on R&D Needs for Improving Civilian Medical Response to Chemical and Biological Terrorism Incidents. Institute of Medicine. 1999. *Chemical and biological terrorism: Research and development to improve civilian medical response.* Washington, DC: National Academy Press.

Kortepeter, M. G. et al. 2001. Bioterrorism. *Journal of Environmental Health, 63(6)*:21–24.

Layne, S. P., T. J. Begelsdijk, C. Kumar, and N. Patel (eds.). 2001. *Firepower in the lab: Automation in the fight against infectious diseases and bioterrorism.* Washington, DC: Joseph Henry Press.

Miller, J., S. Engelberg, and W. Broad. 2001. *Germs: Biological weapons and America's secret war.* New York: Simon and Schuster.

25 Biotechnology: The Future

PURPOSE AND EXPECTED OUTCOMES

Biotechnology is unfolding as a truly legitimate biological revolution whose impact is pervasive in modern society. The outstanding successes recorded so far appear to be the tip of the iceberg. New technologies are being developed while older ones are being perfected or improved. Similarly, new areas of application are being discovered. The biotechnology industry will experience significant changes during the next decade.

In this chapter, you will learn:

1. Some of the new technologies being developed.
2. New applications being considered.
3. Social implications of the development and application of biotechnology that will likely dominate the headlines of the biotech debate.
4. The anticipated involvement of developing economies in biotechnology.
5. Possible trends in the biotech business environment.

The chapter has been divided into five topics: technological advances, Third World involvement, applications, business impact, and social issues.

TECHNOLOGICAL ADVANCES IN THE PIPELINE

The science that drives biotechnology will see significant advances in the next decade. The "big picture" approach to research whereby numerous processes and events are simultaneously monitored in living systems requires the development of more sophisticated equipment and new methodologies to cope with the increasing complexity and challenges of biotechnology research. Automation will be important in the development of instrumentation for high-throughput research and product development.

■ GENOMICS TECHNOLOGIES

Biotechnology will see significant changes in the technologies that are used for research and product development. The Human Genome Project (HGP) has indicated that the next phase of research, proteomics, is more challenging than the previous phase, genomics. Consequently, new technologies and approaches will have to be developed to handle the astonishing amount of information generated by genomics projects.

377

To start with, several genome projects, including the HGP, will need to be completed. Plant genomics will receive significant attention. In 2002, several rice genome projects, both in the public and private sectors, announced the completion of their sequencing projects. The next major crop of international interest is corn. Corn is one of the crops that has been intensively studied in terms of genetics. Along with wheat, corn will receive some attention in the area of genomics research.

In spite of the fact that insects are a major pest of crops, animals, and humans, causing an estimated $26 billion in damages annually to crops and livestock, insects have been neglected in the genome sequencing frenzy (except, of course, the model insect *Drosophila*). Genomic projects have been initiated in some insects such as the malaria mosquito *Anopheles gambiae*. This is of major interest because of the role the mosquito plays in the health problems of people in the tropics. Genomic projects provide information for the comparative genetic study of organisms. Sequencing additional species will provide scientists with tools for such comparative analysis.

■ PROTEOMICS

Technologies for the global approach to studying the genome will be improved or advanced. Proteomics, dubbed "the next big thing," will see the introduction of new technologies and improvement of existing ones for high-throughput analysis. Some of these technologies that already exist but are not mainstream technologies include **isotope-coded affinity tags** (pioneered by Ruedi Aebersold at the Institute for Systems Biology in Seattle, Washington). This technology enables researchers to chemically tag specific proteins in two separate samples with distinct heavy and light radioisotopes. Researchers can then study how protein expression changes with disease by tracking the relative abundance of the tagged proteins, using a mass spectrometer.

Protein chip (like DNA chip) technology has been off to a slow start. The most common approach to its design is using antibodies or antibody fragments. Competition is keen in the industrial and academic arenas to develop the protein array technology. This includes companies like Zyomyx of California (antibody-based chips), which has developed chips to screen 30 cytokines (proteins that play a role in the inflammatory diseases of humans, like arthritis and heart disease). The Phylos Company of Lexington uses antibody fragments, a much more sophisticated protein capture technique, while SomaLogic uses what is considered even more superior technology, aptamers. Protein chips promise to be very powerful diagnostic and research tools. Protein chips promise to be several-fold more useful than their predecessor DNA chips. However, the former has a long way to go yet to attain the stature of DNA microarrays. In general, the next decade will see an explosion in the development and application of microarray technologies to facilitate gene discovery and the use of the tremendous amount of genomic information that is constantly being generated from the numerous genomics projects.

Microfluidics entail chips with networks of sample holders, channels, and reaction chambers to carry out the complex sequence of steps needed to prepare protein samples for high throughput using equipment like mass spectrometers. This technology will improve the speed and sensitivity of proteomic analyses.

Another technology, **differential in gel electrophoresis** (commercialized by Amersham Pharmacia Biotech), provides researchers a tool for the global analysis of proteins. Scientists can view how protein expression changes between two samples. Proteins from two samples are differentially tagged with fluorescent dyes. The two proteins are then mixed together and submitted to a single 2-D electrophoresis. If separate spots show both colors, that indicates that a protein is expressed in both samples.

■ *MARKER-FREE TRANSFORMATION*

One of the criticisms of biotechnology has been in the area of the methodology of transformation in which antibiotic markers are used. As already discussed, critics fear that the antibiotic-resistant genes could escape into the environment to pose health issues by rendering antibiotic therapy ineffective. To address this concern, scientists have started to develop new methodologies to achieve transformation in a marker-free environment. These efforts will continue and likely produce practical methodologies.

■ *APOMIXIS*

Apomixis, the phenomenon of embryo formation in sexually reproducing plants without fertilization, has been quietly making progress in biotechnology. Progeny from apomixis are identical. Consequently, traits transferred into apomictic plants by classical or genetic engineering remain insulated from the effects of genetic recombination associated with the meiotic in sexual reproduction.

The worldwide interest in apomixis research was sparked by two independent research efforts in 1998. Bryan Kindiger, Phillip Sims, and Chet Dewald of the USDA's Agricultural Research Service (ARS) Southern Plains Range Research Station in Woodward, Oklahoma, successfully transferred the apomixis trait by classical methods in eastern gamagrass (*Tripsacum dactyloides*) (Figure 25-1). Another researcher, Wayne Hanna of the ARS Coastal Plains Research Station in Tifton, Georgia, transferred apomictic genes from *Pennissetum squamulatum* to its cultivated relative pearl millet.

Since this groundbreaking work, researchers in various parts of the world including the University of California, Davis; UC Berkeley; and Wageningen in the Netherlands have identified and cloned genes that can induce embryo formation in vegetative tissue. Other researchers, including some from Zurich and Canberra, Australia, have succeeded in inducing the formation of the endosperm (an endosperm—that which nourishes the embryo—is needed to make a seed whole).

Work will continue to help scientists elucidate the molecular aspects of the apomixis and bring it into the mainstream of biotechnology tools for plant manipulation.

FIGURE 25–1
Dr. Bryan Kindiger of USDA-ARS Grazinglands Research Laboratory at El Reno, Oklahoma, examines apomictic species in the field.

■ *PLASTID ENGINEERING*

Plant cells contain extranuclear genomes embodied in plastids, the most abundant being chloroplasts, the site of photosynthesis. Crop yield, essentially, is managing photosynthesis in production. Manipulation of chloroplasts for increased efficiency would help increase the productivity of plants.

Plant biotechnologists have other reasons for manipulating plastids. Plastids are known to yield a high amount of protein. Scientists are focusing on the biopharming of plants—using plants as bioreactors to produce therapeutic proteins. To attain this goal, genetic engineering of plastids needs to become routine like the engineering of their counterparts, the nuclear chromosomes. There are other advantages to engineering plastids. Plastids are maternally inherited. Consequently, transgenic plants with engineered plastids are not subject to the criticism of gene escape through pollen grains. Furthermore, nuclear genes are regulated by their own regulatory sequences, making it difficult to engineer complex traits. On the other hand, plastids, like bacteria, operate on a system where multiple genes are controlled by the same genetic switch (polycistronic), making it more amenable to complex genetic manipulation.

Engineering plastids is a daunting task. So far, significant successes have been accomplished only in the tobacco plant. Monsanto researchers have successfully engineered tobacco chloroplasts to produce human somatotropin. Other crops are being worked on by other research groups.

Another area where plastid transformation holds a bright future is in the engineering of RuBisCo, the principal enzyme in photosynthesis.

BIOINFORMATICS

Computer-based storage and retrieval of genetic sequence data is indispensable in biotechnology research. Bioinformatics has blossomed into a major academic discipline, combining the principles and concepts of computer science and biology. Powerful supercomputers and more sophisticated algorithms will be developed to improve the search and manipulation of the huge databases in the gene and protein banks.

BRIDGING THE BIOTECHNOLOGY DIVIDE: THIRD WORLD PARTICIPATION IN BIOTECHNOLOGY

We often hear about how some of those who could benefit the most from biotechnology through applications in agriculture and medicine are in developing countries. An FAO report in 2002 acknowledged the importance of biotechnology at least as a part of the strategies for working for food security in developing countries. Thanks to golden rice, the role of biotechnology in addressing food security issues of developing countries has been brought into mainstream discussion. Even though golden rice is years away from the food bowls of its intended beneficiaries, indications are that the project is being vigorously pursued to make it a successful endeavor. The next decade will see the introduction of additional GM crops of food value to large populations in the developing world (e.g., cassava, sweet potato). There are various ongoing projects in various parts of the world focusing on the nutritional quality and pest resistance of these crops. An example is the sweet potato program at Tuskegee University, where high-protein transgenic cultivars have been bred. These products will be introduced into developing countries.

As scientists in developing countries become more involved in biotechnology research, their governments will become more favorably disposed to the technology as a legitimate approach to enhancing agricultural productivity.

In the newly industrialized Third World countries, and especially those with long-standing biotechnology programs such as China, India, Taiwan, and Cuba, the next decade will see greater practical application of biotechnology. China invested an estimated $112 million in 1999 in biotechnology research and development. China, in 2002, was ranked number four in the world in terms of acreage cropped to GM products (cotton).

Nations that have the requisite infrastructure for participating in biotechnology (e.g., Brazil, Argentina, Mexico, India, and China) will make greater strides, provided their governments support such programs. Unfortunately, numerous other nations can ill afford to assemble the necessary critical mass of professionals (both scientists and technical personnel) to take advantage of the opportunities in global biotechnology. For such nations, it has been suggested that bipartite and tripartite approaches involving overseas universities and biotechnology companies collaborating to transfer biotechnology expertise would be the approach, at least initially, to participating in the biotechnology opportunities. In all this, progress will likely not be made unless local governments give their blessing and contribute their quota.

APPLICATIONS

In the area of applications of biotechnology to solve problems for society, there will be new products on the market in areas where biotechnology application is already well advanced (e.g., agriculture, medicine, and industry). There are applications that are currently experimental that will become commercialized or at least further advanced toward practical application. The area of "personalized medicine" will be advanced. Also, some applications that received a rocky introduction are likely to be revisited, refined, and reintroduced. Biotechnology, as a rapidly evolving field, will most certainly see new areas of application.

■ AGRICULTURE

In agriculture, "old reliables" like **Bt products** and **herbicide-resistance** crops will continue to dominate on the plant side. New major crops will be added to the pool. Some of the products in the pipeline were listed in Chapter 16. As these products become more acceptable, the acreage devoted to transgenics will steadily increase worldwide. In 2002, China was reported to be the world's fourth-largest grower of transgenic crops (cotton). This trend is likely to continue as more developing nations embrace biotechnology.

Golden rice will make headlines, for better or for worse. Currently, the next phase of fine-tuning the product is intended for increased expression of β-carotene and the breeding of traits into adapted cultivars in intended areas of use. This will take awhile and, as previously discussed, will face additional issues, including acceptability by the projected users.

Other **GM crops** are being developed along similar lines for introduction into developing countries. This includes cassava and sweet potato. Significant announcements about these crops will be expected in the next decade.

Biopharming research and applications will move significantly forward in the next decade. The concept of edible vaccines to use in immunizing people in developing countries is often cited as a key reason why biopharming has a promising future. However, some companies are also focusing on producing pharmaceuticals and industrial proteins in plant systems. Currently, some success has been attained with corn and potatoes. ProdiGene Inc. of Texas has developed a hepatitis B corn that the company hopes has a bright future. The company also produces avidin (a marker protein that helps laboratory workers track what is going on in chemical reactions) in corn at a fraction of the cost. It claims a bushel of corn ($2.50) produces the same amount of avidin that would be produced from a ton of eggs ($1,000). To avoid the problems of Aventis of France with its StarLink™ corn that gave the biotechnology industry a black eye from improper management of crop harvesting, companies are taking steps toward extra precautions in the production of their products. ProdiGene, for example, harvests all its products and does not grow its products for sale. Some companies use the refuge approach. Some estimate that biopharmed drugs could be a $200 billion industry within the next decade. Epicyte is developing human antibodies in plants.

Advances in **reproductive biotechnology** are expected in the next decade. Cloning of mammals will receive significant attention. Because application of the technology to farm animals is less controversial than to humans, cloning of livestock will continue to advance.

■ *MEDICINE*

In medical applications, gene therapy will be revisited. As previously indicated in Chapter 17, there is ongoing research to make the procedure safer. Nonviral gene delivery systems are being developed. Techniques that produce long-term lasting cures are being developed. Genes that are capable of forming triplex helices are amenable to the triplex-forming oligonucleotide technology. Other techniques being investigated include small fragment homologous replacement, viral gene targeting, and chimeraplasty.

Xenotransplantation research continues in both the public and private sectors. One key goal is to overcome post-transplantation organ rejection. Because pigs are the most promising candidate species for xenotransplantation, this animal has received the most attention. A sugar-producing gene has been identified as the culprit for tissue rejection. Pigs lacking the gene for the enzyme galactosyltransferase are unable to synthesize this sugar. Researchers in the public sector include those at Colorado State University and the University of Missouri. In the private sector, Advanced Cell Technology of Massachusetts and Immerge BioTherapeutics of the United States, as well as PPL Therapeutics of Scotland, are actively involved in gene therapy research. The galtransferase gene has been knocked out in fetal cells used to make cloned pigs.

Another area of medical research that will make significant strides is the area of stem cells. The overwhelming potential application of the research will drive scientists to discover new and more publicly acceptable sources of stem cells. Currently, the focus is on adult stem cells. In early 2002, an announcement was made to the effect that such cells had been isolated from monkeys.

Pharmacogenomics will make significant strides in advancing the concept of personalized therapy, thanks to the Human Genome Project.

■ *INDUSTRIAL*

One of the areas of industrial application of biotechnology that will continue to grow is in the development of diagnostic equipment. Such equipment will be needed for research, as well as for other applications. Diagnostic kits for use in the domestic monitoring of personal health (e.g., glucose monitoring) will increase. Diagnostic devices for clinical applications will also be developed to assist physicians in disease diagnosis and the prescription of therapy. Information from the Human Genome Project will also enhance genetic counseling programs.

One focus in the development of diagnostic tools will be monitoring agents of bioterrorism, thanks to the infamous terrorist attacks of September 11, 2001. In the 2003 budget submitted to Congress, the NIH proposed $1.473 million for bioterrorism research activities. Of this, over $500 million was earmarked for drug screening, diagnostics, and animal models and another $300 million for genomics and proteomics research.

Drug discovery efforts will turn out "druggable" proteins to feed the pharmaceutical industry. Some of these will be amenable to bioprocessing. The pharmaceutical industry will develop a new generation of drugs and vaccines to combat hepatitis B, influenza, diabetes, leukemia, and human growth hormone deficiency, among others.

■ *ENVIRONMENTAL*

Environmental applications of biotechnology are both direct and indirect. All the modifications in technology (e.g., marker-free transformation, use of refuges) indirectly benefit the environment. In addition to these efforts, scientists develop technologies that directly benefit the environment (e.g., bioremediation, waste management). New microbes with greater capacity to decompose organic and inorganic wastes will be discovered or engineered. Monitoring devices will also be developed to monitor pollutants.

BUSINESS

In the case of the business of biotechnology, mergers and acquisitions will continue as was discussed in Chapter 16. The proteomics race was alluded to in Chapter 14. The early entrants to the race (e.g., GeneProt) may have a headstart but will soon be challenged by others. Pharmaceutical companies form alliances with small research companies to develop drugs.

Just like Monsanto and other major companies surrendered patent rights for the humanitarian application of their inventions, others will follow suit in an effort to assist poor developing countries in their biotechnology quest. As Third World nations embrace biotechnology, companies will find it profitable to invest in technologies and products that target Third World markets.

Businesses will continue their drive to educate the consuming public to position themselves in a favorable image. They will heed public concerns and modify some of their strategies to avoid repeating previous pitfalls (like the "terminator" program).

SOCIAL ISSUES

Activism against biotechnology will continue as the field develops and becomes more pervasive in its applications in society. Because biotechnology is big business and promises to be even more so in the next decade, activists are likely to intensify their protests against what they perceive as a profit-first mentality and exploitative attitude of businesses. One of the major social issues of the next decade will involve the development and application of stems.

As Third World nations embrace biotechnology and more nations grow GM products, world trade will be significantly impacted. European nations are bound to review and relax their current hard stance against GM products.

Whereas some activism may be irresponsible, biotechnology will benefit from constructive protestations by refining current technologies to make them more environmentally benign and safe.

KEY CONCEPTS

1. The science that drives biotechnology will see significant advances in the next decade.
2. More sophisticated equipment and new methodologies will be developed to cope with the increasing complexity and challenges of biotechnology research.
3. Automation will be important in the development of instrumentation for high-throughput research and product development.
4. New technologies and approaches will have to be developed to handle the astonishing amount of information generated by genomics projects.

5. Technologies for the global approach to studying the genome will be improved or advanced.

6. Work will continue to help scientists elucidate the molecular aspects of the apomixis and to bring it into the mainstream of biotechnology tools for plant manipulation.

7. Bioinformatics has blossomed into a major academic discipline, combining the principles and concepts of computer science and biology. It will continue to play a significant role in biotechnology.

8. As scientists in developing countries become more involved in biotechnology research, their governments will become more favorably disposed to the technology as a legitimate approach to enhancing agricultural productivity.

9. In the area of application of biotechnology to solve problems for society, there will be new products on the market in areas where biotechnology application is already well advanced (e.g., agriculture, medicine, and industry).

10. The concept of edible vaccines to use in immunizing people in developing countries is often cited as a key reason why biopharming has a promising future. However, some companies are also focusing on producing pharmaceuticals and industrial proteins in plant systems.

11. In medical applications, gene therapy will be revisited.

12. Another area of medical research that will make significant strides is the area of stem cells.

13. One of the areas of industrial application of biotechnology that will continue to grow is the development of diagnostic equipment.

14. Monitoring devices will also be developed to monitor pollutants.

15. Activism against biotechnology will continue as the field develops and becomes more pervasive in its applications in society.

OUTCOMES ASSESSMENT

1. Give two significant changes you predict will occur in the area of technology development for biotechnology.

2. In your opinion, do you expect biotechnology to become more pervasive in society or to diminish in importance? Give reasons.

3. Do you expect activism against biotechnology to increase or decrease in the future? Explain.

4. Give two specific areas of application for the benefit of society that you predict will occur in the near future.

INTERNET RESOURCES

1. Future of biotech predicted by 10 experts: *http://biotech.about.com/cs/futuredirections/*
2. Future biotechnology research on space station: *http://www.nas.edu/ssb/btfmenu.htm*
3. Biotech and ethics: blueprint for the future: *http://www.biotech.nwu.edu/nsf/*
4. Pioneer Hibred—future products: *http://www.pioneer.com/biotech/dp_biotech/future.htm*
5. Future of biotech: *http://www.rsc.ca/foodbiotechnology/indexEN.html*

REFERENCES AND SUGGESTED READING

Feber, D. 2001. Gene therapy: Safer and virus-free? *Science, 294:*1638–1642.

Gewolb, J. 2002. Plant scientists see big potential in tiny plastids. *Science, 295:*258–259.

Kaiser, J. 2002. Cloned pigs may help overcome rejection. *Science, 295:*26–27.

Moffat, A. S. 2002. For plants, reproduction without sex may be better. *Science, 294:*2463–2465.

PART VI Additional Resources

Selected Biotechnology Websites

Glossary

Selected Biotechnology Websites

Topic of Interest	Source for Information
Genomics, bioinformatics, and proteomics knowledge base	*http://123genomics.com/*
Access excellence, the National Health Museum. Site for health, bioscience teachers, and educators	*http://www.accessexcellence.org/*
Genetic engineering applications, impacts, and implications	*http://www.biotech-info.net/*
Agriculture and biotechnology consulting services	*http://www.agbios.com/default.asp*
Berkeley drosophila project	*http://www.fruitfly.org/*
Biocomputing hypertext course book	*http://merlin.mbcr.bcm.tmc.edu:8001/ bcd/Curric/welcome.html*
Diversity of information on bioinformatics	*http://bioinformatics.org/*
Biotech business issues	*http://www.biospace.com/*
Information for cell and molecular biologists	*http://www.cellbio.com/*
Animated educational site for general biology	*http://www.cellsalive.com/*
A compendium of Internet accessible and electronic tools for biotechnology, molecular biology, biomodeling, and so on	*http://restools.sdsc.edu/*
Tutorial on BLAST	*http://www.rickhershberger.com/ darwin2000/blast/*
Course and exams on molecular biology	*http://www.genengnews.com/top100.asp*
Animated educational site on genetics with historical perspectives	*http://www.dnaftb.org/dnaftb/*
Information on the genomics of eukaryotic organisms	*http://iubio.bio.indiana.edu:8089/*
Bioinformatics information by Kyoto University	*http://www.genome.ad.jp/*
Provides current information on the science and business of biotech	*http://www.genomeweb.com/*

Links to great genome sites	*http://www.hgmp.mrc.ac.uk/GenomeWeb/*
Provides a guide to molecular sequencing	*http://www.sequenceanalysis.com/*
Internet guide to molecular biology and biotechnology	*http://highveld.com/*
Site for raw data on genomics projects	*http://www.ncbi.nlm.nih.gov/Traces/trace.cgi*
Variety of resources for biotechnology	*http://www.genebrowser.com/*
Plant dictionary	*http://www.hcs.ohio-state.edu/plants.html*
Plant tissue culture information	*http://aggie-horticulture.tamu.edu/ tisscult/tcintro.html*
Service for sequence analysis and protein prediction	*http://cubic.bioc.columbia.edu/predictprotein/*
Repository for processing and distributing 3-D biological molecular structure data	*http://www.rcsb.org/pdb/*
Research tools for biotechnology	*http://www.nih.go.jp/~jun/research/*
Resources for scientists teaching science	*http://instruct1.cit.cornell.edu/courses/ taresources/*
RNA structure database	*http://www.rnabase.org/*
Biology project that provides detailed instruction in the principles and concepts of biology	*http://www.biology.arizona.edu/*

Glossary

Ab initio recognition   The technique of predicting gene structure literally from the beginning by using computer programs and information from the protein coded for the gene.

Activator (of gene)   A protein molecule which stimulates or increases the expression of a given gene, by binding to transcription control sites.

A-DNA   A right-handed helical form of DNA possessing 11 base pairs per turn.

Adenosine Diphosphate (ADP)   A ribonucleoside 5'-diphosphate serving as phosphate-group acceptor in the cell energy cycle.

Adenosine Triphosphate (ATP)   A ribonucleoside 5'-triphosphate serving as a phosphate-group donor in the energy cycle of the cell. (It is the major carrier of chemical energy in all living cells.)

Aerobic respiration   Respiration that occurs in an oxygen-rich environment.

Agar   A complex mixture of polysaccharides obtained from marine red algae.

Agarose   A highly purified form of agar.

Allele   Alternate form of a gene or DNA sequence.

Allosteric enzyme   A regulatory enzyme whose catalytic activity is modulated by the noncovalent binding of a specific metabolite at a site other than the catalytic site of the enzyme.

Alternative mRNA splicing   The post-transcriptional processing phenomenon involving the inclusion or exclusion of different exons to form different mRNA transcripts.

Alu family   A set of dispersed and related genetic sequences, each about 300 base pairs long, in the human genome.

Ames test   A bacterial-based test for carcinogens.

Amino acid   A basic unit or building block of proteins, comprising a free amino (NH_2) end, a free carboxyl (COOH) end, and a side group (R).

Amplified Fragment Length Polymorphism (AFLP)   A type of DNA marker.

Amplify (DNA amplification)   Method to increase the number of copies of a DNA sequence through cloning or Polymerase Chain Reaction (PCR) methodology.

Anabolism   The phase of metabolism concerned with the energy-requiring biosynthesis of cell components from smaller precursor molecules.

Anaerobic   An environment without air or oxygen.

Anneal   The pairing or binding together of complementary DNA or RNA sequences, via hydrogen bonding, to form a double-stranded polynucleotide.

Annotation   Text describing the analysis and various pieces of information associated with DNA sequences, protein sequences, and other materials stored in data bank.

Antibody   An immunoglobulin protein produced by B-lymphocytes of the immune system that bind to a specific antigen molecule.

Anticodon   A nucleotide base triplet in a transfer RNA molecule that pairs with a complementary base triplet, or codon, in a messenger RNA molecule.

Antigen   A foreign substance that elicits an immune response by stimulating the production of antibodies in a host organism.

Antiparallel   The arrangement of molecules that are parallel but oriented in opposite directions (e.g., DNA).

Antisense (DNA sequence)   A strand of DNA that codes for a messenger RNA molecule.

Antisense RNA   A complementary RNA sequence that binds to a naturally occurring (sense) mRNA molecule, thus blocking its translation.

Apomixis   A method of reproduction in which plants produce seed without the process of sexual fertilization.

Aptamers   Oligonucleotide molecules that bind other specific molecules.

Asexual reproduction   Nonsexual (vegetative) means of reproduction.

Autosomes   All chromosomes in a cell except the sex chromosomes.

Bacteriophage   A virus that attaches to and infects bacteria by injecting its DNA into, and multiplying inside of, the host.

Baculovirus   A class of virus that infects lepidopteran insects.

Base pair (bp)   A pair of complementary nitrogenous bases in a DNA molecule.

Base substitution   The replacement of one base within a DNA molecule by another base.

Biocide   Any chemical or chemical compound that is toxic to living things.

Bioinformatics   The knowledge-based theoretical discipline that entails the collection and manipulation of computer-based biological data (from DNA sequencing and others) to make predictions about biological functions or other research objectives.

Bioleaching   The use of microorganisms to recover metals from their ores.

Bioluminescence   The enzyme-catalyzed production of light by living organisms.

Biomass   The total dry matter of all organisms in a particular sample, population, or area.

Bioremediation   The use of organisms to consume or otherwise help remove materials from the environment.

Biotechnology   The use of technologies based on living systems to make products or improve other species.

Callus   A clonal mass of undifferentiated plant cells.

Carbohydrates   A class of carbon-hydrogen-oxygen compounds usually represented chemically by the formula $(CH_2O)n$, where n is three or higher.

Carcinogen   A cancer-causing agent.

Catabolism   The phase of metabolism involved in the breakdown of a complex biological molecule into less-complex components, usually accompanied with the release of energy in the form of ATP.

Catalyst   Any substance that increases the rate of a chemical reaction without being consumed itself in the reaction.

cDNA (complementary DNA)   DNA synthesized from an RNA template using reverse transcriptase which is complementary to the mRNA molecule.

Cell culture   The *in vitro* propagation of cells isolated from living organisms.

Cell differentiation   The process whereby one cell develops into many cells with a modification of the new cells to form specialized structures for particular functions.

Central dogma   The concept that, in nature, genetic information generally flows from DNA to RNA to protein, but not the reverse.

Chimera   An organism consisting of two or more genetically different tissues.

Chromosome walking   A technique for identifying a gene or a chromosomal region of interest by working from a flanking DNA marker identifying successive clones that span a chromosomal region of interest.

Cistron   A DNA sequence that codes for a specific polypeptide. Synonymous with gene.

Clone   An exact genetic replica of a specific gene, cell, or an entire organism.

Cloning   The technique involving mitotic division of a progenitor cell to give rise to a population of identical daughter cells.

Coding sequence   The region of a gene that codes for the amino acid sequence of a protein.

Codon   A triplet of nucleotides that codes for an amino acid or a termination signal.

Coenzyme (cofactor)   An organic molecule that binds to an enzyme and is required for its catalytic activity.

Consensus sequence   The most common nucleotide sequence at each position within a specific DNA molecule in cases where variations in nucleotide sequence occurs.

Constitutive promoter   An unregulated promoter that allows for continual transcription of its associated gene.

Contig   A chromosome map showing the locations of those regions of a chromosome where contiguous DNA segments overlap

Copy number   The number of molecules of a specific plasmid or plastid that is typically present in a single cell.

Corepressor   A small molecule that combines with the repressor to shut down transcription.

Cosuppression   A decrease in the gene expression as result of the artificial insertion and expression of a homologous gene.

Cross-pollination   The transfer of pollen from the anthers of one flower to the stigma of an unidentical flower.

Crossing over   The exchange of chromatid segments between homologous chromatids during meiosis.

Denaturation   The process by which a macromolecule loses its natural conformation, leading to loss of biological activity.

Dominant allele   An allele that is manifest in all heterozygotes.

Down-regulating   The process of causing a given gene to express less of the protein that it normally codes for.

Downstream   The region extending in a 3' direction from a gene.

Electrophoresis   A technique for separating molecules based on the differential mobility in an electric field.

Electroporation   A method for transformation that uses high-voltage pulses of electricity to open pores in cell membranes for foreign DNA to enter the cell.

Endonucleases   A class of enzymes capable of hydrolyzing the interior phosphodiester bonds of DNA or RNA chains.

Enzyme   An organic, protein-based substance produced by living cells to catalyze biochemical reactions.

Epistasis   Interaction between nonallelic genes in which the presence of a certain allele at one locus inhibits the phenotypic expression of the other.

Eukaryote   An organism whose cells possess a nucleus and other membrane-bound organelles.

Exon   The segment of a eukaryotic DNA that is transcribed into an mRNA.

Explant   The material used to initiate tissue culture.

Feedstock   Raw material used for the production of chemicals.

Fermentation   The enzyme-catalyzed reaction in which molecules are broken under anaerobic conditions.

Fertilization   The fusion of two gametes of the opposite sex to form a zygote.

Forward mutation   A mutation from the wild type to the mutant.

Frameshift   A displacement of the reading frame in a DNA or RNA molecule as result of an addition or deletion of one or more nucleotides to or from the DNA or RNA molecule.

Functional genomics   The science devoted to understanding the function of genes or DNA sequences in an organism.

Gamete   A germ or reproductive cell comprising a single copy of each chromosome in an organism.

Gene   A sequence of nucleotides in the genome of an organism to which a specific function can be assigned.

Gene cloning   The process of synthesizing multiple copies of a particular DNA sequence using another organism as a host.

Gene expression   The process of producing a protein from its DNA- and mRNA-coding sequences.

Gene pool   The sum total of all the alleles of all genes of all individuals in a particular breeding population.

Gene probe (genetic probe or DNA probe)   A short, specific, artificially produced segment of DNA that is complementary to the desired gene and is used to detect the presence of specific genes (or shorter DNA segments) within a chromosome.

Gene silencing   The suppression of gene expression.

Gene splicing   The enzymatic joining of one gene to another or the removal of introns and joining of exons during mRNA synthesis.

Gene therapy   The insertion of genes into selected cells in the body in order to accomplish a specific therapeutic effect.

Genetic code   The sequence of three nitrogeneous bases in a DNA molecule that codes for an amino acid or protein.

Genetic engineering   The genetic manipulation of an organism that involves the DNA directly at the molecular level.

Genetic linkage map   A chromosome map showing the linear order of the genes associated with the chromosome.

Genetic map   A diagram showing the relative sequence and position of specific genes along a chromosome.

Genetic marker   A gene associated with a genetic event.

Genetics   The science of the study of heredity.

Genome   The genetic complement (entire hereditary material) contained in the chromosomes of a given organism.

Genomics   The scientific study of genes and their role in an organism's structure and function.

Genotype   The genetic makeup of an individual.

Genus   A group of structurally related species.

Germ line   Inherited material that comes from the eggs or sperm and is passed on to offspring.

Glycolysis   A metabolic process in which sugars are broken down into smaller compounds with the release of energy.

Haploid   A cell with one complete set of chromosomes.

Heredity   Transfer of genetic information from parents to their offspring.

Heritability (broad sense)   The fraction of phenotypic variation that is due to genetics.

Hybrid   The offspring of two parents differing in at least one genetic characteristic.

Hybridization   The hydrogen bonding of complementary DNA and/or RNA sequences to form a duplex molecule. Also, the mating of two unrelated individuals to produce a hybrid.

Immunoassay   The method of identifying and quantifying a substance that involves the use of antibodies.

Imprinting   A process in which certain genes within an organism's cells are inactivated very early in the development of the organism.

In situ   In the natural or original position.

In vitro   Living in test tubes (in glass) or an artificial environment.

In vivo   In the living organism.

Intron   A non-coding DNA sequence (intervening sequence) within a coding sequence or gene that is removed during post-transcriptional processing.

Isozymes (isoenzymes)   Multiple forms of an enzyme that differ in their properties, such as substrate specificity and maximum activity.

Jumping genes   Genes that change positions within the genome.

Karyotype   A chart depicting the chromosome complement of a cell characterized according to size and configuration.

Kb   An abbreviation for 1,000 (kilo) base pairs of deoxyribonucleic acid (DNA).

Knockout The inactivation of specific genes.

Lac operon An operon in *E. coli* that codes for three enzymes involved in lactose metabolism.

Library (DNA) A collection of cloned DNA usually from a specific organism.

Ligation The formation of a phosphodiester bond to link two adjacent bases separated by a nick in one strand of a double helix of DNA.

Linkage A measure of the degree to which alleles of two genes assort independently during meiosis.

Linkage group A set of gene loci that can be placed in a linear order according to linkage relationships.

Linkage map A chromosome map showing the linear order or the genes associated with the chromosome, based on the frequency of recombination in the offspring's genome.

Locus The position of a gene on a chromosome.

Lysis The process of cell disintegration or membrane rupturing.

Macromolecules Large molecules in the cell with molecular weights ranging from a few thousand to hundreds of millions.

Major Histocompatibility Complex (MHC) A chromosomal region (approximately 3,000 Kb) which encodes for three classes of transmembrane proteins, MHC I, II, and III proteins, that play a critical role in the success of organ transplantation.

Map distance A number proportional to the frequency of recombination between two genes.

Mapping The process of determining the physical location of a gene or genetic marker on a chromosome.

Meiosis A cellular process in which the nucleus undergoes two successive divisions following a single replication of the chromosomes to produce four haploid nuclei.

Messenger RNA (mRNA) The class of RNA molecules that copies the genetic information from DNA, in the nucleus, and carries it to ribosomes for translation into protein.

Microarray (DNA) An ordered set of DNA molecules of known sequences on a piece of glass or other material.

Micropropagation A technique for clonal propagation of a plant using tissue culture.

Microsatellite DNA A portion of the DNA containing numerous sequential repeats of a specific small DNA sequence.

Mitosis A nuclear division that produces two daughter nuclei with the same number of chromosomes.

Model organism An organism that is utilized in scientific experimentation to produce information that is then applied to more complex organisms.

Monosaccharides The chemical building blocks of carbohydrates with an empirical formula of $(CH_2O)_n$.

Mutagen A chemical or physical agent that is capable of producing a genetic mutation in a living organism.

Mutant A product of a mutation or heritable genetic change.

Mutation Any change that alters the sequence of the nucleotide bases in the DNA of a cell or an organism.

Nick A break in one strand of a double-stranded DNA molecule.

Nonsense codon A triplet of nucleotides that does not code for an amino acid.

Nonsense mutation A mutation that converts a codon that specifies an amino acid into one that does not specify any amino acid.

Northern blot A technique used to identify and locate mRNA sequences that are complementary to a piece of DNA called a probe.

Northern blotting (or hybridization) A method for transferring RNA fragments from an agarose gel to a nitrocellulose filter, following electrophoresis, for further analysis by hybridization with appropriate probes.

Nucleic acid A macromolecule containing a phosphate group, a sugar group, and nitrogeneous bases.

Nucleoside   A building block of DNA and RNA, consisting of a nitrogenous base linked to a five-carbon sugar.

Nucleotide   A building block of DNA and RNA, consisting of a nitrogenous base, a five-carbon sugar, and a phosphate group.

Oligomer   A relatively short molecular chain consisting of repeating units.

Oligonucleotide   A DNA polymer consisting of only a few nucleotides.

Oligosaccharide   A relatively short molecular chain consisting of 10 to 100 simple sugar units.

Open reading frame   A long DNA sequence that is uninterrupted by a stop codon and encodes part or all of a protein.

Operon   A group of functionally related structural genes located adjacent to each other, transcribed into a single mRNA, and controlled by the same operator.

Organelles   Membrane-bound structures with distinct functions that occur in eukaryotic cells.

Organogenesis   The capacity of nonmeristematic tissue to produce organs *de novo*.

Origin of replication (ori)   The nucleotide sequence that marks the initiation point for DNA synthesis.

Oxidative phosphorylation   The enzymatic addition of a phosphate group to ADP to produce ATP coupled to electron transport from a substrate to molecular oxygen.

Peptide   Two or more amino acids covalently joined by peptide bonds.

Peptide bond   A covalent bond between the α-amino group of one amino acid and the α-carboxyl group of another amino acid.

Photosynthesis   The process in green plants of converting carbon dioxide and water into sugar using light energy as the power source.

Phytochrome   A reversible protein pigment in plants that is associated with the absorption of light that serves to direct the course of plant growth, development, and differentiation.

Phytoremediation   The use of certain plants to remove contaminants or pollutants from either soils or the environment.

Plastid   A self-replicating organelle inside a plant cell that is not a part of the reproduction cell genome.

Point mutation   A mutation involving a change of only one nucleotide in a DNA molecule.

Polyacrylamide Gel Electrophoresis (PAGE)   A technique of electrophoresis in which molecules are separated on the basis of size and charge in a polyacrylamide gel.

Polycistronic   Coding regions representing more than one gene in the mRNA.

Polygenic   System of traits determined by many genes, each having a slight effect on the expression of the trait.

Polylinker   A short DNA sequence in a cloning vector that contains several restriction enzyme recognition sites.

Polymer   A large molecular unit composed of repeated subunits.

Polymerase   An enzyme that catalyzes the assembly of similar or identical subunits into a larger unit.

Polymerase Chain Reaction (PCR)   A reaction that uses the enzyme DNA polymerase to catalyze the amplification of a DNA strand through repeated cycles of DNA synthesis.

Polymorphism   The presence of several forms of a trait or gene in a population.

Polynucleotide   A DNA polymer composed of multiple nucleotides.

Polypeptide (protein)   A polymer composed of multiple amino acid units linked by peptide bonds.

Polyploidy   A condition in which a cell, tissue, or organism has more than the $2n$ chromosomes per nucleus.

Polysaccharide   A long-chain molecule composed of multiple units of monosaccharides.

Position effect   A change in the expression of a gene as a result of its translocation to a new site in the genome.

Positional cloning   A technique used by researchers to zero in on the gene(s) responsible for a given trait or disease.

Post-transcriptional processing (modification) of RNAs   The enzyme-catalyzed processing or structural modifications of recently transcribed RNA molecules to become functionally finished products.

Post-translational modification of protein   Enzymatic processing of a polypeptide chain after a recent translation from its mRNA that may include the addition of carbohydrate moieties to the protein or the removal of a portion of the polypeptide chain in order to produce a functional protein in the correct environment.

Pribnow box   The consensus sequence TATAATG centered about 10 base pairs before the starting point of bacterial genes, which is a part of the promoter and is especially important in binding RNA polymerase.

Primer (DNA)   A short sequence of DNA that is paired with one strand of the template DNA in the Polymerase Chain Reaction (PCR) technique.

Promoter   A nucleotide sequence in the DNA molecule that occurs upstream from the transcription start site that contains the recognition or binding sites for RNA polymerase and a number of proteins that regulate the rate of transcription of the adjacent gene.

Prosthetic group   A heat-stable metal ion or an organic group that is covalently bonded to the apoenzyme protein, and is required for enzyme function.

Protein engineering   The selective, deliberate (re)designing and synthesis of proteins for specific functions.

Protoplast   A part of the cell that includes the cell membrane and all of the intracellular components, but excludes the cell wall.

Quaternary structure   The three-dimensional structure of an oligomeric protein.

Reading frame   A series of triplet codons beginning from a specific nucleotide.

Recessive gene   A gene whose phenotype is expressed only when both copies of the gene are mutated or missing.

Recognition sequence (site)   A nucleotide sequence that is recognized by a restriction endonuclease.

Recombinant   A cell that results from a recombination of genes.

Recombinant DNA   The technology of cutting and recombining DNA fragments from different sources.

Recombination   The formation of new gene combinations involving the joining of genes, sets of genes, or parts of genes through independent assortment or crossing over, or through laboratory manipulation.

Replicon   A chromosomal region containing the DNA sequences necessary to initiate DNA replication processes.

Reporter gene   A specific gene that is inserted into the DNA of a cell to alert researchers that a specific event, such as gene expression, has occurred.

Repressor   A DNA-binding protein in prokaryotes that blocks gene transcription by binding to the operator.

Restriction endonuclease (enzyme)   A class of endonucleases that cleaves DNA after recognizing a specific sequence.

Retrovirus   An RNA virus that utilizes the enzyme reverse transcriptase to reverse copy its genome into a DNA intermediate, which integrates into the host cell chromosome.

Reverse genetics   Using linkage analysis and polymorphic markers to isolate a disease gene in the absence of a known metabolic defect, then using the DNA sequence of the cloned gene to predict the amino acid sequence of its encoded protein.

Selectable marker   A gene with a readily assayable expression that is included in genetic manipulation to alert researchers that an event such as transformation or transfection with a vector containing the marker gene has occurred.

Self-pollination   The transfer of pollen from one plant to another of the same genetic makeup.

Sense   Normal or forward orientation of a DNA sequence in a genome.

Sequence-tagged site (STS)   A unique (single-copy) DNA sequence used as a marker in the mapping of a chromosome.

Sequencing (of DNA molecules)   The process of obtaining the sequential order of nucleotides in the DNA backbone.

Shotgun cloning   A technique for isolating a gene by producing a restriction digest of the genome, placing each fragment into a bacterium, and then screening the numerous transformed bacteria to locate those that contain the desired gene.

Shotgun sequencing (whole-genome shotgun sequencing)   A technique for sequencing DNA whereby an organism's genome is first fragmented, and then randomly selected pieces of the DNA are individually sequenced and reassembled by piecing together overlapping ends.

Shuttle vector   A vector capable of replicating in two unrelated species.

Silent mutation   A mutation that causes no detectable change in the biological characteristics of a gene's product.

Site-directed mutagenesis   A technique for inducing single mutations in a cell's DNA at a predetermined region.

Somaclonal variation   The genetic variation that arises in cells in culture.

Southern blot analysis   A technique for transferring DNA fragments from a gel to a filter.

Stem cells   Certain cells that can grow or differentiate into different cells or tissues of the body of an organism.

Structural gene   A gene that codes for an RNA or protein product other than a regulator molecule.

Structural genomics   The study of the genome that focuses on the DNA sequence and other features of the subunits that comprise those sequences.

TATA box   An adenine- and thymine-rich promoter sequence located 25 to 30 bp upstream of a gene, which is the binding site of RNA polymerase.

T-DNA (transfer DNA, tumor-DNA)   The transforming region of DNA in the Ti plasmid of *Agrobacterium tumefaciens.*

Tissue culture   The growth and maintenance of cells or tissues from an organism *in vitro* under sterile conditions.

Totipotency   The ability of a cell to grow and differentiate into all of the types of cells and tissues present in an organism's body.

Trait   A characteristic of an organism, which manifests itself in the phenotype.

Transcription   The process of copying the genetic information contained in the template strand of the DNA to produce an mRNA strand.

Transduction   The transfer of DNA sequences from one bacterium to another via lysogenic infection by a bacteriophage.

Transfection   The uptake and expression of a foreign DNA sequence by cultured eukaryotic cells.

Transformant   A cell or individual that carries a foreign DNA sequence in its genome.

Transformation   The process of transferring a foreign piece of DNA into a cell.

Transgene   A foreign gene or genetic material that is transferred into the genome of a cell of another organism using recombinant DNA technology.

Transgenic   An organism in which a foreign DNA gene has been incorporated into its genome.

Transposon   A DNA sequence that is capable of replicating and inserting a copy of itself at a new location in the genome.

Upstream   The region extending in a 5′ direction from a gene.

Vector   The agent used in recombinant DNA research to carry new genes into cells.

Wild type   The natural form of an organism that exists.

Xenotransplantation   The implantation of parts of an organism to another or a different species.

Zygote   The product of the fertilization of an egg and a sperm.

Index